Make It a Green Peace!

Make It a Green Peace!

The Rise of Countercultural Environmentalism

FRANK ZELKO

OXFORD
UNIVERSITY PRESS

OXFORD
UNIVERSITY PRESS

Oxford University Press is a department of the University of Oxford.
It furthers the University's objective of excellence in research,
scholarship, and education by publishing worldwide.

Oxford New York
Auckland Cape Town Dar es Salaam Hong Kong Karachi
Kuala Lumpur Madrid Melbourne Mexico City Nairobi
New Delhi Shanghai Taipei Toronto

With offices in
Argentina Austria Brazil Chile Czech Republic France Greece
Guatemala Hungary Italy Japan Poland Portugal Singapore
South Korea Switzerland Thailand Turkey Ukraine Vietnam

Oxford is a registered trade mark of Oxford University Press
in the UK and certain other countries.

Published in the United States of America by
Oxford University Press
198 Madison Avenue, New York, NY 10016

Library of Congress Cataloging-in-Publication Data
Zelko, Frank S.
Make it a green peace! : the rise of countercultural
environmentalism / Frank Zelko.
p. cm.
Includes bibliographical references and index.
ISBN 978-0-19-994708-9
1. Greenpeace International—History.
2. Environmental protection—Political aspects.
3. Antinuclear movement—History.
4. Environmentalism—History.
5. Counterculture—History. I. Title.
TD169.Z45 2013
333.72—dc23 2012042260

3 5 7 9 8 6 4 2

Printed in the United States of America
on acid-free paper

For Ana and Stefan

CONTENTS

ACKNOWLEDGMENTS

You know a project has taken too long when several of the people who deserve the most thanks did not live to see the final product. This book could not have been written without the enthusiastic and cheerful cooperation of numerous former Greenpeace activists, many of whom not only allowed me to interview them and pore over various documents in their basements and attics but also extended their warm hospitality, including food and drink and sometimes even a place to sleep. My cosmic gratitude goes out to the late Dorothy Stowe, Ben Metcalfe, Jim and Marie Bohlen, Bob Hunter, and David McTaggart. An earthly but no less heartfelt thank you to Patrick Moore, Paul Watson, Rod Marining, Rex Weyler, Robert O. Taunt III, Steve Sawyer, Kelly Rigg, David Tussman, Michael M'Gonigle, and numerous others who were willing to share their past with me, either through conversation, personal papers, or both.

The Greenpeace offices in Washington, DC, Seattle, San Francisco, Toronto, and Hamburg, in addition to the Greenpeace International office in Amsterdam were kind enough to provide me with access to their archives and libraries. I would also like to thank the archivists and librarians who helped me find useful materials at the City of Vancouver Archives, University of British Columbia Special Collections, Archives Canada in Ottawa, the Commonweal Collection at the University of Bradford, and the Sophia Smith Collection at Smith College.

Researching an organization whose archives and former members are scattered across several continents and located in some of the world's more expensive cities was possible only through generous financial support at various stages of the project. I am exceedingly grateful to the following institutions for their vital assistance: the Fulbright Program, the Aspen Institute Non-profit Sector Research Fund, the German Academic Exchange Service, the German Historical Institute in Washington, DC, the University of Queensland New Faculty Research Fund, the Max Planck Institute for History in Göttingen, the Gerda

Henkel Foundation, the Lattie F. Coor Faculty Development Award at the University of Vermont, and the Rachel Carson Center for Environment and Society at Ludwig-Maximilians University in Munich.

Some sections of the book have appeared elsewhere in different form. I am grateful to the following editors and the publishers for allowing me to share my analysis and arguments in articles and for permitting me to republish them here: *Society and Animals* (Brill, 2012); *BC Studies: The British Columbian Quarterly* (2004); and *Historians and Nature: Comparative Approaches to Environmental History*, edited by Ursula Lehmkuhl and Hermann Wellenreuther and published in 2006 by Berg (an imprint of Bloomsbury). Many thanks as well to Greenpeace for granting me permission to reprint numerous photographs from their vast photo archive.

Several of my valued colleagues read either all or part of the manuscript at various stages and offered insightful criticism and suggestions. My thanks to Karl Brooks, Kurk Dorsey, Arn Keeling, Kristine Kern, Kieko Matteson, Christof Mauch, Karen Oslund, Adam Rome, Carl Strikwerda, James Turner, Graeme Wynn, and especially Donald Worster. Don was a brilliant and generous mentor during my days as a doctoral student at the University of Kansas, and he remains an inspirational colleague and friend. His scholarship inspired my initial interest in environmental history and is responsible for much of what I have subsequently learned. I am also grateful for the friendship and intellectual stimulation of numerous fellow graduate students who came to Lawrence to work with Don, including Jay Antle, Kevin Armitage, Lisa Brady, Kip Curtis, Sterling Evans, Michael French, Nancy Jackson, Matthew Logan, Neil Maher, Bruce Stadfeld McIvor, James Pritchard, Amy Schwartz, and Paul Sutter. Susan Ferber of Oxford University Press has been an exemplary editor, cutting and polishing a sprawling manuscript into a tighter, more focused, and much-improved book. I am also immensely grateful to Graham Burnett for sharing his superb book on whale science while it was still in manuscript form. Many thanks as well to Maribel Novo for transcribing numerous interviews and for many years of intercontinental friendship.

As a result of a somewhat peripatetic lifestyle, I have lived in four different countries while researching, writing, rewriting, and editing this book. Such an unmoored existence has its challenges. That the experience has been rewarding is due to the friendship, good company, and kindness of numerous people. As a long-term victim of alphabetical discrimination, I thank them here in reverse alphabetical order: Renee Worringer, Amani Whitfield, Kathy Truax, Olaf Tarmas, Ken Tanzer, Lisa Steffen, Felice Stadler, Christabelle Sethna, Nadine Requardt, Uwe Reising, Amanda Rees, Janette Philp, Rusty Monhollon, Abby McGowan, Kelly McCullough, David Massell, Laurie Kutner, Adrian Ivakhiv, Terrance Hayes, Katja Hartmann, Geoff Ginn, Andrew Gentes, Sean Field,

Sarah Ferber, Nina Ehresmann, Charles Closmann, David Christian, Kathy Carolin, Andrew Bonnell, Gesa Becher, and Jodi Bailey.

Eddie Zelko has frequently ridden to the rescue when uncooperative computers have threatened to derail my life. Over the years he has offered just the right blend of brotherly patience, good cheer, and skepticism. Like me, Eddie is the beneficiary of my parents' hard-won middle-class life in the Melbourne suburbs. War, political upheaval, and poverty denied them the benefits of an extensive formal education. Their lives were shaped instead by the rigors of emigration and factory jobs, which in turn made possible a fulsome education for their sons. Whether I have made the most of my parents' toil is perhaps disputable. However, my gratitude, respect, and love for them are not. I dedicate this book to Ana and Stefan Zelko.

Make It a Green Peace!

Introduction

Most of us probably don't know how it feels to unexpectedly have our head clamped in the jaws of a live killer whale. Not an angry or hungry killer whale, but a friendly killer whale in an aquarium. Bob Hunter, Vancouver's most famous hippie intellectual and one of the founders of Greenpeace, knew what it felt like. It happened to him in 1974 while he was visiting the Vancouver Aquarium at the behest of a whale scientist who was hoping to convince Greenpeace to mount a campaign against whaling. The experience changed his life: "I had been through marathon t-group therapy sessions and emotionally exhausting workshops with the great Gestalt therapist Fritz Perls, but neither experience had been so far out of the framework of my understanding that it left me as shaken as I was now." Hunter quit his job, divorced his wife, and devoted his life to saving "the serene superbeings in the sea," those exquisitely adapted creatures that "had mastered nature by becoming one with the tides and the temperatures long before man had even learned to scramble from the shelter of the caves."[1] Hunter's cetacean-inspired epiphany fired him with fierce conviction: the abominable practice of whaling had to end. This was not a view shared by all environmentalists.

In the early 1970s, the U.S. Congress House Subcommittee on Fisheries and Wildlife Conservation held a series of hearings on the subject of marine mammal protection. Among those who testified were representatives of America's oldest and most established wilderness protection groups, such as the Sierra Club, the Audubon Society, and the National Wildlife Federation. Although it was important to ensure that the world's populations of whales and seals remained as healthy as possible, these organizations argued, they did not support a policy of absolute protection. As long as the survival of the species was ensured, they believed, it was legitimate to use its "surplus" members for the benefit of people. In his testimony before the subcommittee, Thomas Kimball of the National Wildlife Federation employed phrases such as "renewable resources," "stewardship," and "professional wildlife management." The "harvesting of surplus wildlife populations," his organization maintained, was an "important management tool if the continuing long-range well being of an animal population is the ultimate objective."[2]

Bob Hunter would have none of this. A few years after these hearings, he and his fellow Greenpeacers came across a fleet of Soviet whaling boats off the coast of California. They leaped into several motorized inflatable dinghies and skimmed across the open ocean, zipping between whalers' harpoons and fleeing pods of sperm whales, in effect acting as human shields for the defenseless giants. Not long after that, these same individuals were scrambling across ice floes off the coast of Newfoundland, throwing their bodies over harp seal pups to save them from club-wielding seal hunters. Whaling and sealing, these ardent activists insisted, were not merely issues of wildlife conservation or resource stewardship. Rather, they were ecologically destructive and morally reprehensible acts that represented humanity's ignorance and thoughtless cruelty toward other sentient life-forms.

How should we understand these activists, whose impassioned antics under the Greenpeace banner challenged the heretofore staid conservation-oriented discourse of wildlife protection—and the standard repertoire of environmentalism in general—and supplanted them with a form of nonviolent protest and countercultural holism that has influenced environmentalism ever since? Hunter had felt the firm clamp of orca teeth on the back of his skull, but even this dramatic experience cannot fully explain his commitment.[3] What else inspired him and his fellow activists to take such drastic, self-imperiling actions to protect other species? How did they come to hold such uncompromising views? In an effort to answer such questions, this book explores the complex roots of Greenpeace, tracing the development of the organization from its emergence amid the various protest movements of the 1950s and 1960s to the end of its volatile, dramatic, and at times quirky first decade in 1980.

Since the beginning of the 1970s, no single organization has done more than Greenpeace to bolster and reshape environmental protest around the world. Its founders were the first environmentalists to adopt the Gandhian nonviolent protest strategies employed by the peace and civil rights movements. They combined this with the Quaker notion of "bearing witness"—the idea that a crime or atrocity can be challenged by observing it and reporting it to others—and hitched it to a media strategy heavily influenced by Marshall McLuhan, the Canadian communications scholar who developed such enduring concepts and aphorisms as "the global village" and "the medium is the message." In addition, Greenpeace's founders were self-conscious internationalists, a stance motivated both by a form of postnationalist romanticism that envisioned a world without borders and by the ecological imperative that nature did not recognize the artificial boundaries of nation-states.

Greenpeace also made environmentalism look cool. Its vivid and confrontational protest style resonated with the antiwar demonstrators of the 1960s and 1970s, while its iconic imagery and links with popular musicians challenged older stereotypes that associated environmentalism with middle-aged Sierra Club hikers in corduroys and cardigans. The new Greenpeace-inspired environmentalists wore tie-dyed T-shirts and long hair, smoked dope and dropped acid, and fomented a consciousness revolution that sought nothing less than a radical change in Western culture. For better or worse, this hip, edgy, occasionally somewhat flaky image continued to linger long after Greenpeace had largely abandoned its more eccentric countercultural traits. The fact that Greenpeace became for many people a kind of synecdoche of environmentalism, therefore, means that in some quarters environmentalism continues to be tainted by its association with the sixties counterculture.

By the early 1980s, Greenpeace had grown into an international environmental powerhouse centered in Europe, with a complex hierarchical—some might say "corporate"—structure and branch offices in numerous countries. Today, it is one of the planet's most recognized environmental groups. Its logo is almost as familiar as those of Coca-Cola and McDonald's.[4] Its present-day activities involve a variety of campaigns, from lobbying governments and intergovernmental agencies, such as the International Whaling Commission, to sponsoring the production of new technologies, like environmentally friendly refrigerators and automobiles. While its prominence and influence are undeniable, the institution that Greenpeace has become is not necessarily what its founders had in mind. Throughout its early history, there were moments when Greenpeace could have taken different paths. Some might have led to its demise, while others may have allowed it to develop as more of a grassroots social movement. Despite its unpredictable evolution and the internecine struggles that gave rise to its present form, it has retained the direct-action style that first set it apart in the 1970s and imbued its activists with unique élan.

Drawing on a wide-ranging set of sources, from newspaper articles, meeting minutes, internal correspondence, and numerous interviews with former Greenpeacers, to philosophical writings, manifestos, and personal accounts by prominent and lesser-known thinkers, including Bob Hunter, this book investigates the diverse ideologies and outlooks that gave Greenpeace its distinct character from its founding forward. How did its origins shape the path the organization took? To what extent has it lived up to the vision and ideals of its founders? And how have its multifaceted origins proved both inspiring and problematic in its evolution and present incarnation?

By situating Greenpeace within the context of the postwar peace movement and the sixties counterculture and examining its spectacular rise on the world scene through simultaneously quixotic and muscular media-savvy campaigns, this book seeks to provide a much deeper and more comprehensive understanding of the organization's brand of radical, direct-action environmentalism than other scholarly and journalistic works. Previous studies of Greenpeace can be broken down into several broad categories: members' memoirs and autobiographies, which seem to be something of a cottage industry among former participants and provide an indication of the unique role that Greenpeace has occupied in the environmental movement;[5] official and semi-official organizational histories;[6] media studies that analyze Greenpeace's skillful use of mass communications and these campaigns' effects, in turn, on the organization itself;[7] and sociological analyses of Greenpeace's strategies and structure.[8] There are also several journalistic accounts that describe how Greenpeace functions and some of the more prominent campaigns in which it has been involved.[9] As far as historical scholarship is concerned, there is an almost inexplicable incongruity between the minor consideration given the organization in the historiography of environmentalism and the significant role it has played in shaping environmental activism in the United States and abroad for the past four decades.[10]

The dearth of serious historical scholarship on Greenpeace means that certain important elements in the history of environmentalism have received little attention. Most prominent among these is the influence of the twentieth-century peace movement. Chapter 1, therefore, explores the "peace" half of Greenpeace through the lives of two of the organization's founding couples, Irving and Dorothy Stowe from Rhode Island and Jim and Marie Bohlen from Pennsylvania. The Stowes and Bohlens had strong connections with various pacifist and antiwar organizations, particularly those inspired by Quakers. These groups foreshadowed Greenpeace's protest tactics and strategies; in fact, Greenpeace's first campaign closely mimicked Quaker-organized antinuclear protests from the 1950s.

Chapters 2 and 3 introduce the protest culture that the Stowes and Bohlens found when they moved to Canada in the late 1960s to escape, in their eyes, the clutches of U.S. militarism. Much to their surprise, the city in which they settled—Vancouver—was simmering with alternative politics and grassroots activism, from antiwar groups to antipollution organizations. The Stowes and Bohlens, seasoned activists in their forties and fifties, brought to local peace and ecology movements the tactics and values they had learned from decades of activism. In turn, they found themselves interacting with younger activists—Canadians and Americans—whose ideas and lifestyles presented interesting challenges and opportunities. Many of these individuals would play a key role in the emergence of Greenpeace, chief among them Bob Hunter. Hunter was only in his late twenties but had already published three books—one novel and two ambitious works of cultural criticism—and was a columnist for the *Vancouver Sun*. A critical examination of Hunter's texts and columns reveals a fertile mind with a penchant for grand theory. Apart from his considerable role in shaping Greenpeace's innovative and controversial media strategies, Hunter did more than anyone to inject Greenpeace with the spirit and values of the sixties counterculture and holistic ecology.

Chapter 4 narrates and examines Greenpeace's first campaign—an attempt to disrupt an underground nuclear explosion by sailing into the testing area, a remote region of the Aleutian Islands southwest of Alaska. The maiden voyage was important in forging Greenpeace's identity and revealed fundamental tensions among the group's many founders. The most obvious of these was the split between the older generation of peace movement protestors who were inclined toward a sober and respectable form of scientific rationalism and a group of younger activists who embraced various countercultural beliefs and values. The participants labeled this dichotomy the "mechanics versus the mystics," and it would remain a fundamental cleavage within the organization throughout the 1970s. The Aleutian voyage also inspired the birth of one of Greenpeace's core myths—the idea that they were the "warriors of the rainbow," a reference to a Native American prophecy that foretold the coming of a band of earth warriors who would save the world from environmental destruction.

From 1972 to 1974, Greenpeace directed its attention toward French nuclear testing in the South Pacific, a campaign taken up in chapters 5 and 6. Ben Metcalfe and David McTaggart, the two most prominent figures during this period of Greenpeace's history, were in many respects completely unrepresentative of the nascent organization's internal culture. Metcalfe was a wily and cynical journalist who felt that the manipulation of public opinion by elites offered the only hope for substantive social and political change. Despite this—or perhaps because of it—he proved to be a more than adequate exponent of "mind bombing," a media theory that Bob Hunter had derived from Marshall McLuhan. McTaggart was a conservative forty-year-old resort developer with no prior connection to either the peace movement or the counterculture. Yet he successfully sailed his little ketch all the way from New Zealand to the test site near Tahiti twice, in the process proving to be a considerable headache to the French military. He went on to play a major role in the formation and running of Greenpeace International.

The South Pacific campaign represented an early attempt at global environmental activism. It required coordinating activists in several countries and led to the establishment of the first Greenpeace group outside Canada—in New Zealand. While McTaggart was taking on the French Navy, other Greenpeacers were organizing protests in New York, London, and Paris, before eventually setting up camp at the United Nations Conference on the Human Environment in Stockholm, the UN's first major conference on international environmental issues. Thus the South Pacific campaign marked Greenpeace's entry into the arena that political scientist Paul Wapner has called "world civic politics": a level of politics where the promotion of broad cultural sensibilities represents a mechanism of authority that is able to shape human behavior.[11]

In 1975, Greenpeace underwent a dramatic change in its campaign focus, philosophy, and membership base. Until then, it could best be described as an antinuclear group with an environmental emphasis. By deciding to mount a direct-action campaign against whaling, however, the organization embraced a biocentric philosophy that challenged the widespread idea that humans were the supreme beings on the face of the planet. They vividly illustrated their commitment to this notion by placing their bodies between pods of fleeing whales and the explosive harpoons of the Soviet whaling fleet. In the process, the activists captured a series of stunning images that would make them renowned throughout the world, giving them entrée into the world's most lucrative environmental "market"—the United States.

Chapters 7, 8, and 9 explore this spectacular and timely entrance into the antiwhaling movement. In addition to offering a historical account of changing attitudes toward whales, this section explores the rather surprising and bizarre connections between cetacean research, Cold War bioscience, neuroscience, and sixties counterculture. Greenpeace and other antiwhaling activists based their opposition to whaling on the controversial notion that whales and dolphins represented a form of higher intelligence, a "mind in the waters," to use the title of a famous book on the subject.[12] However, by arguing that all whales deserved to be saved because of their supreme intelligence, Greenpeace was unwittingly invoking the great-chain-of-being worldview that ranked a species' worth according to whether it possessed traits humans value: intelligence, advanced communication skills, and a theory of mind. Such a view was at odds with the ethics of holistic ecology and biocentrism, in which nature has an intrinsic value independent of humans. The tension between this great-chain-of-being approach and Greenpeace's professed biocentrism would became particularly evident when various Native American tribes—the inspiration behind Greenpeace's "warriors of the rainbow" image—demanded that they be allowed to continue their "traditional" practice of hunting whales.

In 1976, with new Greenpeace branches springing up throughout North America, the organization's attention shifted from the Pacific to the Atlantic. Once again, its goal was to protect marine mammals, this time the adorable harp seal pups born each winter on ice floes off the coast of Newfoundland and Quebec. Chapters 10 and 11 focus on this campaign, an early example of the kind of imbroglio that environmentalists have become entangled in countless times over the past few decades, in which their interest in saving a species or preserving a habitat conflicts with the needs of local working people for whom that species or habitat constitutes a resource. The antisealing protests attracted a substantial number of animal rights activists to Greenpeace. From an animal rights perspective, there can be no question of compromise when it comes to the killing of whales or seals—abolition is the only goal worth pursuing. Therefore, those within Greenpeace who were willing to take a more pragmatic, ecological approach to sealing found that they had almost as much trouble with some of their own supporters as they did with the sealers. Despite these problems, Greenpeace's leaders—Bob Hunter in particular—made a concerted effort to work with the mostly impoverished Newfoundlanders who constituted the sealing industry's labor force. The fact that they failed was due as much to the intransigence of the sealers and their supporters within the Canadian government as to Greenpeace's own shortcomings. At the same time, the questionable actions of some of its activists forced Greenpeace to define the acceptable boundaries of its direct-action approach.

By 1978, Greenpeace was beginning to experience severe growing pains. The original Vancouver group, who considered themselves the leaders of the rapidly expanding, loosely controlled organization, was facing a mountain of debt from several years of nonstop campaigning. Meanwhile, the various American offices, particularly the relatively wealthy San Francisco branch, began to chafe at what they perceived as Vancouver's authoritarian leadership. The controversial antisealing campaign had caused a significant decline in Canadian donations to the Vancouver group, forcing them to rely on the San Francisco office to underwrite their operations. When Vancouver tried to tighten its control over the various Greenpeace branches worldwide, they met considerable resistance from the Americans. The result was an acrimonious legal battle in which the Vancouver branch of Greenpeace sued its brethren in San Francisco. The nascent European groups, led by Vancouver native David McTaggart, took advantage of the opportunity to consolidate their power. The final chapter describes the complex series of deals that gave rise to the European offices' emergence as the leaders of Greenpeace, and the shift of power from the Pacific coast of North America to the countries bordering the North Sea. One result was that Amsterdam, Hamburg, and London replaced Vancouver and San Francisco as Greenpeace's most important offices, which has continued to the present day. Another was McTaggart's attempt to strip the new organization of some of the more flamboyant countercultural tendencies it had inherited from the Vancouver hippies, an effort that was only partially successful.

The Greenpeace story is worth telling merely for its abundant drama, pathos, and absurd moments of comic relief. Beyond that, it provides many insights into environmentalism, social movements, and the history of protest in the twentieth century. Greenpeace never became the revolutionary, world-changing movement that its idealistic founders hoped it would; yet, there is no doubt that it has successfully and enduringly highlighted environmental problems in ways that no other group has managed. In the process, it revealed some of the cracks and fissures in the broad structural constraints—such as global capitalism and the mechanistic and instrumental view of nature—that influence peoples' thoughts and actions in the modern world. As some of the organization's founders quipped during their more sanguine moments, this was quite an achievement for a bunch of peaceniks and hippies from a medium-sized city on the west coast of Canada.

1

Speak Truth to Power

Whatever else people might think of Greenpeace as an organization, few would deny that it has a brilliant name. Simple, elegant, and richly evocative, it rolls easily off the tongue and fits comfortably into a one column headline. In the words of a Canadian journalist writing in 1977, "the name shimmers with images of tranquil pastures and Bertrand Russell, lotus petals and Mahatma Gandhi."[1] From a historian's perspective, however, the name's simple elegance can sometimes be misleading. For most of its history, Greenpeace has been perceived primarily as an environmental organization. "Green" not only comes first but is also emphasized in the pronunciation: We say *Green*peace, rather than Green*peace*. However, this emphasis on "green" at the expense of "peace" obscures some important aspects of the organization's early history.

Greenpeace became a predominantly environmental organization in the mid-1970s when it embarked on its famous antiwhaling campaign. This was also the first campaign to receive widespread coverage in the United States. For most Americans, therefore, Greenpeace first appeared on the scene as a Canadian environmental protection organization operating on the high seas. The antiwhaling campaign was soon followed by the high-profile protests against the annual harp seal slaughter off the coast of eastern Canada. Greenpeace, therefore, was never as "American" as the Sierra Club or the Audubon Society, and the organization's international focus further compounded this perception. A closer examination of Greenpeace's origins, however, reveals a different picture. The organization may have started life in Canada, but to a large extent, its activist roots lie south of the 49th parallel. This becomes especially clear when one situates Greenpeace within the history of the post–World War II American peace movement, particularly the wing of the movement that employed nonviolent action as a vital component of its protest repertoire. From this perspective, the Greenpeace story is part of a broader history of global protest and activism that includes anticolonial struggles, particularly in India, and various pacifist and antiwar movements throughout the world. The people who can arguably lay greatest claim to be Greenpeace's "founders," Irving and Dorothy Stowe and Jim and Marie Bohlen, were veterans of these movements. These Americans were deeply immersed in the culture and tradition of the postwar peace movement. Not surprisingly, the organization they helped create bore the indelible stamp of that tradition. An examination of their lives, and of the peace movement of which they were a part, reveals that while Greenpeace may have been born in British Columbia, its origins are more cosmopolitan than such a birthplace suggests.

There seems to be a certain percentage of the population—conservatively around 10 percent—for whom loquaciousness is a genetic condition. If this 10 percent were broken down into the category of people most likely to start making a speech at any given time, then Irving Stowe would probably have fallen into the ninety-ninth percentile.[2] Fellow Greenpeacer Bob Hunter, no slouch in the art of rhetoric, remarked that Stowe would make a speech if you said, "Good morning, Irving." Stowe's wife, Dorothy, recalled that it was not uncommon for Irving to deliver an impromptu speech if a speaker failed to turn up for an engagement, and his off-the-cuff remarks were often articulate, logical, and trenchant. On the flip side, however, there are some who remember how Stowe's speeches, impromptu or otherwise, could degenerate into interminable rants.[3]

Stowe's passionate need to air his views was not limited to the spoken word. He was also a prolific and indefatigable letter writer. A restless sleeper, Stowe would usually be awake between 2 and 5 a.m. and would make use of his insomnia by shooting protest letters off to politicians, government officials, editors—whoever had rankled his moral sensibilities. One night he might pen a lengthy letter to President Eisenhower or the secretary of defense, criticizing the decision to employ nuclear-powered Polaris submarines; the next would be devoted to a series of terse notes to various newspaper editors decrying their obsequious political commentaries. Stowe had strong opinions on many subjects, including race relations, labor unions, urban planning, pollution, vegetarianism, and smoking. Had he confined himself to rhetoric, he might be dismissed as just another self-righteous, opinionated crank. But to his credit, Stowe followed through on many of the issues that concerned him, joining groups that were fighting for a particular cause. If the appropriate group did not yet exist, he formed it himself. It was this unrelenting critical drive and commitment to activism that would lead Stowe, along with his like-minded wife, Dorothy, to play a major role in the founding of a unique new organization, one that brought together important elements of the peace and environmental movements and combined them with the philosophy of nonviolent direct action.

Born in 1915 into a middle-class, Jewish family in Providence, Rhode Island, Irving Stowe was the product of a culture that produced some of the most gifted intellects and committed activists in twentieth-century America. It was from this social milieu, forged in the crucible of the northeastern urban centers, that figures such as Noam Chomsky, Barry Commoner, Howard Zinn, Irving Howe, and Arthur Miller, as well as countless lesser-known writers, intellectuals, scientists, and activists, emerged. The rich Talmudic literary tradition and the Jewish emphasis on education fostered a critical intellectual and activist culture that became closely associated with, and a major contributor to the development of, the political Left in the United States. Stowe grew up within this mostly secular Jewish milieu in Providence, attending Brown University, where he majored in economics, before going on to study law at Yale.[4]

In 1951, Stowe was introduced to Dorothy Rabinowitz, a union activist and social worker who, like Irving, was a product of Providence's Jewish middle class. After attending Pembroke College, the women's college at Brown, Dorothy helped organize a social workers union in Rhode Island and became its first president. It took Irving a couple of years to convince Dorothy to marry him, and they wed in 1953. By this time, Irving, who had never been very interested in Judaism, had joined the Unitarian church, so a traditional Jewish wedding was not something he nor, for that matter, Dorothy particularly desired.[5] They compromised by having a rabbi preside over what was otherwise a civil ceremony. One of Irving's great passions was jazz. With his self-confident and gregarious nature, he was never afraid to approach famous people and became quite friendly with many of the musicians he admired. George Shearing, the blind British jazz pianist, was the best man at their wedding. The dinner celebration took place at the premises of the local branch of the National Association for the Advancement of Colored People, another organization in which Irving had become active.[6]

Throughout the fifties, Irving and Dorothy continued to be involved in a variety of causes. In addition to those already mentioned, Irving joined the Mental Health Association, whose bylaws and constitution he helped draft, and became a legal representative for injured merchant seamen, suing several major oil corporations on behalf of seamen who had been injured due to the companies' neglect. He was also the chairman of the Rhode Island branch of the American Veterans Committee, a left-wing version of the American Legion. Stowe was a keen supporter of the United Nations, seeing it as the best mechanism for insuring an international and lasting peace. On occasion he received threatening letters from overheated patriots who saw such "One Worlders" as part of an international communist conspiracy.[7]

With their activism in a panoply of progressive causes, the Stowes were in many ways the archetypal activist couple of the McCarthy era—middle-class, urban, highly educated, left-wing, secular Jews. But of all the causes they joined, the Quaker-inspired peace movement would have the greatest impact upon their lives.

The history of the peace movement in the United States owes much to the work of dissenting religious groups, particularly Quakers. One of the fundamental tenets of Quakerism was pacifism. Since every person had direct access to God, everyone was a potential channel of truth, no matter how misguided he might seem at any given moment. Violence against people, therefore, only served to suppress love, truth, and freedom. This testimony led the Friends to oppose all wars and preparations for wars. Many refused to pay war taxes or to be conscripted, actions that frequently incurred the wrath of non-Quakers during wartime. Another form of protest was the notion of "bearing witness," which involved registering one's disapproval of an activity and putting moral pressure on the perpetrators simply through one's presence at the scene. When the United States entered the First World War in 1917, several prominent Quakers, including the writer and philosopher Rufus Jones and Henry Cadbury, a Haverford College professor and descendent of the eponymous chocolate-manufacturing dynasty, formed the American Friends Service Committee (AFSC) in an effort to bring greater organizational focus to Quaker pacifism. The AFSC aimed to harness traditional Quaker sentiments and channel them into promoting peace and justice. Within six months of its formation, the committee trained over a hundred men and women for civilian relief work in Europe. It also created a social order committee, which contained study groups focusing on broad issues such as the structural causes of poverty and the problems of industrialization and the American military buildup. In the interwar years, the AFSC broadened its social activism, feeding the children of striking Appalachian miners and helping to organize agricultural and craft cooperatives among sharecroppers and workers who were victims of the Great Depression. Although its critique of poverty, corporate power, and militarism placed the AFSC in the camp of the ideological Left, it went to great lengths to ensure that it remained politically nonpartisan.[8]

Quakers, with their pacifist roots, believe that political protest must always be nonviolent. However, since nonviolent action encompasses so many possible methods and activities, it is difficult to define the term with any precision.[9] In his classic study of the topic, Gene Sharp proposed the following broad definition:

> Nonviolent action is a generic term covering dozens of specific methods of protest, noncooperation and intervention, in all of which the actionists conduct the conflict by doing—or refusing to do—certain things without physical violence. As a technique, therefore, nonviolent action is not passive. It is *not* inaction. It is *action* that is nonviolent.[10]

Nonviolent action, therefore, is not synonymous with pacifism. Nor is it identical to religious and philosophical systems emphasizing nonviolence as a matter of moral principle. All pacifists, it is safe to say, practice nonviolence in their protest, but not all nonviolent protestors are pacifists. What for some people is the ultimate fulfillment of a belief system is for others merely a useful technique—a means to an end.

Nonviolent action, according to Sharp, is based on an implicit view of the nature of political power: "That governments depend on people, that power is pluralistic, and that political power is fragile because it depends on many groups for reinforcement of its power sources." Groups that advocate political violence, on the other hand, tend to take the view that political power is monolithic and stems from a few dominant people or institutions.[11] Intuitively, therefore, one would think that nonviolent action would most likely succeed—and would be least likely to meet violent resistance—in open and democratic societies such as the United States. Not surprisingly, therefore, there is a rich tradition of nonviolent action in American history, from the American colonists' refusal to pay taxes to the British, to Henry David Thoreau advocating civil disobedience in the face of "immoral" laws and institutions such as slavery. There are also, however, numerous examples of nonviolent action being successfully employed in totalitarian states, with the so-called Velvet Revolutions against communism in Eastern Europe constituting a prominent example.

Sharp identifies three broad categories of nonviolent action: The first is nonviolent protest and persuasion. This mostly involves the use of symbolic actions to express disapproval and dissent. Examples include such oft-used techniques as marches, parades, and vigils. Sharp's second category is noncooperation. This includes the withdrawal or withholding of social, economic, or political cooperation through strategies such as strikes, economic boycotts, and a refusal to participate in elections. The third class is referred to as nonviolent intervention and involves activities such as sit-ins, nonviolent obstruction, nonviolent invasion, and setting up a parallel government.[12] One of the defining features of postwar social movements was the frequent and innovative use of nonviolent intervention, particularly among civil rights and antiwar protestors, a development that owed much to the Indian struggle against British colonialism led by Mohandas Gandhi.

The person who was most responsible for introducing Gandhi's philosophy of nonviolent protest, or *Satyagraha*, to North America was a Quaker named Richard Gregg. Satyagraha, which derived from the Hindi term for "truth-force," evolved from Gandhi's long struggle for Indian independence and became the foundation for most of the nonviolent direct action movements of the twentieth century. Its key principles included refusing to return the assault of an opponent, refraining from insulting opponents, not resisting arrest, and behaving in an exemplary manner if imprisoned. From these precepts Gandhi derived an escalating program of nonviolent protest measures, from negotiation and arbitration through to agitation, strikes, civil disobedience, usurping the functions of government, and, finally, setting up a parallel government. Few in America had heard of Satyagraha before the mid-1930s when Gregg, after returning from a four-year stint at an Indian spiritual retreat, published his landmark *The Power of Nonviolence*. Gregg argued that nonviolent resistance was not only a morally superior form of protest, but also a more effective means of creating social change and resisting aggression. Although his book was widely read, Gregg's ideas were apparently too radical for their time; none of the major peace groups attempted to use them to block U.S. involvement in the Second World War. Nevertheless, Gregg played a crucial role in inculcating the North American peace movement with Gandhian principles and strategies, particularly with his emphasis on the performative aspects of nonviolent action, which could elicit sympathy from onlookers and guilt and shame from opponents.[13]

The U.S. decision to drop atomic bombs on Hiroshima and Nagasaki was met with widespread public approval at the time. Irving Stowe and Dorothy Rabinowitz were shocked and appalled by the event, but they were in the minority. A Roper poll conducted in the fall of 1945 showed that 53.5 percent of those surveyed thought it had been the best action under the circumstances and, astoundingly, a further 22.7 percent felt that the United States should have quickly dropped many more of the bombs before Japan had a chance to surrender. However, as historian Paul Boyer has shown in his analysis of the cultural response to the atom bomb, such feelings were soon replaced by a general mood of fear as the United States began a series of tests at Bikini Atoll in 1946.[14]

Throughout the 1950s, peace groups such as the War Resisters League (WRL), Peacemakers, the AFSC, and others continuously staged protests and engaged in educational and organizational activities to highlight the dangers of nuclear war and to draw people's attention to the connections between militarism, poverty, racism, colonialism, and capitalism.[15] This activity was particularly important during the early 1950s when, in Boyer's words, "the obsessive post-Hiroshima awareness of the horror of the atomic bomb had given way to an interval of diminished cultural attention and uneasy acquiescence in the goal of maintaining atomic superiority over the Russians." During the difficult McCarthy years, the peace groups fanned the embers of the movement, keeping it alive until a change in the public mood provided an opportunity for it to flourish once more. This change duly arrived in the mid-1950s, when a further series of tests on Bikini Atoll forced the emergency evacuation of people on the nearby Marshall Islands, as well as sickening and eventually killing several Japanese fishermen who had not been warned to stay out of the area. When radioactive rain fell on the Midwest in 1955, the consequences of nuclear testing finally hit home in a palpable way.[16]

In June of 1955, a small group of protestors from WRL and Peacemakers, as well as several other groups, staged the nation's first civil defense protest in New York City. Pacifists had long ridiculed civil defense plans as being, in the words of the veteran peace activist A. J. Muste, "an integral part of the total preparation for nuclear war." Prominent peace activists such as Dave Dellinger, Bayard Rustin, Jim Peck, and Muste were among those arrested for refusing to take shelter during the annual air raid drill. The action was soon replicated in Philadelphia, Chicago, and Boston. Over the next decade, it became a regular part of the pacifist protest repertoire.[17] Nonviolent direct action also became an integral part of the civil rights struggle, and there was considerable overlap between the two movements in terms of their values, tactics, and personnel. As early as 1947, for example, peace activists from A. J. Muste's venerable pacifist organization, the Fellowship of Reconciliation, assisted the Congress of Racial Equality in a two-week "Journey of Reconciliation" through the upper South. The action represented the first biracial attempt to test the recent Supreme Court decisions against segregation in interstate travel. For men such as Muste, Peck, and Dwight Macdonald, the struggle against racial oppression constituted a logical extension of their radical pacifist philosophy. For them, the fight for civil rights and nuclear disarmament were both integral parts of the broader struggle for human dignity and survival.[18]

Among the numerous groups to emerge from the peace movement in the postwar era, the one that most closely prefigured Greenpeace was the Committee for Nonviolent Action (CNVA). Established in 1957 by a group of Quakers who were looking for a more dynamic form of protest than the AFSC was willing to risk, the new organization brought together representatives of all the major pacifist groups with the purpose of conducting nonviolent direct actions that were too difficult for local groups to carry out on their own. Their first action was a civil disobedience campaign at a Nevada nuclear weapons testing facility in 1957 in which eleven pacifists were arrested for crossing over into a prohibited area.[19] The *New York Times* commented that this action "marked the unusual employment in this country of the 'civil disobedience' tactics made famous by M.K Gandhi."[20] According to Gene Sharp's taxonomy of nonviolent actions, such protests, which he refers to as "nonviolent invasions," constitute a significant step in the evolution of a more radical form of nonviolent protest, entailing civil disobedience and the risk of severe repression.[21]

In 1958, CNVA initiated one of its most effective and, in retrospect, influential actions when it organized a yacht, the *Golden Rule,* to sail into the Eniwetok test zone in the Pacific. The *Golden Rule*'s captain, Albert Bigelow, had been the captain of a destroyer escort in the U.S. Navy during World War II but had become a pacifist after Hiroshima and Nagasaki. After the war, he became the housing commissioner of Massachusetts but was continuously "seeking some sort of unified life-philosophy or religion" that expressed his deeply felt religious pacifism. In 1952, he resigned his commission in the Naval Reserve a month before he would have become eligible for his pension. Then in 1955, his family hosted a pair of "Hiroshima maidens"—young women who had been disfigured during the atomic blast and whom Norman Cousins and other leading pacifists had brought to the United States for plastic surgery. At that time, Bigelow, after many years of flirting with Quakerism, decided to join the Society of Friends. He immediately became an important and influential figure in the peace movement, befriending leaders such as A. J. Muste and Bayard Rustin and becoming involved in groups such as the Committee for Nonviolent Action. In 1957, Bigelow was among several leading pacifists who were arrested for trespassing onto the grounds of an Atomic Energy Commission (AEC) test site in Nevada as part of a CNVA action. In February 1958, he announced, via the pages of *Liberation* magazine, that he intended to sail into the U.S. atomic testing zone.[22]

Bigelow's inspiration was British Quaker Harold Steele who, in the spring of 1957, came up with the idea of sailing a yacht into the British nuclear test site on Christmas Island (Kiritimati) in the middle of the Pacific. He managed to get as far as Tokyo but was unable to organize a ship and a crew. Bigelow and other Quakers were inspired by Steele's efforts and decided to mount their own voyage to the Marshall Islands in order to oppose the series of blasts, known as "Hardtack," which the U.S. Atomic Energy Commission had scheduled in 1958. It would be a symbolic act based on Gandhian principles of nonviolence. Their protest, it was hoped, would act like "a magnifying glass to focus the rays of conscience" on the immorality of nuclear weapons.[23]

Bigelow and his crew of three set sail from California in their thirty-foot ketch and made it to Hawaii before encountering resistance from the AEC. At the time, there were no laws that specifically barred people from entering the atomic proving grounds; it was simply assumed that military warnings and the threat of radiation sickness would keep people from deliberately sailing into the site. However, once it became clear that the crew of the *Golden Rule* was determined to sail from Hawaii toward the Marshall Islands, the AEC embarked on a series of tactics, which a U.S. Court of Appeals would later find illegal, to prevent the boat from leaving Honolulu. They issued, without public hearing, a regulation that made it a crime for U.S. citizens to enter the test area and then arrested Bigelow and his crew for stating that they intended to violate that regulation. The crew of the *Golden Rule* refused bail on principle and were kept in a Honolulu prison for most of the summer while awaiting trial.[24] While Bigelow tried to shift the focus of the trial onto the morality of nuclear testing, arguing that it was comparable to Nazi medical experiments and that it constituted "contamination without representation," the prosecution argued that Bigelow's violation of the AEC's newly created regulation would cause "irreparable damage" to the security of the United States and the "free world." In his final judgment, Hawaiian Federal Court Judge Jon Wiig questioned the very idea of civil disobedience. Quoting from a 1911 U.S. Supreme Court judgment, he stated: "If a party can make himself a judge of the validity of orders which have been issued, and, by his act of disobedience, set them aside then the courts are impotent." He gave the crew a sixty-day suspended jail sentence and a one-year probation period, which they promptly violated by announcing their intention to continue their journey to the test site. The local media, meanwhile, accused the crew of being communist stooges, while the mainland newspapers largely ignored the event altogether.[25]

Bigelow's efforts may have received minimal coverage outside Hawaii, but they became legendary among Quakers and pacifists nationwide. Protestors formed picket lines outside federal and AEC buildings in various cities during Bigelow's trial, bearing placards reading: "No contamination without representation" and "Free the *Golden Rule*." One man was inspired enough to try to emulate the *Golden Rule* and to complete the voyage it had been forced to cut short. Earle Reynolds was a physical anthropologist from Mississippi who had worked for the AEC after the war. For four years he had studied the effects of radiation on human growth in Hiroshima and Nagasaki. While in Japan, he also examined the American military's rationale for using atomic weapons against the Japanese. He came to the conclusion that neither bomb could be justified from a strictly military point of view, let alone a moral one, and that the Nagasaki bomb in particular was a callous and opportunistic act on the part of the military to test one more bomb "in the field" before the war was over. Upon investigating the American weapons testing program, Reynolds was shocked to discover that the U.S. government had refused to allow the World Court to rule on the legality of using the Marshall Islands as a nuclear proving ground. The more he read about the issue, he recalled, "the more unhappily amazed I became at what my government has been doing, in my name."[26]

In 1958, after completing a three-year, round-the-world voyage in their yacht, the *Phoenix*, Reynolds and his family arrived in Hawaii just as the *Golden Rule* crew was about to go on trial. Reynolds had never considered himself an activist, let alone the kind of person who would break the law to prove a point, but a meeting with Bigelow and his crew made a powerful impression on him. He found that the Quakers possessed an honesty and integrity that was beyond question and they had no reservations about their actions. "This country could use a lot more men like them," he recorded in his journal. Once it became clear that the *Golden Rule* would not be allowed to sail to the Marshall Islands, Reynolds and his family decided to go in its stead. Listing their official destination as Hiroshima, they managed to sail into the test zone near Bikini before being arrested by the U.S. Coast Guard and flown back to Hawaii. The arrest flouted a number of U.S. and international laws—lack of a warrant, failure to be told the reason for arrest, illegal transportation to a foreign soil, and several others.[27] Reynolds, like Bigelow and others in the peace movement, was naïve enough to believe, or at least hope, that these legal technicalities, combined with the moral superiority of his position, would be enough to exonerate him of any guilt, as well as enabling him to strike a blow against the nuclear testing program. He was genuinely shocked, therefore, that his government would embark on a series of ad hoc maneuvers of dubious legality in an effort to stop his protest.

In 1960, CNVA organized Polaris Action, a series of protests over several months at the Connecticut docks where the nuclear-armed Polaris submarines were being constructed. Polaris Action's nonviolent direct-action style strikingly foreshadowed the tactics that Greenpeace would adopt more than a decade later. The activists—many of them young men who self-identified as beatniks—repeatedly tried to enter the submarines docks. They also paddled boats bearing peace messages into the path of launching vessels. On several occasions, protestors attempted to board the submarines, and two of them managed to succeed by swimming through the frigid November waters and climbing up the submarine's guide ropes. This event garnered considerable media attention but also earned the activists a stiff nineteen-month jail term. Such innovative and daring direct action tactics were supplemented by an educational campaign aimed at dissuading locals from participating in the manufacture of the submarines.[28]

Polaris Action did not exactly transform the United States into a nation of pacifists, but it was nonetheless one of the most high profile of the early antinuclear nonviolent direct-action protests. After a while, some of the townspeople in New London began to express sympathy for the demonstrators. A small number of Polaris workers even offered to quit their jobs if CNVA could find them alternative employment. The continuous media coverage of the protests ensured that public awareness of the issue remained high, particularly in New England, where a pacifist sloop called the *Satyagraha* dropped anchor at ports throughout the Northeast to reinforce CNVA's actions. Using tactics that Greenpeace would successfully employ in future decades, the *Satyagraha* sailed along the coast, stopping at port towns along the way to educate people about nuclear issues in general and on the Polaris submarine in particular. At each port of call, new people would be enlisted into CNVA's ranks, providing hundreds of college students with their first taste of civil disobedience. Throughout the late 1950s and early 1960s, CNVA continued to organize actions that helped attract new people to the radical peace movement. In 1960 and 1961, they sponsored the San Francisco to Moscow Walk for Peace, an action that involved thousands of people throughout the United States and Europe who would join the march as it came through their hometown or region.[29]

Polaris Action and the voyages of the *Golden Rule, Phoenix,* and *Satyagraha* can be seen as the direct forerunners of Greenpeace. Their tactics inspired the first Greenpeace campaign and their values, with their emphasis on nonviolence and "bearing witness," would also become Greenpeace's core values. The CNVA protests were both a culmination of several decades worth of evolution within the peace movement and a springboard for the social movements of the 1960s. In varying proportions, they blended Quaker pacifism, the Marxist critique of capitalism, anarchism's suspicion of centralized power, an emerging countercultural sensibility, and Satyagraha into a forceful critique of Cold War militarism and a striking set of protest strategies. Many of these ideas and tactics were later adopted, in various forms, by antiwar, civil rights, and environmental groups, and often it was people such as Irving Stowe who conveyed them to a new generation of protestors.

For Irving Stowe, as for so many others, the development of the atom bomb was a seminal event. The bomb cast a great shadow over humanity and had an enormous political and cultural impact throughout the world. For those in the peace movement, nuclear weapons made war unthinkable, inspiring a sense of urgency that helps explain the more radical actions of the postwar era.[30] Nevertheless, the majority of people managed to push such fears to the back of their minds. Those in industrialized democracies such as the United States took comfort in the economic prosperity that characterized the postwar era, withdrawing into the protective shell of middle-class domesticity.[31] However, a small number were not prepared to sit back and entrust the world's fate to politicians and generals. Suffice to say that Irving and Dorothy Stowe were among those who decided that they could not sit passively by while the world seemed to move inexorably toward nuclear warfare.[32]

A New Deal liberal, Stowe felt betrayed by the postwar conduct of the United States. The willingness to use the atomic bomb, along with the Truman administration's hard line toward the Soviet Union and its crackdown against organized labor, angered Stowe and led him to search for alternative organizations through which he could better express his political sentiments. He also became angry at the Unitarian church, which he had joined because of what he felt were its enlightened antiwar views, when some of its ministers supported the idea of universal military training. Disillusioned by liberalism but not attracted to the hard Left, Stowe searched for a respectable way to express his discontent. The Quakers, with their adherence to pacifism, were an acceptable option for a Rhode Island Tax Commission lawyer, and Irving began to attend the local Friends' meetings.[33]

Stowe's Quakerism was cultural rather than religious. He and Dorothy attended Friends' meetings because they believed they offered a set of values that not only opposed militarism but also presented a structural critique of its causes and a well-defined set of protest strategies. In 1955, the American Friends Service Committee published a lengthy pamphlet entitled *Speak Truth to Power*, which offered a pacifist critique of U.S. foreign and domestic policy and sketched out a path to a peaceful society based on nonviolent principles. The publication was widely read in the pacifist community, and Irving and Dorothy studied it closely and were strongly influenced by its analysis. The AFSC argued that a massive buildup of nuclear weapons was extremely unlikely to produce a lasting peace since history has shown that, eventually, anger and resentment will outstrip a people's fear of war. Instead, America's actions only confirmed Soviet doctrine about the imperialist and war-mongering intentions of the United States. Apart from antagonizing the Soviet Union, U.S. foreign policy had also caused the steady erosion of core American values, as well as precipitating the McCarthyism of the early 1950s. "Proceeding from the false assumption that whatever is anti-communist is therefore democratic, many Americans have supported or acquiesced in measures that have generally been considered the central characteristics of totalitarianism," such as spying on fellow citizens, anonymous denunciations, and prosecution for beliefs rather than acts. The result was "the growing confusion of our thought and language until we no longer feel any astonishment at the use of a phrase like 'the free world' to include all nations, however dictatorial, and colonies, however exploited, that are not under Soviet control."[34] The moral bankruptcy of America's policy was further illustrated by the rhetoric of military strategists who talked of the possible necessity of, some day, deploying "our atomic arsenal on China" while hardly any voices "cry out in the moral wilderness."[35]

The root cause of this malaise, according to the AFSC, was not some illusory clash of ideologies between East and West, but rather the effort to master nature and the "glorification of material things" that had come to characterize modern industrial societies, regardless of their political underpinnings. Humankind's conquest of nature, "far from giving [us] mastery over [our] world . . . has apparently brought with it only the fulfillment of terrifying prophecy" in the form of a potential armageddon. And "Man's . . . failure to set any limits on his material needs" was a form of "idolatry lead[ing] him to lust for power, to disregard human personality, to ignore God, and to accept violence or any other means of achieving his ends."[36]

Rather than continuing down a road that was leading American society toward a greater degree of social control and military preparedness, the AFSC advocated a foreign policy based on the Quaker creed of nonviolence. The committee insisted that this was not merely a utopian vision. Although "the choice of non-violence involves a radical change in men, it does not require perfection." The ultimate example of the effective use of nonviolence on a large scale, the AFSC argued, was Gandhi's campaign for Indian independence in which "multitudes of men and women, without being raised to individual sainthood, were able to make an entirely new response to injustice and humiliation." Fortunately, it would not be necessary for a majority of Americans to adopt the pacifist credo in order to realize a nonviolent foreign policy. Anticommunist extremists, for example, represented only a small minority in American society, but "their positions actually set the poles and pull the whole range of public discussion toward them." In the same way, the AFSC argued, a minority of pacifists could, through their leadership and example, make nonviolence "politically relevant in a very practical way." Even in the extreme case of an enemy invasion, nonviolent resistance could still be effective in "undermin[ing] the morale of the enemy and remov[ing] his will to conquer."[37]

For the majority of Americans whose opinions were shaped by the Cold War rhetoric emanating from the Pentagon and State Department, the AFSC's ideas probably seemed quaint, if not dangerously naïve. Undoubtedly, *Speak Truth to Power* contained many contradictions and problematic analogies. Could one really compare Gandhi's India, where a distant foreign power imposed an exploitative colonial regime over millions of Indians, with Eisenhower's America? Apart from minority groups such as African Americans, few Americans would have felt exploited and disenfranchised in the way that Indians had under British rule. The AFSC's contentions that the emerging military-industrial complex was "oppressing" Americans and turning the United States into an "imperialist" nation may have rung true to academics of a Marxist bent, but it would not have had much resonance for ordinary Americans. *Speak Truth to Power*'s critique of mindless consumerism—"the glorification of material things"—sounded gloomy and pessimistic to a society enthralled by the endless gadgets and leisure activities that were becoming available to an ever greater number of Americans. Furthermore, as peace movement historian Charles DeBenedetti observed, radical pacifism "seemed frankly anomalous in a society that had gained so much from war."[38]

Nevertheless, for many pacifists and those who had perhaps flirted with the idea of embracing a nonviolent philosophy, *Speak Truth to Power* tied together many of the loose ends of pacifist thought and appeared to offer a trenchant critique of U.S. militarism and an attractive solution to the nation's political and social ills. Admittedly, it preached mostly to the converted, but it had a galvanizing effect on the Quaker and pacifist communities. The fact that it was published at the time when more radical forms of nonviolent direct action began to flourish among these groups is more than mere coincidence. Furthermore, its critique of humankind's drive to dominate nature and of the crass materialism that was a hallmark of 1950s America presaged the values that would characterize the early environmental movement and pointed toward the links between unfettered industrial and military development and environmental destruction.

Irving and Dorothy Stowe were not the most prominent peace movement leaders during the 1950s, but they were well connected within progressive circles in the Northeast. Their good friends included Harvey and Jessie Lloyd O'Connor, whose leftist credentials were impeccable. Harvey played a prominent role in the Seattle General Strike of 1919 and had worked as a journalist and editor for numerous left-wing publications, while Jessie was a lifelong community activist and radical journalist (her mother, Lola Maverick, was the founder of the U.S. section of the Women's International League for Peace and Freedom). The two had lived in Moscow during the late 1920s and early 1930s, writing for the English language *Moscow Daily News*.[39] The Stowes, therefore, moved among leftist royalty and participated in many of the protest activities that took place on the northeastern seaboard throughout the 1950s. For example, they were involved in CNVA's Polaris Action and March for Peace and were part of a group of antinuclear demonstrators picketing a speech by President Eisenhower at a military base at Fort Adams, Rhode Island. Irving also protested against civil defense exercises in Providence, refusing to take shelter when the warning sirens sounded, and he and Dorothy were also active in the civil rights movement.

The Stowes were part of a small but significant element of the American population that questioned the path that their nation was taking and attempted to divert it. A 1959 poll of its readers by *Liberation* magazine provides a profile of this group's social origins and political views. Not unexpectedly, the prototype was a middle-class intellectual, with most readers highly educated, professional (the largest single occupational category was "teacher"), and generally left wing in their social views. When given a choice between "capitalism," "socialism," "mixed economy," "communism," or "cooperative communities," about half chose the last option, with the remainder split evenly between "socialism" and "mixed economy," while capitalism and communism rated 2 and 1 percent respectively. Most were members of pacifist and civil rights organizations, with WRL, Peacemakers, NAACP, CORE, and AFSC the most prominent among them. They read magazines and journals like *Peace News*, the *Nation, Dissent*, the *Progressive*, and *Peacemaker*, and they opposed loyalty oaths and the spanking of children and defended homosexuality and interracial marriage.[40] They were staunchly, indeed radically, egalitarian.[41] Clearly, such people constituted a very small minority of the American population, a fact that helps explain why radical pacifism never became the mass movement its proponents had envisioned. Quakerism, democratic socialism, and Satyagraha, despite their philosophical compatibility with many core American values, held little appeal for postwar Americans in the thrall of consumerism and living in fear of a nuclear attack. In Charles DeBenedetti's words, although the peace movement "lives in the application of certain symbols esteemed by the larger culture . . . [it] stands as a minority reform in America because it constitutes a subculture opposed to the country's dominant power culture and power realities."[42]

The Stowes were not among the most radical people within the peace movement, but in one respect they were quite literally willing to go further than most. Deeply dismayed by their nation's military program, which they felt was leading the world inexorably down the path toward nuclear war, Irving and Dorothy began to think about leaving the United States, a difficult, if not drastic step for citizens of a nation that had always prided itself on its tolerance and acceptance of oppressed peoples from around the world. The final straw proved to be the discovery, in 1959, of radioactive strontium-90 in mother's milk. This was followed by the appearance of countless articles in the popular media dealing with the health effects of fallout, with scientists forecasting drastic increases in leukemia and genetic deformities.[43] Popular films and novels such as Neville Shute's *On the Beach* (1957), Helen Clarkson's *The Last Day* (1959), and Stanley Kubrick's *Dr. Strangelove* (1964) reflected and fanned these fears and helped propagate an increasingly gloomy cultural mood in which the nightmare scenario of a nuclear world war began to seem increasingly inevitable, particularly to people such as the Stowes, who were predisposed toward such pessimistic predictions.[44]

Not surprisingly, such fears led people to explore potential postnuclear holocaust survival strategies. One of the notions that gained some currency was that the inhabitants of the southern hemisphere stood a better chance than those in the north. By 1961, the Stowes were convinced that the United States would inevitably become involved in a nuclear war with the Soviet Union. This genuine fear, combined with their belief that children raised in the United States had a higher chance of contracting a deadly disease because of nuclear fallout, prompted Irving and Dorothy to seriously consider emigration. Their new home would be New Zealand. To many Americans, such drastic action would have been incomprehensible, not to mention cowardly and unpatriotic. To those outside the radical pacifist community, the Stowes must have looked, at best, like stubborn idealists, at worst, like chronic misfits. It took them some time to gain the New Zealand government's permission—apparently Dorothy's membership in the Women's International League for Peace and Freedom was a red flag for the immigration officials—but in 1961, the Stowes settled in Auckland, where they were to remain for the next five years. Neither Irving nor Dorothy would ever live in the United States again.[45]

Soon after arriving in New Zealand, Irving found work at the law faculty at the University of Auckland while Dorothy was employed as a social worker. They also changed their name from Strasmich to Stowe because, as Dorothy put it in a letter to Jessie O'Connor, "we became sick to death with spelling Strasmich and not being understood and also being asked if we were 'Dalmatians.'" They chose the name Stowe in honor of Harriet Beecher Stowe, the nineteenth-century abolitionist and author of *Uncle Tom's Cabin*.[46] Just as they had in Rhode

Island, they devoted most of their spare time to attending Quaker meetings and various protests. Irving continued to write letters and helped organize peace marches and pickets outside the U.S. consulate. He also placed advertisements in newspapers condemning America's escalating involvement in the Vietnam War, an issue that was to increasingly occupy his time and influence the family's lifestyle.[47]

In May of 1965, responding to U.S. pressure to fulfill its obligation as part of the Australia, New Zealand, United States Security Treaty, the New Zealand government decided to send troops to Vietnam. The Stowes were outraged. The long tentacles of U.S. militarism, they felt, had again become entangled in their lives, creating a morally intolerable situation and once more forcing them to uproot their family and search for a new home. Australia, too, had joined the Vietnam War effort by this time and so was out of the question, as was apartheid South Africa, while Europe was a potential theater of nuclear war. The only realistic option was Canada. Irving had stopped over in Vancouver once on one of his flights between New Zealand and the United States and had been impressed with the spectacular city nestled between the Pacific Ocean and the mountains of British Columbia. So, in 1966, the Stowes left New Zealand and settled in a city a mere thirty miles from the U.S. border.[48]

In the mid-1960s, Vancouver was still a long way from the cosmopolitan Pacific Rim city of today. Its size, British heritage, and temperate climate made it not too dissimilar from Auckland. Despite its apparent provincialism, however, the city was beginning to experience the initial rumblings of rebellion and discontent that, in a few short years, would turn it into the countercultural capital of Canada. It was in Vancouver that the Stowes, and Irving in particular, would give full reign to their activist urges. After settling into a house near the University of British Columbia campus, Dorothy once again found employment as a social worker while Irving nominally registered himself as an "estate planner." If Irving ever had any intentions of building a legal career in Canada, they soon fell by the wayside. He would instead devote himself to protest and activism on a full-time basis.[49]

Like the United States, Canada had an active antinuclear movement, though one heavily influenced by its British and American counterparts. For example, the Canadian Committee for Nuclear Disarmament (CCND), founded in Edmonton in 1962, was modeled on the original Committee for Nuclear Disarmament, which had been established in Great Britain in 1959. CCND's student counterpart, the Combined Universities CCND, also played an important role in raising people's consciousness about nuclear weapons. In the late 1960s, it would spawn the Canadian version of the radical U.S. student group, Students for a Democratic Society. The Canadian movement particularly focused on pressuring the major political parties to avoid cooperating with U.S. nuclear policy in order to prevent Canada from becoming a pawn in the Cold War. An early high point was a large seventy-two-hour Parliament Hill protest in Ottawa in 1961, which garnered significant media attention throughout Canada. As in the United States, Quakers were heavily involved in Canadian peace activism, spreading the same philosophy of nonviolent direct action that American groups had helped pioneer in the 1950s.[50]

By the mid-1960s, the Canadian peace movement, like its U.S. counterpart, was devoting much of its attention to the Vietnam War. Although Canada was not officially involved in the war, its government was cooperating closely with the Pentagon, and Canadian firms and universities were profiting handsomely from their role in the manufacturing of weapons, including napalm and Agent Orange. The arrival of thousands of American draft resisters from the mid-1960s onward gave further impetus to the Canadian antiwar movement and stoked the anti-American sentiments that had long been an undercurrent of Canadian history.[51]

As soon as they arrived in Vancouver, the Stowes joined the local anti-Vietnam War movement. They picketed the U.S. consulate and wrote countless letters to Canadian officials and newspapers decrying U.S. militarism and urging Canadians to oppose the war. They also joined the Committee to Aid American War Objectors, a group formed by American expatriates to help newly arrived draft resisters settle in Vancouver. Among the founders of this committee were a couple from Pennsylvania, Jim and Marie Bohlen, whom the Stowes quickly befriended and with whom they would lay the foundations for the coalition of activists that would eventually form Greenpeace.[52]

Although committed to many of the ideals of the Quakers, Jim Bohlen never actually joined the Society of Friends. Ironically, the very feature of Quakerism that he found least attractive was one that appealed strongly to Irving Stowe. "I went to several of their meetings," recalled Bohlen, "but I came away less than impressed because I found that there were a lot of people who went to the meetings just so they could hear themselves speak. They had a sort of egotistical need to be heard. That's the way a Quaker meeting works. You go into a room and sit down, nobody says anything or reads anything until someone feels compelled to speak. And some of the stuff you hear is very mundane."[53] Bohlen was neither a compulsive speaker nor a patient enough listener to endure people's rants or banal observations. Nevertheless, he found that the Quakers possessed an unshakeable spirit and great emotional strength and endurance. Bohlen's feelings about Quakerism were similar to the attitude he had held toward his communist friends in New York City during his youth: He admired them and sought out their company, without ever taking the next step and joining the group. A skeptic at heart, Bohlen was wary of associating himself too closely with any particular group or ideology lest he be tainted by the inexplicable or irrational behavior that passion for a cause sometimes produces. It was, however, a trait he managed to overcome from time to time, often in ways that surprised him.

Bohlen was born in 1926 and spent most of his childhood alternating between the Bronx and the Hungarian and Jewish areas of Greenwich Village. Because his parents' career paths forced them to live separately for much of the time, he was raised by his grandparents and other relatives for most of his childhood. His teenage years were even more unsettled: He lived at various places up and down the eastern seaboard, moving from the Bronx to his grandparents' farmhouse in New Jersey; then living with his father in Philadelphia; in Miami Beach with his mother; and from 1940, when his parents finally managed to bring their lives together again, on a farm in Bucks County, Pennsylvania. In 1943, after graduating from high school, he began studying aeronautical engineering at New York University.[54]

In 1944, when he turned eighteen, Bohlen decided to preempt his likely induction into the U.S. military, and the potential frontline service it entailed, and volunteered for the navy. He spent most of what remained of the war in Wisconsin and California, training to become a radio technician, before being assigned to a salvage boat in the Pacific a few weeks before the war's end. He was in the middle of the Pacific Ocean when the atomic bombs were dropped on Hiroshima and Nagasaki. Unlike Irving Stowe, he recalled feeling gleeful at the news that the Japanese were to surrender and that the million or so American troops whose lives were going to be sacrificed in an invasion of Japan would be spared. Later in his life, Bohlen came to view this figure as military propaganda designed to make the use of atomic weapons more acceptable, but at the time, his feelings reflected those of the average American.[55]

Since he had entered the war very late and was young, single, and of low rank, Bohlen was one of the last discharged from active duty and remained in the navy for almost a year after the Japanese surrender. He learned to deep-sea dive and was involved in many grisly salvage operations around Japan and the Aleutian Islands, where he discovered how cold and treacherous the waters of the North Pacific could be.[56]

After returning from the war, Bohlen married Ann Arndt, an art student from Allentown, Pennsylvania, and returned to NYU to complete his engineering degree. When he graduated in 1949, he found himself in a job market that was overcrowded with engineers. Initially he had to settle for a taxi-driving job in Manhattan before his father was able to get him a drafting job at the firm where he worked as a senior engineer. Unsatisfied with the position, Bohlen worked until he saved up some money, and then in 1950, he and Ann took a yearlong trip to Europe.

Bohlen's experience reflects how closely professional careers in research science and engineering were connected to the emerging U.S. military-industrial complex. Bohlen returned from Europe as the United States entered the Korean War, and suddenly the demand for research engineers skyrocketed. He took a job with a firm that produced a new material—fiberglass reinforced plastic (FRP)—and by 1954, he was a custom products sales engineer for Lumm Laminates on Long Island, a company whose work was largely based on military contracts. Another of Lumm's employees was the engineer and futurist, Buckminster Fuller, with whom Bohlen became good friends. Bohlen worked with Fuller in the development of FRP geodesic domes that were designed to protect U.S. radar installations in the Arctic. Bohlen became quite fascinated with the design, so much so that in later years he would design and build his own house on Denman Island, British Columbia, in the form of two large geodesic domes.[57]

Throughout the 1950s, Bohlen followed a rather typical middle-class life path: a college education, marriage and two children, a corporate job secured through his father's influence, and an attempt at running his own business. Yet during this time, Bohlen was also questioning the values that undergirded that lifestyle and that compelled him to strive for the trappings of middle-class success. Oddly enough, according to Bohlen, the trigger for this reflection was not the nuclear arms race or McCarthyism, but rather the works of the maverick American author Henry Miller. Bohlen had first picked up a copy of Miller's infamous *Tropic of Cancer*, which was banned in the United States due to its explicit sexual content, while in Paris during the 1950s. He managed to smuggle a copy into the country by, appropriately enough, secreting it in his dirty laundry. Apart from the thrill of reading a forbidden novel, Bohlen was deeply persuaded by Miller's "vigorous denunciation of middle class values." Miller's thesis, as Bohlen understood it, was not merely that middle-class values produced a bland, self-satisfied, and uncritical society, but that the rampant materialism they promoted, "if allowed to proliferate, will provoke the fracturing of the human community, and ultimately the extirpation of society." Bohlen read all of Miller's work that he could get his hands on and became, in his own words, "a fanatical and devout Millerite."[58]

Although he did not think of himself as a beatnik, Bohlen followed an intellectual trail that was not too dissimilar from the one that Allen Ginsberg and Jack Kerouac were embarking on during the late 1950s and early 1960s. Like the beatniks, Bohlen found the pursuit of the American dream of middle-class comfort and security to be an increasingly meaningless experience. His moral discomfort was exacerbated by the fact that he was enmeshed deeply in America's burgeoning military-related industries. His search for a more meaningful life led Bohlen to the Zen Buddhist philosophy of the Japanese philosopher and monk Daisetz Teitaro Suzuki (1870–1966).[59] Zen, which became popular among the early beatniks and heavily influenced writers such as Ginsberg and Gary Snyder, teaches that subordinating one's life to the struggle for material success is counterproductive and leads to social anomie and spiritual emptiness. Bohlen also embraced Gandhi and Satyagraha, and by the late 1950s he was beginning to lead something of a double life: he continued to pursue a decidedly middle-class lifestyle, while devoting his spare time to the task of integrating the antimaterialist and antimilitarist philosophical constructs of Miller, Suzuki, and Gandhi "into an action-oriented lifestyle."[60]

Part of Bohlen's search for an alternative set of values led him to the Quakers in his local Pennsylvania community. Although, like the Stowes, he could not identify with the more religious aspects of Quakerism, he found the idea of nonviolent action and the antiwar and antimaterialist message in texts such as *Speak Truth to Power* very appealing. Thus he felt he could admire and learn from the Quakers, even join in their protests, without having to subscribe to all their rituals and beliefs.[61]

During one of his bouts of unemployment, Bohlen volunteered to attend a civil defense workshop in Florida. Part of the workshop's goal was to discuss various strategies of self-protection in the event of a nuclear attack. After being shown a horrific slide show of some of the victims of Hiroshima and Nagasaki, Bohlen and the other participants were told that the simple "duck and cover" strategies that the government and military had been advocating throughout the 1950s were still the best means that civil defense experts could come up with to protect people against a hydrogen bomb blast. All people needed to do, experts promised, was dig a trench, climb in, and cover themselves with a door. Bohlen was furious that the U.S. government was continuing to plan for a nuclear war while deluding its citizens into thinking that they could save themselves through such absurdly inadequate means. Whatever reticence he may have had about becoming a full-fledged antinuclear activist was eradicated by his experience at the workshop. He stormed out halfway through the proceedings, flew back to Pennsylvania, returned the funding he had received to attend, and became "determined . . . to actively oppose nuclear weapons of mass destruction until they were wiped off the face of the earth."[62]

Just as it had been for the Stowes, the discovery of traces of the radioactive isotope strontium-90 in mother's milk throughout North America in the late 1950s was important to Bohlen's radicalization. It was also a seminal moment for the modern environmental movement. It is hard to imagine a more powerful symbol of impurity than the thought of innocent and helpless infants threatened by the insidious contamination of their mothers' milk. Scientists such as Barry Commoner began to investigate the potential health consequences of nuclear testing, consequences that had been downplayed and little discussed by the Atomic Energy Commission. Along with Rachel Carson's investigation into the damage caused by the copious and indiscriminate use of modern pesticides, the threat of leukemia and long-term genetic damage posed by nuclear fallout were key elements in the emergence of the public concern about environmental contamination.[63]

Demonstrating concern about nuclear proliferation and chemical pollution was one thing, but in the early 1960s, Bohlen was not yet ready to completely abandon the deeply ingrained bourgeois values with which he had been raised. Once again, however, he discovered that he could not find work that would enable his family to continue its accustomed lifestyle without being directly involved in the military-industrial complex. Reluctantly, he swallowed his misgivings and took an engineering job with Hercules Powder Company, which had a lucrative government contract to design and build guided missile motors. The only consolation for Bohlen was that he managed to get himself assigned to a small team of engineers and technicians whose task it was to explore potential nonmilitary uses for the missile technology. Nevertheless, the contradiction between Bohlen's career and family responsibilities, on the one hand, and his increasingly antiestablishment values, on the other, became ever more difficult to reconcile. He was spending his weekends at antiwar protests organized by Quakers and other peace groups, and his weekdays working for a company whose chief source of profit came from supplying the nuclear weapons industry. One of the main casualties was his marriage, which collapsed, in part, under the weight of this contradiction and the frustration that resulted from it.[64]

Bohlen met his second wife, Marie Nonnast, a well-known illustrator of children's books, at a Quaker-organized antinuclear testing protest in front of Philadelphia's City Hall. Marie's first marriage, from which she had a teenage son, had also recently ended. They found that they shared similar values and began to devote a greater amount of time to attending protests and to trying to live their lives with fewer middle-class materialist trappings. They strongly opposed the U.S. involvement in the Vietnam War and attended many protests in Philadelphia, New York, and Washington, D.C., finding themselves "elated, both by the prospect of contributing to the anti-war effort, and by being in the company of like-minded souls."[65] In 1965, as the U.S. involvement in Vietnam began to escalate, the Hercules Powder Company put its civilian research on hold and ordered all its employees to concentrate their efforts on the development of an antipersonnel shoulder-fired rocket. The warheads were to be filled with small pieces of razor blades designed to tear and penetrate deeply into human flesh, making treatment and recovery almost impossible. This was more than Bohlen could stomach, and he resigned from the company in mid-1965.[66]

In 1966, Marie's son, Paul, graduated from high school and became eligible for the draft. He decided that if he were drafted, he would go to Canada. One of the stipulations of the relationship between the United States and Canada is that individuals who move from one country to another cannot be prosecuted for crimes committed in their native land unless those crimes are mutually recognized. Since Canada did not draft men into its armed forces, it did not extradite legally resident U.S. immigrants who had evaded the draft. In fact, Canada viewed draft resisters as political refugees who faced prosecution if they were returned to the United States. Nevertheless, it was still a difficult choice for a young man to make. Apart from the moral shame associated with draft evasion, the resisters also had to effectively start a new life in Canada without much hope of returning home—even to visit family—because they risked arrest and imprisonment.[67]

In 1967, the Bohlens moved to Vancouver, where Jim found employment at a government-run forest products laboratory on the campus of the University of British Columbia. Like Irving and Dorothy Stowe, Jim and Marie immediately became involved in local antiwar activities. One of the first things they noticed was the number of restless and often directionless young American men drifting about town. Vancouver was the main destination of West Coast draft resisters, but there was little in the way of a support network once they arrived. In response to this, the Bohlens, along with some other American expatriates they had met at UBC, formed the Committee to Aid War Objectors. The organization developed a network of sympathetic contacts with whom draft resisters could stay when they first arrived in Canada, as well as offering them several other services, such as job and housing bulletins, a mailbox system, magazine and newspaper subscriptions, and counseling. For the next couple of years, the Bohlens hosted a continuous succession of newly arrived draft resisters.[68]

While most Canadians disapproved of the U.S. involvement in Vietnam, the prevailing attitude seemed that Canada should not become directly involved either way. Many Canadians, however, were oblivious to the fact that their country's industries and workers were manufacturing arms and equipment that were vital to the American military. Those in the Canadian peace movement were keen not only to voice their opposition to the war but also to highlight Canada's complicity. Since the Second World War, the two governments had closely collaborated in their defense systems, setting up the Joint Board of Defense in 1947 and an integrated North American Air Defense system in the 1950s. In 1959, the two nations signed a defense-sharing agreement under which each country would spend an equal amount of their defense dollars in the other and the United States would drop its requirement that all military contracts go exclusively to American firms. As a result, Canadian military contractors profited immensely from the war, with U.S. defense expenditure in Canada increasing from $25 million in 1964 to $317 million in 1966.[69]

On college campuses, Canadian students discovered that some forty universities across the country were being funded to conduct research that had direct military applications to the U.S. war effort in Vietnam. From 1966 onward, antiwar demonstrations erupted throughout Canada, often initiated by student groups, many led by American draft resisters studying at Canadian universities.[70] At UBC, a hodgepodge group of old and new leftists, called the Internationalists, organized several antiwar marches in downtown Vancouver. Another more broadly ecumenical group, End the Arms Race, also sponsored antiwar demonstrations in town. Jim and Marie attended one of these marches soon after arriving in Vancouver in 1967. They were hoping to find a group of Quakers with whom they could join in regular nonviolent protests, and about halfway down the line of marchers they spotted a Quaker banner held aloft by Irving and Dorothy Stowe. The Bohlens went over and introduced themselves, and the nucleus of the Greenpeace coalition was born.[71]

Irving and Dorothy Stowe and Jim and Marie Bohlen were as much the products of the Cold War as Joseph McCarthy, Richard Nixon, and their supporters on the right of American politics. People on both sides of the ideological divide were shaped by the horrors of the Second World War and the Holocaust and by the ever-present fear of nuclear war. For various reasons—the nature of their upbringing, their social milieu, their geographical origin, and a host of undoubtedly complex psychological factors—the Stowes and the Bohlens gravitated toward a worldview in which salvation resided in international cooperation and mutual understanding rather than military strength and strategic alliances. The things that many Americans saw as the nation's strengths—its unsurpassed military capability, its wealthy and powerful corporations, and its unrestrained consumerism—were the very factors, they believed, responsible for fueling the Cold War abroad and creating a moribund culture at home. For the Bohlens and the Stowes, the radical pacifism of groups such as the Committee for Nonviolent Action and the War Resister's League represented the best hope for the creation of a peaceful and just society and a meaningful and vibrant culture. The Quaker notion of "bearing witness" and Gandhi's Satyagraha provided radical pacifists with a set of tools that allowed them to express their opposition to the dominant power structure in a way that was consistent with their belief in the power of nonviolent action. Not surprisingly, therefore, the organization that the Stowes and Bohlens founded in Vancouver was deeply imbued with the radical pacifist values they brought with them from their experience in the American peace movement, values that to this day remain at the core of Greenpeace's philosophy.

2

The Enemies of Anarchy

Greenpeace's paternity has always been a subject of dispute. Longtime Greenpeacers like to joke that if you walk into a dingy fishermen's pub anywhere between Anchorage and San Francisco, you are likely to find some old geezer who claims to have been a founder of Greenpeace. While it is very clear today that Greenpeace is an organization, with all the positive and negative connotations that come with the term, this was not the case initially. During its early years, Greenpeace was a loosely organized network of activists and journalists, more a social movement than an organization. Its origins, therefore, can best be analyzed as a process rather than a singular "birth" or "founding." Thus the joke about a Greenpeace founder in every West Coast pub turns out to be only a half-joke. The first Greenpeace campaign—which occurred before the organization's formal existence—involved dozens, if not hundreds, of people. While some, such as Irving Stowe and Jim Bohlen, worked tirelessly on the campaign for over a year, others may have simply set up a support group or a welcoming committee in an Alaskan fishing village or a coastal logging camp in British Columbia. Given the organization's fluid origins and subsequent fame, one can perhaps understand why so many people might succumb to the "founder" temptation. After all, where is the cutoff point between a founder and a mere peripheral supporter?

Determining whether Greenpeace had many founders, a handful, or none at all is a contentious activity. Nevertheless, it is clear that some of the early members did more founding than others. Irving Stowe and Jim Bohlen certainly fall into that category, as does Bob Hunter. All three had a considerable—and distinctive—impact on the nascent Greenpeace organization and helped shape its activist style and culture. It is important, therefore, to understand how they formed their environmental ideas and the context in which this occurred.

To fully understand the emergence of postwar environmentalism requires viewing it as part of a broader history of ecological ideas. By the late 1960s, the word "ecology" had morphed from a term that referred to a branch of biology that studied the way various species interacted with one another in nature into a worldview denoting a certain cultural and political stance. This was an unusual development for a scientific discipline that evolved in the rarified air of university biology departments. As Donald Worster notes in his history of ecological ideas, few other academic disciplines have entered the public lexicon as catchwords denoting a particular worldview or political party. No one has yet proclaimed himself or herself to be a "deep entomologist" in the same way that people have embraced deep ecology.[1] Nevertheless, the history of ecology and environmentalism should not be conflated: the former certainly influenced the latter, but they are not synonymous, despite the fact that many 1960s and 1970s activists used the term "ecologist" to refer to people that we today would call "environmentalists."

According to historian of science Sharon Kingsland, ecology emerged as an adjunct of early twentieth-century utilitarian conservation: "The same imperatives and the same need to rationalize resource use that supported the conservation movement also supported research in ecology." Thus early twentieth-century ecological science "was part of an effort to control life and to apply rational methods to a complex set of problems generated by the American desire to migrate into and adapt to new landscapes." In this sense, it was a "quintessentially Progressive Era science" whose practitioners "sought to render natural processes more predictable" so that further landscape transformation could be better controlled.[2] As the goals and needs of the American state shifted into Cold War mode, so too did the research agenda of ecology. The result, exquisitely ironic given subsequent events, was that the Atomic Energy Commission became the major institutional supporter of ecology in the immediate postwar era. The irony would be slightly less perverse if the AEC had hired ecologists to critique nuclear power or to assess its risks. However, the commission was not particularly interested in what ecologists had to say about such matters. Rather, as historian Stephen Bocking points out, they were sponsoring "research supportive of the overall objective of furthering nuclear technology." Thus many influential ecologists obtained funding in nuclear research by embracing, or at least not challenging, this prevailing consensus within the American nuclear establishment.[3]

So how was ecology transformed from an obscure academic science—one that was deeply enmeshed in the project of controlling nature to benefit the state—into a radical, antistatist worldview? According to Worster, this was largely the result of the mid-twentieth-century scientific advances that accelerated environmental despoliation. The ultimate symbol of this great leap forward in humankind's destructive potential was, of course, the atomic bomb. The bomb, according to Worster, "cast doubt on the entire project of the domination of nature that had been at the heart of modern history. It raised doubts about the moral legitimacy of science, about the tumultuous pace of technology, and about the Enlightenment dream of replacing religious faith with human rationality as a guide both to material welfare and to virtue." Nothing else, Worster argues, can adequately explain the onset of environmentalism and ecology's role within it. There had been no drastic changes in human behavior, no radical alterations in the capitalist system that could otherwise explain ecology's sudden emergence as a discourse of environmental redemption.[4]

While external factors such as the rise of the military-industrial complex certainly played a major role in ecology's rise to prominence, they do not tell the entire story. From an epistemological perspective, ecology was primed for rebellion. Ecologists, after all, focused on the interconnections between species; they studied the natural world as a system, which predisposed them toward a holistic worldview, and many of them adopted normative models in which nature tended toward a balanced and stable state.[5] Disturbing a harmonious, balanced system was inherently problematic. And by the middle of the twentieth century, the rate of human disturbance had reached the point where the entire intricately interconnected and delicately balanced web of relationships that constituted "nature" appeared in danger of collapse. At a certain point, therefore, ecology slipped from the realm of "objective" science into the world of ethics and politics. If nature was supposed to exist in a state of balance, and humans were endangering that state, then surely it was time for our species to show some self-restraint: to reflect upon the way our actions affected nature and think about ways of reducing our impact. This kind of "moralistic ecology"—"moralistic" in that it displayed a concern for moral issues, as opposed to the self-righteous, finger-wagging sense of the term (although that element was not entirely absent)—emerged in the 1960s and soon came to dominate the general public's understanding of ecology. Bolstered by the work of crusading scientists such as Rachel Carson and Barry Commoner, it appealed to people who were critical of modern industrial society and its institutions, particularly those who were predisposed toward cultural and philosophical holism.[6] And it is here that ecology—as a worldview more than as a discrete science—intersects with some of the other holistic discourses—such as Zen Buddhism, New Age spiritualism, and Native American mythology—which influenced the cultural radicalism of the 1960s.[7]

The rise of the modern ecological worldview represents a momentous cultural shift in humanity's understanding of its place in nature. In historian Michael Bess's words, it constitutes "the reinvention of nature as a bounded and fragile space, requiring intensive human nurturing and protection." While the roots of this shift date to the eighteenth century, its manifestation as a mass phenomenon, "a central feature in the cultural landscape of modernity," occurred only after the Second World War.[8] This new view, which by the 1970s was commonly referred to as "environmentalism," was characterized by four key ideas about nature and how humans needed to live within it. First, nature was finite and could easily be overexploited and damaged by careless human actions. Second, everything within the natural world was interconnected, and disturbances in one place could have unanticipated consequences elsewhere. Third, in extreme cases, and the twentieth century was nothing if not extreme, such disturbances could cause systemic ruptures, radically altering ecosystems and severely damaging their ecological integrity, in some cases even rendering them uninhabitable. Finally, the potential consequences of human impact on the rest of nature were so severe that nothing short of a global socioeconomic transformation would be required.[9]

Until the early 1960s, Jim Bohlen had not taken much interest in issues such as wilderness preservation. Most of his activist hours were devoted to the antinuclear cause. The only environmental problems that had concerned him were those, such as the discovery of strontium-90 in milk, that were a direct product of nuclear weapons. Marie, on the other hand, was an active Sierra Club member who had been interested in wilderness preservation, both locally and nationally, for many years. Through Marie's enthusiasm for, and knowledge of, the wilderness areas of the mid-Atlantic states, Bohlen, whose views of nature had largely been shaped by his career in scientific research and engineering, began to gravitate toward biocentrism. It would be an exaggeration to say that he was suddenly seized by a John Muir-like passion for wild places. Nevertheless, he came to appreciate the wilderness for more than just its recreational or aesthetic potential. It was a view that complemented his interest in Zen Buddhism and the values he had inherited from the Quakers and radical pacifists he counted among his friends. Given their activism in the peace movement, it seemed only natural that Jim and Marie would gravitate toward the more activist wing of the Sierra Club.[10]

The Sierra Club's Atlantic chapter, of which the Bohlens were members, was part of this wing, a fact that was reflected by its strong support for Club president David Brower's controversial efforts to instill the organization with a greater degree of political activism.[11] For people such as the Bohlens, Brower was a heroic figure who was attempting to convert a conservative, tradition-bound organization into a more progressive and politically aggressive force for environmental protection. Not surprisingly, therefore, the Bohlens' first taste of environmental activism came during a Sierra Club campaign to halt the development of a flood-control project on the Delaware River. Like all rivers, the Delaware was subject to periodic flooding, causing damage to farms and factories along its banks. In response, the Army Corps of Engineers developed a plan to pump water upstream into an enormous artificial lake, thereby allowing it to control the river's flow. The construction, in the Sierra Club's opinion, would inundate thousands of acres of fertile agricultural land and eastern hardwood forest, as well as disturb fish breeding and the general ecological well-being of the river and its watershed. Through a concerted petition and letter-writing campaign, the Club's Atlantic chapter raised enough public opposition to help defeat the plan. This success spurred the Bohlens to join other environmental battles. Furthermore, their experience in the antinuclear and anti-Vietnam War movements prompted them to think about how the tactics of such movements could be employed in environmental protests.[12]

When Jim and Marie Bohlen arrived in Vancouver in 1967, there was no local Sierra Club chapter, nor was there an equivalent Canadian organization. For middle-class Americans who had taken the existence of multiple wilderness preservation organizations for granted, the dearth of such groups in Canada came as a considerable disappointment. British Columbia in particular, with its cut-and-run timber industry, must have seemed like a throwback to the late nineteenth-century American West. In Canada, unlike much of the United States, jurisdiction over natural resources lay largely in the hands of the provincial governments. For American environmentalists accustomed to a system where the federal government curbed the more rapacious instincts of the western states, having the British Columbian government in charge of its own forests was comparable to allowing the state legislature in Laramie to determine the fate of all the trees in Wyoming. Terry Simmons, an American graduate student at Vancouver's Simon Fraser University (SFU) and one of the founders of the British Columbia chapter of the Sierra Club, wrote that being a conservationist in British Columbia was like "being a civil rights worker in Alabama." Unlike in the United States, where the Sierra Club could sue the Department of the Interior for failing to enforce conservation laws, in British Columbia people still needed the permission of the government before they could sue it. The result was a "conservationist's nightmare" with few legal impediments to exploitive economic development.[13]

To newly arrived Americans who had been involved with wilderness issues in their home country, the dearth of environmental organizations in Canada came as something of a shock. Simmons had gone to Simon Fraser University, a new futuristic campus perched atop a mountain a few miles east of Vancouver, after completing an anthropology degree at the University of California at Santa Cruz in 1967. Unlike many others of his generation, he was not fleeing the draft (the effects of a motorcycle accident had made him ineligible). In fact, he did not even consider himself an opponent of the war in Vietnam. He was, however, intensely interested in conservation issues and felt that British Columbia could use a good dose of American-style wilderness protection and natural resource conservation.[14]

In the late 1960s, the waves of ecological consciousness from the United States were spilling over the border into Canada, where many had also read such popular authors as Rachel Carson and Barry Commoner. As a result, some British Columbians became increasingly determined to address their government's lack of concern with environmental issues. In December 1968, a group of social activists, UBC and SFU academics, and the leader of the local Transit Union, met in a house in Coquitlam, just outside Vancouver, to discuss plans to oppose Premier Bennett's recent approval of strip mining in the Kootenay Valley in southeast British Columbia. The following month, they held an open meeting at SFU that drew over two hundred academics, students, and members of the local counterculture. One of the participants, a thirty-four-year-old engineer named John Stigant, opened the meeting by tipping a barrel of oily water from Vancouver's Burrard Inlet onto the floor and declaring, "We are the filthiest animals on the planet." The group went on to discuss the various environmental problems they saw plaguing the province, particularly emphasizing urban pollution and health issues such as automobile emissions, pesticide use, and deteriorating water quality. They decided to form a new "ecology" society to publicize environmental problems and pressure governments and industry to abandon environmentally destructive practices.[15]

One of those who took part in the group's early activities was Bob Hunter, who was developing his own ideas about ecology and society and was eager for the new group to become a vanguard for an ecological and cultural revolution. He suggested that they call themselves the Society for the Prevention of Environmental Collapse and urged his readers to back the organization. There were "several reasons for supporting this group," he declared. "The main reason is that our civilization has gone into a tail-spin and the human race appears to be heading for one of its periodic smash-ups—perhaps its last."[16] For the group's more academic and pragmatic members, however, Hunter's exhortations were a little too dramatic. They decided to keep the acronym, SPEC, but to replace it with the more sober, though somewhat uninspired, title the Scientific Pollution and Environmental Control Society. They became particularly active in opposing widespread pesticide use, offshore oil drilling, air and water pollution, and strip-mining and tried to form an apolitical coalition whose goal would be "the preservation and development of a quality environment through the stimulation of public interest, and consultation and cooperation with industry, government, labour, and academic communities."[17] SPEC groups rapidly spread to most major Canadian cities, where they concentrated primarily on the kind of urban-industrial issues—such as smog and water pollution—that were typical of various new American organizations, such as the Environmental Defense Fund, that also sprang up in the late 1960s. This suggests that Canadians had a greater affinity for this predominantly urban-oriented environmentalism than they did for wilderness preservation of the Sierra Club variety.[18]

Terry Simmons watched SPEC emerge with great curiosity and for a time flirted with the idea of joining the organization. But the group's urban bias and its disinclination to seriously address traditional wilderness preservation and resource conservation issues led Simmons to conclude that they were simply not up to the task of imposing an American-style order on the province's ad hoc and destructive natural resource exploitation. So, he turned to the only organization that he felt had the tradition and resources to tackle the deeply entrenched pioneer attitude that many British Columbians held toward their wilderness areas. He contacted the Sierra Club in California, of which he was still a member-at-large, and asked them for a list of other members in the Vancouver area. The Club informed him that it had some sixty members living in the region. Simmons contacted as many of them as he could and set up a meeting at SFU in July 1969. Among those on his list were Jim and Marie Bohlen, who brought along their friends Irving and Dorothy Stowe. The participants, the majority of whom were Americans, were in full agreement with Simmons: only an organization such as the Sierra Club was capable of saving British Columbia's landscape from decimation. So they decided to set up the BC Sierra Club, the first branch outside the United States, and registered themselves under the BC Society's Act in September 1969.[19]

Up until the founding of the BC Sierra Club, the Bohlens and the Stowes had been devoting most of their activism to various Quaker-related antiwar activities. There were, however, some notable exceptions. Jim's first act of environmental direct action occurred several months before the inaugural Sierra Club meeting. One spring day, he was sitting outside his laboratory near the UBC campus, eating his lunch, and watching migrating birds make their way north up the Pacific flyway, the major bird migration route on the west coast of North America. Many birds were landing and resting in a large, old western red cedar under which Bohlen was picnicking. Suddenly, a man wielding a chainsaw approached the tree. "He asked me to move," Bohlen recalled, because

> he needed to cut down the tree in order to make way for a temporary trailer addition to the laboratory. I was appalled. Not knowing what else to do, I stood up and spread-eagled myself, with my back against the tree trunk. I told the logger, somewhat passionately, this tree will *not* be cut down. The logger was so startled by my behaviour that he retreated . . . It gave me a new feeling and I liked it.[20]

Bohlen's spontaneous "tree-hugging" episode was followed by a more organized example of civil disobedience. A friend of Bohlen's and a fellow Sierra Club member, Bill Chalmers, had formed a group called the Save the Natural Beach Committee to prevent the Vancouver Board of Parks and Recreation from constructing a road along a pristine beach near the UBC campus. Bohlen, Chalmers, and several other respectable, middle-aged environmental activists, as well as numerous UBC students, turned up at the site on the morning that construction was due to begin and formed a human chain to impede the bulldozers. The action was successful and the project was eventually shelved.[21] It was only a small protest dealing with an event of purely local significance, yet one should not underestimate its importance. The philosophy of Satyagraha and the tactics of radical pacifism and civil rights were now being used in the service of environmentalism. According to Gene Sharp's typology of nonviolent action, environmentalists had hitherto been reluctant to move beyond the first stage in the protest hierarchy. Letter writing, lobbying, disseminating information, and, infrequently, a protest march had always been the movement's modus operandi. The Save the Natural Beach Committee's act of defiance, which Sharp would define as "nonviolent obstruction," was several rungs higher on the nonviolent action ladder.[22]

Irving Stowe, meanwhile, was reveling in the role of full-time activist. By 1969, he had helped establish a network of grassroots organizations throughout Vancouver, tackling everything from town planning to nuclear weapons testing. His friendship with Bohlen had led him to take a far greater interest in environmental issues than he had previously displayed, though he still maintained his commitment to a plethora of other causes. He established the Take Back the Earth Committee, a group that pushed for a sensible approach to urban planning that would "avoid the rampant growth problems that have plagued so many American cities."[23] They organized demonstrations in downtown Vancouver in an effort to prevent the construction of a bridge from the city to Kitsilano, a development they felt would lead to further automobile traffic in the inner city. Another of Stowe's groups, United For Survival, was founded in an effort to establish an umbrella organization for various progressive organizations throughout British Columbia. Stowe's bitterness toward the United States largely dictated the tone of the group's rhetoric. Canada, he argued, was being taken over by the United States and would soon effectively cease to exist as an independent nation. "The majority of voters," he declared, "still think of 'survival' in terms of KEEPING THEIR STANDARD OF LIVING." One way in which Canada could maintain the more desirable aspects of its high standard of living, while avoiding some of the ecological and cultural destruction emanating from the United States, was to promote the establishment of a new trading bloc, a "Common Market for National Survival," which would set strict regulations about trading with nations that possessed nuclear weapons or promoted ecologically destructive developments such as nuclear power plants and big dam projects. The Take Back the Earth Committee and United For Survival spanned both ends of Stowe's "think globally, act locally" philosophy, while also displaying his more bombastic tendencies and his increasingly impractical and impassioned anti-Americanism.[24]

Throughout 1970 and 1971, Stowe wrote a column titled "Greenpeace Is Beautiful" for Vancouver's underground newspaper, the *Georgia Straight*. The column allowed him to publicly vent his feelings about the evils of the United States and the travesties of capitalism, as well as enabling him to pontificate on the virtues of his own brand of environmentalism. At fifty-five years of age, Stowe was old enough to be the father of just about everyone associated with the newspaper, and his stout build and huge bushy black beard led one journalist to describe him as "a Jewish Santa Claus."[25] Despite his age, Stowe got along well with the young hippies and radicals who ran, or simply hung out at, the *Straight*. Politically, Stowe liked to describe himself as a small "c" conservative: "I want to conserve our environment, our resources, our people," he told a journalist, "that's why I call myself conservative. The men who want to go on testing nuclear weapons are the real radicals." However, few people would have described Stowe as a "conservative" in the mainstream definition of the term. Bob Hunter, for example, was convinced that Stowe was a Maoist. Even Bohlen, who had a similar background and political outlook to Stowe, felt that Irving had perhaps "done something in the States that wasn't quite kosher," so strong were his anti-American sentiments.[26]

By the late 1960s, Irving's political outlook put him squarely in the camp of the New Left. He quoted the German philosopher Herbert Marcuse, the so-called father of the New Left, with approval and advocated Tom Hayden's concept of the revolutionary collective, which emphasized power from below and a distrust of elite organizations.[27] He frequently echoed the views expressed in the chief New Left organ, *Ramparts*, arguing that environmental problems such as air and water pollution did not simply involve one-dimensional solutions but would require "many radical changes in our lives." Like the New Left, he dismissed the "soft" environmentalism of the government and the mass media, whose solutions to environmental problems involved tightening regulations and employing technological Band-Aids, as mere distractions. "The traditional conservation and anti-pollution approaches have failed," insisted Stowe. Indeed, "they were doomed to failure, because they naively assumed that government and industry could be pressured into protecting people's health and lives." Politicians, he argued, "welcome people's growing interest in 'cleaning up the environment' as a happy sign that they will be litter-picking instead of protesting the USA war against Southeast Asia and Canada's role as accomplice." But if people were serious about saving their environment, they had to adopt a "people's action approach" that recognized that "government and industry are the problem." "Pollution, garbage, waste, [and] the rape and destruction of Canada's resources and wilderness areas," Stowe thundered, are "symptoms of a world-wide ECOLOGICAL CRISIS." Avoiding this crisis would require "revolutionary changes in attitudes and lifestyle that will make establishment politicians wish that people were back on the streets protesting only war!" According to Stowe, people needed to expose and abandon all the establishment values and slogans—"profits," "efficiency," "cheapness," "GNP," "progress"—that had been used to manipulate them and examine and choose their actions toward others and the environment in accordance with what Stowe referred to as a "survival ethic," a more anthropocentric version of Aldo Leopold's "Land Ethic." "Do your actions," he challenged his readers, "enhance the health, continued existence, and freedom of your brothers and sisters and yourselves and preserve the environment?"[28]

Like the New Left, Stowe was comfortable with the neo-Marxist rhetoric of third-world dependency theorists such as Frantz Fanon. "The domination of the Western World over the Third World," he resolutely declared, "capped off by the political and economic domination of the US over both, and crowned by the domination of corporate enterprise over the public interest is simply and tersely set down in the political economy of resource exploitation." For Stowe, therefore, developing a "full ecological awareness," required that people recognize the interrelationship between exploitation, war, Western lifestyles, and ecological destruction. Stowe was certain that an end to the destructive ecological practices of the modern world would only come about if people actively fought for it. The "system," he insisted, "cannot be reformed by trying to operate within its rules (the game is rigged in favor of keeping the ruling class rich and in power)." The only way that people "can make the present death-oriented system impotent is by using their power of *collective non-cooperation*." This would involve "not eating the system's artificial foods, not buying the system's gadgets, and rejecting 'education' for establishment jobs." Pushing this line of thinking to the limit, Stowe argued that there was "only a difference in degree between those people who in Nazi Germany were willing to work in the gas crematoria and people who are willing to work for Dow Chemicals." Just working for a paycheck was no longer acceptable, Stowe insisted. Instead, "a job to produce a standard of LIVING must be a job which contributes to life and health and the protection of the environment."[29]

While Irving Stowe represented a quasi-New Left influence on Greenpeace, Bob Hunter was the organization's countercultural prophet.[30] Like more celebrated chroniclers of the counterculture, such as Theodore Roszak and Charles Reich, Hunter attempted to analyze the counterculture and outline its future path.[31] Like Reich, he viewed history as part of the unfolding of a classical Hegelian dialectic: growth in human consciousness and self-knowledge challenges older ways of conceiving of the world; the old consciousness perishes and is replaced by the consciousness that undermined it, but the older consciousness refuses to fade away entirely, and eventually a new synthesis emerges that shifts human self-knowledge into a completely new realm. Although Reich's book was titled *The Greening of America*, ecology was far more central to Hunter's vision of consciousness revolution than it was to Reich's. Furthermore, Hunter was not content to remain on the counterculture's intellectual sidelines. Instead, he became a committed activist and a crucial figure in the evolution of Greenpeace's philosophy and tactics.

Hunter, who was born in 1941, grew up on the outskirts of Winnipeg and was raised largely by his French-Canadian mother, who worked at a nearby restaurant, earning just enough to keep the family above the poverty line. Although a bright child, Hunter hated the regimentation and discipline of school. He was also something of a rebel: when he received a bursary to study fine arts at university, he chose to publicly burn it at his high school graduation ceremony. Instead of attending university, Hunter decided that he would live life to the fullest and become a writer. Looking back on his impetuous and naïve actions, Hunter was convinced that it was his decision not to attend university that ultimately gave him the confidence to become a writer. Many of his friends had similar ambitions and went to college only to learn "what insignificant little gnats they were compared to the literary giants." Despite his rebelliousness and romanticism, Hunter had trouble leaving the security of his Winnipeg environment. He only just managed to get out before his twenty-first birthday, which he celebrated on a Yugoslav freighter in the middle of the Atlantic. The ship was due to pass through Havana at the height of the Cuban missile crisis, but changed its course at the last moment, thus denying Hunter his big break in journalism.[32]

Hunter then spent several months traveling through Europe and living in England, where he met his first wife, Zoe, an aspiring actress who worked at the same library as Hunter. Up to that point, Hunter's literary influences were largely confined to science fiction, which he read voraciously, and occasional forays into Nietzsche and beatnik literature, particularly Jack Kerouac. Not surprisingly, his political views at the time were largely filtered through the lens of science fiction literature, and he had little time for the mundane machinations of everyday politics. Zoe, however, was an antinuclear activist and was involved in the Campaign for Nuclear Disarmament (CND), the influential antinuclear protest group led by renowned figures such as Bertrand Russell. CND organized the famous Aldermaston marches, the first of which occurred in April 1958, when thousands of protestors walked from London to the nuclear weapons facility in Aldermaston, fifty-two miles away. The march became an annual event, and in 1963, Zoe persuaded Hunter to cut short their Welsh honeymoon in order to take part in the protest, the first significant political action of Hunter's life.[33]

Hunter returned to Winnipeg in late 1963 with Zoe, who was pregnant, and began to work for one of the city's newspapers. He immersed himself in the local beat scene, wearing the obligatory black turtleneck sweater and engaging in long smoky conversations with other would-be philosophers and serious writers. He also began to work on *Erebus*, his first published novel. Semi-autobiographical in nature, *Erebus* was about the life of a rebellious and directionless young man working in a Winnipeg slaughterhouse in the early 1960s.[34] Raw, disturbing, and studded with brutal violence on and off the killing floor, the novel attempts to use the slaughterhouse as a metaphor for civilization, contrasting the foul and bloody abattoir with the solitude and beauty the narrator experiences on his camping trips to a remote island in Lake Winnipeg. Written about, and for, the emerging sixties generation, the book is ruggedly naturalistic and revels in brutal honesty and self-criticism. Although the novel's tone is often bleak, it is interspersed with humor and ends on an optimistic note: the cynical protagonist finds fulfillment and satisfaction as a teacher at a progressive school set up by his idealistic and recently blinded best friend. By the time the book was published, Hunter and his family had moved to Vancouver, where he was working as an assistant editor for the *Sun*. When the *Sun*'s editor heard that *Erebus* had been nominated for the Governor General's Award, one of Canada's most esteemed literary prizes, he gave Hunter his own column. Hunter quickly established a following among the paper's readers, many of whom viewed him, with admiration or disapproval, as the leading voice of the counterculture in Vancouver.[35]

Hunter's column gave him the freedom to dabble in whatever interested him. For the most part, he focused his attentions on the alternative side of town. Despite participating in the Aldermaston march, Hunter had nothing like the activist credentials of people such as Irving Stowe and Jim Bohlen. Nevertheless, he began pushing numerous causes and made appearances at various protests. At one point, he even planned to run for mayor of Vancouver, until his stoned-out hippie lawyer (and future Greenpeace director), Hamish Bruce, forgot to file the requisite paperwork on time. But his real interest at this stage was cultural analysis. He wanted to be an intellectual leader of the coming revolution, both chronicling its emergence and shaping its future. This revolution, as he saw it, would not be of the traditional political or social kind but rather would take place largely in people's heads. It would, in short, be a "consciousness revolution."

Although Hunter considered himself on the political Left, he was not drawn to Marxism; in fact, Isaac Asimov's *Foundation* trilogy was a far greater influence than *Das Kapital*. He was also a keen student of the Gestalt therapist Fritz Perls and of the Canadian philosopher Marshall McLuhan, in addition to being well-acquainted with the major authors who appealed to the sixties counterculture, such as Aldous Huxley, Alan Watts, Teilhard de Chardin, Gary Snyder, Paul Ehrlich, Erich Fromm, and Jacques Ellul. His column, and a lengthy strike at the newspaper in the late 1960s, afforded him the opportunity to read widely and travel throughout the United States, where he met influential figures such as Theodore Roszak and Fred Hampton, the leader of the Chicago Black Panthers.[36]

After Hunter wrote a favorable column about Gestalt therapy, Fritz Perls invited him to lunch and offered him free therapy. Hunter found Perls's holistic approach to psychology deeply convincing, particularly his claim that most people had very little self-awareness. This, according to Perls, meant that they could neither properly understand themselves nor empathize with others. Much of Gestalt therapy involved individual sessions with Perls, who would sit with his patients and make them project themselves onto an empty chair, encouraging them to engage in a dialogue with their projection. Rather than simply analyzing the content of the dialogue, Perls would ask patients if they were aware that they were tapping their foot while speaking or if they were conscious of the quality of their own voice and the fear or anger it contained. Gradually, patients would come to a deeper self-awareness and understanding of their psychological makeup. For Perls, awareness was the key to everything. Rather than always focusing on change as a method of improving one's psychological well-being, the thing to do was to simply be aware of one's problems—fully, deeply aware—and with that awareness came the ability to discover and solve problems.[37]

In 1970, Hunter published the first of two books that tried to explain the "consciousness revolution" that he saw taking place among the youth of the world. *The Enemies of Anarchy* was Hunter's attempt to outline a "gestalt sociology" that defined the problems of the world in largely psychological terms and traced the outlines of the coming consciousness revolution that would constitute humankind's last effort to save itself from certain extinction. Using a deft, if somewhat cheap, rhetorical ploy, Hunter defined "organized anarchy" as the institutionalization of selfish individualism, which had permeated society and dominated all modern nations, regardless of their political ideology or social organization. The "anarchists," therefore, were the bedrock institutions of modern society: the government, the church, the corporation, the university, and the people who ran them and worked for them. That is, the vast majority of people were anarchists, in that they contributed to the perpetuation of a system of aggressive individualism and unfettered acquisitiveness that was rapidly pushing the planet to the brink of environmental collapse. Pollution, for example, was "a direct result of unchecked anarchic industrial survival."[38]

The salvation of mankind, therefore, lay in adopting "attitudes and techniques which are anti-anarchic—which move, that is, *against* organized anarchy." Such a movement was already taking place, not as a result of education, but rather as a spontaneous self-adjustment within certain members of the population. This homeostatic process of modification was occurring largely, though not exclusively, among the youth of Western society and was most advanced in the United States, where the process of organized anarchy had developed the furthest. The chief agents of this effort to break down institutionalized individualism, according to Hunter, were ecology, humanistic psychology, a greatly heightened appreciation of oriental philosophy and religion, psychotropic drugs, rock music, and the increasing application of computers to a wide array of problems. In short, the sixties counterculture was the force for long-term stability and survival, while mainstream society was leading the world down the anarchic path to ruin. In Gestalt terms, the counterculture was effective not because it was trying to revolt, but because it was projecting itself onto Fritz Perls's empty chair and gradually attaining self-awareness on a mass scale.[39]

The solution to organized anarchy was what Hunter referred to as "large-scale integration," a process whereby individualism in all its negative forms would be rendered meaningless and the monolithic structures created to serve it would be demolished. Just as cancer cells worked for themselves at the expense of other cells, organized anarchy caused humans and institutions at all levels, from the individual to the nation-state, to work only for their own interests, regardless of the consequences for others. "The exploitive model of cancer," therefore, has become "the model on which our progress and development is based." Large-scale integration, through heightened human awareness and a more precise definition of human beings' role in the natural world, would weave together the torn fabric of society, breaking down its anarchic structures and leading to a state of "organic and highly defined organization," which Hunter called "convergence." Unlike the exploitative logic of cancer that characterized organized anarchy, a state of "convergence" would function like a healthy human body in which the cells of each organ, while working to maintain that organ, did not do so in competition with, or opposition to, the cells of all the other organs. As integration gradually broke down institutionalized individualism, people would once again view the world in a more holistic fashion and organize themselves and their institutions to function in ways that did not do continuous harm to society and the natural world.[40]

The society that would emerge from this process would be one in which "techniques are standardized and universally applied" and where

> the economic systems of each country will emerge as variations of the same post industrial theme. . . . Health standards, environmental controls, population controls, and so on, will be, of necessity, worldwide. . . . Planning will become absolutely essential in all areas. Organization—and the organizations will all be basically the same—will replace social order. . . . Global management will become a reality . . . and individuals everywhere will be organized into integrated units and the relationships between these units will be programmed—orchestrated, if you like—by large planning agencies using electronic data processing systems.[41]

Perhaps somewhat paradoxically, the projected end product of Hunter's "large-scale integration" and "convergence" sounds like Sweden if it were run by

IBM managers on acid. The struggle to achieve this enlightened managerial utopia would not take place in the street or in the factory, or even on the campus. There would be no rebellions, riots, or coups d'état, all of which had only brought "illusory change." It would instead be a psychological, inward-looking revolution guided by ecological awareness, Eastern mysticism, technological advances, and mind-altering drugs. These would allow people to break the shackles of institutionalized individualism and Cartesian dualism and to once again see the world whole.[42]

According to Hunter, the counterculture generation was already well on the road to developing a more holistic worldview. In the second of his books on the consciousness revolution, *The Storming of the Mind,* Hunter argued that the sixties counterculture was fostering a radically new perception of reality: "Since the critical problems which now threaten our existence can only be understood in terms of gestalts, or wholes, or flows, or synergistic effects," he declared, "the eye of the new generation is the only one capable of even seeing the problem . . . [since] those who have been trained to see things as being separate and unrelated . . . cannot readily see whole systems or flows." Instead, they remain mired in the Cartesian dogma which insists that humankind and nature are separate entities and must be viewed as such, leading to what Hunter, drawing on Marcuse, called the "operational mode," whereby reality consists only of that which can be "measured and related to a set of operations or functions." This means that science "has effectively divorced itself from any generalized awareness of larger processes which cannot as yet be brought into the framework provided by the concept of length." The equivalent in the social sciences has been the behaviorist reduction of the concept of nature "to a kind of quantifiable object—at best, an *object d'art,* at worst, a bio-electronic perpetual motion machine."[43]

While mind-expanding drugs, rock music, Eastern religion, and gestalt were all major contributors to the development of a more holistic consciousness, the single most important factor, according to Hunter, was ecological awareness. In fact, he argued, one could just as well say "the new holistic consciousness is basically an *ecological* consciousness," which breaks through the constraints of the operational mode and "leads us inevitably to a holistic philosophy." Ecology, Hunter explained,

> teaches us to recognize gestalts, to understand synergy, to appreciate the extent to which we are only one facet of an environment. Already, it has opened the path leading to a theology of the earth, thus short-circuiting Christianity and pointing the way back toward the kind of unfragmented, harmonious mental space understood perfectly by the ancient Chinese and—more intuitively—by primitive peoples the world over.

> Ecological consciousness, in short, is the common denominator of the real revolution which is just now beginning inside the gate of the comfortable concentration camp fashioned by technique. It is the root whose growth will make the difference, in the future, between freedom and unfreedom, stagnation and flowering.[44]

Ecology, as Hunter conceived it, was more than just another science. Rather, it approximated something like a spiritual awakening. It was not merely emerging in response to environmental collapse, as many people supposed, but was the chief agent in the breakdown of organized anarchy. Unlike cultural and political resistance, ecology was "post-historic," confronting not just social and political realities but "the penultimate question of how man must order his existence in relation to his planet." For every housewife who joins an antipollution group, Hunter enthused,

> one more human being has been added to the numbers of those who have had, or are about to have, some first-hand experience with the limitations of technological "rationality." It was faith in that "rational" order which initially turned her off to the possibility of regularities in nature which could not be quantified, categorized, and bottled by science. Now, as her faith in "rationality" crumbles under the impact of oil slick on beaches and poison in the air her children must breathe, she is that much less turned-off. . . . It is a small beginning but it has tremendous reach. Once one's unqualified faith in the rationality of science has been shattered, one is no longer a true slave to the technological order. At least a small part of one's psyche has been liberated.[45]

Since environmental problems affected the entire planet, ecology had the potential to be a far greater unifier than all previous liberation movements. The cries of Black Power, for example, may have fallen on deaf ears, "but the cries of Green Power will not," and if environmental deterioration is not at least slowed in the near future, wrote Hunter, anticipating Earth First! and the Sea Shepherd Conservation Society, "the non-violent phase will pass and into the forefront will move the Green Panthers or their equivalent."[46]

Clearly, Hunter's views on ecology were influenced by the popular ecology of the day, from Rachel Carson and Barry Commoner, to Paul Sears' and Paul Shepard's vision of ecology as a "subversive science."[47] But for Hunter, ecology was infused with a spiritual, almost transcendental quality that gave it the potential to become a "secularized religion," and it seems that his mind was groping toward the deep ecological philosophy that writers such as Arn Naess and Gary Snyder were in the process of outlining at around that time. Humans must learn to expand their circle of ethics to embrace the natural world, Hunter wrote in one of his newspaper columns.

> Brotherly love is nothing more than a preference for our own kind. If it does not flow over the boundaries into a love for all forms of life, it remains racist in the ultimate sense. . . . Justice which does not include all forms of life, from tadpoles to whales, is no justice at all . . . We humans are all elitists. We would exclude other forms of life from our trade-offs. . . . If we do not return quickly to a state of grace, of harmony with the earth and reverence for all life, non-human as well as human, then we will in all likelihood perish.[48]

Another aspect of Hunter's thinking, one that set him apart from the likes of Reich and Roszak, was a greater openness to technological advances, particularly in the sphere of mass communications. In light of his critique of rationalism and the "operational mode," his embrace of technology seems paradoxical. However, as historians Fred Turner and Andrew Kirk have argued, the idea that the sixties counterculture was antithetical to the technological developments of the Cold War era is simplistic and, in certain cases, incorrect. Tech-savvy countercultural networkers such as Stewart Brand—the founder of the famous *Whole Earth Catalog*—believed that the same technology that could be oppressive in one context could be liberating in another. For example, the field of cybernetics described a world that was becoming rapidly interconnected by interlinked information patterns. As Turner notes, this brought a sense of comfort to those who were inclined toward holistic thought: "in the invisible play of information, many thought they could see the possibility of global harmony."[49]

Among certain scientifically literate members of the counterculture, therefore, technology offered all sorts of liberatory possibilities. It is no surprise that a voracious reader such as Hunter picked up the on this vibe; his devotion to McLuhan and his fascination with science fiction suggest he had long been predisposed to it. While skeptics such as Ellul, Roszak, and Reich saw runaway technology as a prime cause of environmental collapse, bemoaning the "technocracy" that arose to facilitate its development and implementation, Hunter had a more optimistic view. Certain technological advances, by constantly compelling people to redefine their roles in the world, sped up the process of "convergence." In this regard, he was heavily influenced by McLuhan, whom he referred to as "our greatest prophet" and who saw technology as the locomotive of human history. The domestication of large animals, for example, freed people from the backbreaking labor of transporting heavy goods, thereby forcing them to redefine themselves as something other than beasts of burden. At the other extreme, the invention of the computer, with its superior capacity for linear thinking, gave people the opportunity to concentrate on more creative and innovative tasks. People were not so much replaced by machines as "displaced," forced to seek new roles and functions, to discover new resources, and to dig ever deeper into their potential. "The collective journey to the centre of the self is therefore something of a forced march, during which we are herded along by our own creations—which keep snapping at our heels like sheep dogs."[50] This is not to suggest that Hunter's attitude to science and technology was unproblematically celebratory. Rather, he urged activists to master new forms of technology, particularly those associated with mass communications, and use them for their own ends.

Like McLuhan, therefore, Hunter saw technology as a catalytic force in the evolution of human consciousness. The mass communications systems of the late twentieth century, he argued, would have a tremendous impact in reshaping society and altering human self-awareness. For McLuhan, the new electronic media constituted a kind of globalized central nervous system, though one that was worn on the body's exterior and, therefore, was constantly exposed to the outside world. In his study of McLuhan's influence on Greenpeace, Stephen Dale provides an elegant summary of McLuhan's central thesis:

> He saw the planet as a "global village" in which events and impressions from around the world are absorbed by microphones and camera lenses and teletype machines in the same way that sense impressions are gathered by the eyes, ears, and skin of the human body. All of this information is analysed, contextualized, and interpreted in the brain of mass media image factories and sent back out through that electronic nervous system to the various appendages of human society around the globe. According to McLuhan, this creates a "unified field of experience." People in opposite corners of the Earth see their own experiences placed in the context of a manufactured composite image of global events and are told how to react to the outside world, in much the same way as the limbs of a human body are instructed to react to impressions gathered through the eyes, ears, nose, or some other part of the anatomy. The electronic media made it possible to imagine a world that thinks—and acts—in unison.[51]

For Hunter, the mass media and McLuhan's "global village" constituted the medium through which the new holistic, ecologically inspired human consciousness was to be propagated on a massive scale. Quoting McLuhan, he argued that "if literate western man were really interested in preserving the most creative aspects of his civilization, he would not cower in his ivory tower bemoaning change but would plunge himself into the vortex of electric technology and, by understanding it, dictate his new environment—turn ivory tower into control tower." While revolutionaries of the past had required armed struggle as a means of achieving their ends, the modern mass communications system provided a "delivery system" through which the agents of the new consciousness could "bomb" the enemy's mind. Throughout the postwar era, according to Hunter, the mass media had been contributing, though largely unintentionally, to the breakdown of organized anarchy in the McLuhanesque fashion outlined above. By the late 1960s, however, the process was becoming far swifter and more deliberate for "behind the cameras now are not the old consciousness technicians, but the new ones." And if the pen was a thousand times mightier than the sword, then, Hunter estimated, television was at least a million times more powerful. Building on his decidedly military "mind bombs" metaphor, Hunter argued that television could be "targeted with complete accuracy to strike at a point precisely two inches behind the victim's eyes. No bullet flies so fast, so far, with such unerring accuracy. Not even a hydrogen bomb can affect so many people at once."[52]

Bob Hunter was the single most influential figure in the first decade of Greenpeace's history. Initially, however, he was a junior partner in a coalition that was dominated by Irving Stowe and Jim Bohlen. While there was a certain amount of overlap between Hunter's countercultural view of environmentalism and Irving Stowe's New Left approach, there was also a considerable degree of tension between them. In Hunter's Hegelian view of history, the best prospect for revolutionary change lay in radically altering people's perception of the world and their place within it. The most promising means of achieving this lay in harnessing the tools of mass communications technology to disseminate the ideas and values of the counterculture. Such an approach downplayed the need for traditional political tactics, such as lobbying and pressuring politicians. In Hunter's opinion, engaging in the political process, with its drawn-out procedures and its tendency toward compromise, was like wading through a giant, strength-sapping bog. Rather than becoming mired in political battles that always favored the forces of the status quo, revolutionaries would be better off bypassing politics as much as possible. Consciousness change, after all, would not be achieved with mere placards and petitions. Rather, it was LSD, the *I Ching*, and, above all, the camera that represented mankind's best hope for a postindustrial, ecologically sustainable future.

To Irving Stowe and Jim Bohlen, Hunter's theories seemed a little too grandiose and impractical. It was naïve, they felt, to believe that the Cold War, consumerism, and the politics that sustained them would simply melt away once enough people started to think outside mainstream cultural parameters. Furthermore, although both men were liberal in their social views, neither felt comfortable with the more libertarian side of the counterculture, particularly the emphasis on mind-altering drugs. Despite their differences, however, the values that united these three men were far greater than those that divided them. Bohlen's activist approach to wilderness preservation, Stowe's New Left critique of the military-industrial complex, and Hunter's countercultural analysis all reflected a deep suspicion of notions of progress, growth, and security that mainstream society took for granted. All three embraced the holistic philosophy of popular ecology and viewed the environment as the most pressing issue facing humankind. Each, in his own way, embraced ideas that questioned the values and norms of mainstream Western society. Although nobody could have predicted this a decade earlier, late 1960s Vancouver turned out to be a propitious place for people who wanted to enact a few radical ideas.

3

The Canadian Crucible

British Columbia seems an unlikely birthplace for a radical new form of environmentalism. A vast, rugged, sparsely populated land, its political culture was almost entirely shaped by the resource-extractive industries that dominated its economy into the 1970s. From the early 1950s onward, the Social Credit (Socred) government, under the leadership of the demagogic W. A. C. Bennett, aggressively promoted a virulent form of state capitalism aimed at wringing the utmost from the province's vast reserves of timber and mineral wealth. The population, by and large, approved. Bennett was able to forge a stable electoral majority based on the support of big business, rural conservatives, petit-bourgeois shopkeepers, and the antisocialist middle classes in Vancouver and Victoria. Nor did Socred's major opponents—the province's powerful labor unions—have a more benign view of humankind's relationship with the natural world. They sought a more equitable distribution of the province's resource wealth rather than resource conservation or wilderness preservation. Moreover, the distant federal government had little influence upon British Columbia's management of its public land. Even if Ottawa had possessed powers akin to those of Washington, D.C. over its vast western hinterland, there is little in the history of Canadian conservation to suggest that things would have been substantially different. Until the late 1960s, there were no influential environmental organizations, such as the Sierra Club, active in the province.

Yet hardy plants sometimes spring forth from seemingly barren soils. By the late 1960s, British Columbia as a whole may not have been ripe for the appearance of a group such as Greenpeace, but Vancouver was different. Throughout the decade, the picturesque provincial city had been shaped by a series of movements and events, many of them distant and bearing no direct connection to life in British Columbia, which laid the foundations for the emergence of Greenpeace. These included the rise of the counterculture, the Vietnam War, American nuclear testing in the Aleutian Islands, and a growing anti-Americanism on the part of many Canadians. These issues, along with various demographic and social changes that increasingly put the city at odds with the population in the province's hinterland, created an environment that was conducive to the growth of the vibrant oppositional subculture from which Greenpeace emerged. Many of the people who played influential roles in Greenpeace's early history were the products of this milieu. This chapter will introduce them and examine how their lives were shaped by this distinctive subculture.

By the late 1960s, the most conspicuous elements of the Vancouver counterculture congregated around the inner-city quarters of Gastown and Kitsilano. With their proximity to the local beaches, their stock of old, cheap, and occasionally abandoned houses, and a multitude of cafés and stores catering to various alternative lifestyles, these areas were a magnet for dropouts, alternative lifestylers, and disaffected youth. It was not long before Vancouver developed a reputation as a haven for the counterculture, and this, combined with its mild climate (by Canadian standards), turned the city into a mecca for countercultural experimentation. Vancouver, in short, became the Canadian equivalent of San Francisco, and Kitsilano rapidly developed into a local version of Haight-Ashbury.[1]

Naturally, the relatively sudden emergence of a counterculture ghetto provoked considerable alarm among the more conservative citizens of Vancouver, who tended to lump street gangs, vagrants, and hippies together, using the worst elements of the former two to define the latter.[2] One government study, however, estimated that genuine "lifestyle Hippies" constituted only 15–20 percent of the of the Kitsilano population. The rest were mostly school dropouts, runaway teenagers with family problems, people with mental illness, and university and high school students exploring the counterculture during vacation. Although many people recognized that a heavy concentration of such groups could cause severe social problems, few were concerned that any kind of organized movement would emerge from the scene. One committee report in 1967 found "no concrete evidence to support the implication of Hippies in any political party or movement" and expected the scene to exhaust itself rapidly. Even the more overtly political arm of the alternative scene, the Vancouver Liberation Front, described by a local official as a "cloak for an assortment of radical groups from Hippies and Yippies to Trotskyites, Marxists, and anarchists," was still seen as too disorganized and immature to affect any kind of serious political or social movement.[3]

Those who dismissed the political potential of the local counterculture, however, soon received a rude awakening. Despite the high proportion of transient youth and the putatively apolitical nature of the counterculture scene, there were small but significant pockets of activists within this subculture who were politically engaged and who organized protests and demonstrations. Frequently, these protests bore the stamp of the guerilla theater pioneered by a group of American activists who called themselves Yippies. One of the peculiar features of the Vancouver's alternative scene was that, to a greater degree than most other cities, it tended to blend the radicalism of the New Left with the symbols and lifestyle of the counterculture, a trend that was evidenced by the popularity of Yippie-style tactics.

The two most influential figures in the American Yippie movement were Jerry Rubin and Abbie Hoffman. Both came from middle-class Jewish families: Rubin from Cincinnati and Hoffman from Worcester, Massachusetts. By the mid-1960s, both had become disillusioned with the American Left's earnest and, in their eyes, rather staid radicalism. At an antiwar rally in March 1966, Rubin told the assembled crowd that in order to reach people "who have never heard our ideas before we are going to have to become specialists in propaganda and communication." Radicals, in other words, needed to learn to manipulate the tools of mass communication and the symbols of mass society if they were serious about changing America. Mere language, Rubin insisted in 1968, "does not radicalize people—what changes people is the emotional involvement of action. What breaks through apathy and complacency are confrontation and actions." Rubin, therefore, supported "everything which puts people into motion, which creates disruption and controversy, which creates chaos and rebirth . . . people who burn draft cards . . . burn dollar bills . . . say FUCK on television . . . freaky, crazy, irrational, sexy, angry, irreligious, childish, mad people."[4]

Hoffman, like Rubin, also found the standard political tools of the Left—the manifestos, meetings, and marches—to be tedious and increasingly ineffectual. Instead, he advocated various forms of street theater that drew attention to issues through the playful manipulation of symbols or the parodying of opponents. Rather than earnestly proclaiming that the government should "end the war" or "fight poverty," activists should draw mass media attention to these issues by staging weird, absurd, and colorful protests. In early 1967, Hoffman put his theory into practice when he and some friends joined the official tour of the New York Stock Exchange. They behaved as normal tourists until they reached the balcony overlooking the exchange, whereupon they ran to the railing and began throwing handfuls of dollar bills to the stockbrokers massed below. The brokers reacted just as Hoffman had hoped, clamoring wildly to catch the falling notes and crawling around on all fours to pick up any that had slipped through to the floor. Inspired by Marshall McLuhan's work on how mass media was changing society, Hoffman felt that the most important role of the activist was to engage in activities that promoted consciousness change rather than agitating for political change via the standard protest repertoire of the Left. Activists needed to realize that for youth in particular, the understanding of reality came not through actual experiences with everyday life, but rather from the images that television beamed en masse into people's homes. Hoffman's ideas clearly influenced Bob Hunter, Greenpeace's resident countercultural intellectual.[5]

Apart from whatever influence his ideas may have had on Hunter and others within the Vancouver counterculture, Jerry Rubin also played a direct role in inspiring and galvanizing the local Yippie movement. In October 1968, Rubin was invited to appear at the University of British Columbia by a left-wing student group that was attempting to publicize the university's involvement in military-related research. After a rousing speech, Rubin succeeded in inciting a large group of students and sundry radicals, accompanied by a pig, to invade the faculty club. The "occupation" spilled out onto the campus lawn, where some 200 people spent the night, with the number swelling considerably over the next two days. The event was carried out in classic Yippie fashion, with the protestors proclaiming the occupation a "Festival of Life North" (a reference to the "Yippie Festival of Life" that had taken place in Chicago that summer) and electing a "mayor" for their ephemeral shantytown. Musicians, street theater actors, and a mime troop also joined the festival, which broke up after three days when protestors negotiated a deal with the university administration in which they promised to peacefully end the protest if the university agreed to drop all its charges against the people involved.[6]

Among the protestors was Rod Marining, a nineteen-year-old street theater activist from North Vancouver. A tall, slightly goofy-looking teenager with a mane of wavy chestnut hair and two different-colored eyes, Marining led a street theater group called the Rocky Rococo Company, a band of itinerant amateur thespians who were willing to perform at various protests in exchange for three gallons of wine. Marining's mother had allowed the sons of American friends who were evading the draft to stay at their house in North Vancouver, and Marining, not surprisingly, became highly critical of America's involvement in the Vietnam War and any Canadian complicity with it. He had also developed a concern for environmental problems, an interest that had in part been sparked by Vancouver's rapid expansion into the wilderness areas along its fringes. Marining grew up in such an area and was particularly incensed when his favorite childhood frog pond in East Vancouver was demolished to make way for a McDonald's restaurant. He was among those who gave a speech at the UBC occupation, after which he was elected the "non-leader" of the Northern Lunatic Fringe of Yippie![7]

The Vancouver Liberation Front (VLF) was another band of Yippie-inspired street theater activists, though one that was more confrontational than Marining's group. The VLF's antics included performing an exorcism on a Vancouver police station and clashing with the police at one of its protests at a local beach.[8] Among their members was a burly teenage sailor from eastern Canada named Paul Watson. Watson, or "Captain Watson," as he preferred to be called in his post-Greenpeace days, has become a controversial, indeed polarizing, figure within the environmental movement, founding and running the radical Sea

Shepherd Conservation Society, whose acts of "ecotage" include ramming fully manned whaling and fishing boats at sea. He is also notorious for his frequent acts of historical revisionism, making it difficult to construct an accurate biographical portrait. Bob Hunter once described him as a knee-jerk radical who had a North Vietnamese flag stitched to his army jacket and wore "Red Power buttons, Black Power buttons, and just about any kind of anti-establishment button that could be imagined." Watson denies this, arguing that Hunter was confusing him with Rod Marining. However, others who knew Watson at the time tend to lean toward Hunter's version, and Watson's active participation in the armed standoff between the Oglala Sioux and the U.S. government at Wounded Knee in 1973 lends credence to their views. In any case, Watson does not deny that he was associated with the VLF, which remained one of Vancouver's more influential Yippie-inspired activist groups throughout the late 1960s and early 1970s.[9]

The Vancouver counterculture's most influential organ was an alternative newspaper called the *Georgia Straight*. The *Straight* was founded by a group of beatnik writers in the mid-1960s. One of its founders, Dan McLeod, who continues to run the paper as of this writing, recalls that it was modeled after the American alternative newspapers that were springing up throughout the United States and banding together under the umbrella of the Underground Press Syndicate. Whereas most of the American papers tended to fall into two distinct categories—either serious *Ramparts*-style political journals or more spiritual, psychedelic, hippie publications—the *Georgia Straight*'s founders, feeling that Vancouver was too small to sustain a variety of alternative publications, decided that they would try to blend the two styles. This decision no doubt contributed to the blurring of the line between the counterculture and New Left-style activism that characterized the city's subculture. The result was a colorful, anarchic publication that dealt with a veritable potpourri of subjects. It was not uncommon to find columns by Vancouver Liberation Front activists arguing for the overthrow of the state side-by-side with articles about Hari Krishna or nude volleyball tournaments. In addition, the *Straight* had a strong commitment to ecology and regularly reported on environmental issues throughout the province and on the broader implications of adopting an ecologically centered lifestyle. It would also become a conduit for some of the more radical ideas of early Greenpeacers.[10]

The site where Vancouver's alternative scene met the city's mainstream was Stanley Park, a sprawling urban green space adjacent to Vancouver's downtown. Throughout the late 1960s and early 1970s, the park's grassy hillocks and sandy beaches were gathering places for all manner of demonstrators, allowing old-time peaceniks to mingle with Yippies and providing newcomers with the opportunity to experience a rich array of protest cultures in a relaxed, almost picnic-like atmosphere. The spring and summer of 1970, for example, saw the park play host to Festival for Survival, a demonstration against urban development and pollution in Vancouver sponsored by the BC Sierra Club, SPEC, the Unitarian church, and various student groups. The event was organized by a group called the Society to Advance Vancouver's Environment (SAVE), which counted Irving Stowe among its founders. A *Georgia Straight* journalist was struck by the similarity the festival bore to various park protests in the United States and by the way that many people who had never demonstrated before rubbed shoulders with Yippie pranksters and "old time disarmament marchers and protest type people."[11]

In early 1970, Vancouver's city government announced that it had given the luxury hotel chain, Four Seasons, permission to construct a huge hotel, replete with several towers, at the entrance to Stanley Park. Though technically private property, the area had always been accessible to the public, and the decision provoked a storm of controversy. The issue soon developed into a crusade for various activists who were determined to frame it as a battle between private development and public space, in addition to portraying it as an act of environmental vandalism. Rod Marining, who worked as the daily horoscope editor at one of the city's major newspapers, the *Vancouver Province*, had read numerous wire stories about the People's Park protest in Berkeley the previous year and was inspired to imitate the action. Along with his street theater group and several dozen friends, Marining set up a camp near the proposed Four Seasons construction site. He was joined by members of the more confrontational Vancouver Liberation Front, among them Paul Watson. Knowing that the construction site would be abandoned over a long weekend during the spring, the protestors used the opportunity to lead an audacious, and ultimately successful, direct action. Borrowing equipment from various sympathizers—including a bulldozer from a construction company, sod and saplings from a nursery, and a wheelbarrow from Irving Stowe—they removed the bolts from the fence and, with the help of some 300 fellow protestors, pulled the entire structure down in just a few minutes. The activists quickly covered the roads and construction areas with sod, into which they planted the saplings, before setting up their tents on the site and proclaiming it "All Seasons Park."[12]

Using his contacts in the local media, Marining was able to release his version of events to the press before the Four Seasons' management or the city government could react. Before the weekend was over, thousands of street kids and hippies had flooded into the park from Kitsilano and Gastown. On Monday morning, the conservative, prodevelopment mayor, Tom Campbell, and his police chief arrived to find that the construction site had been completely transformed and colonized by the very people whose indolence and lack of organization they were constantly criticizing. Campbell threatened to send in riot police and have the protestors arrested, but Marining and his supporters refused to back down. For the next several weeks, a standoff between the police and the site's occupiers ensued. Eventually, a father of one of the protestor's agreed to purchase the property for $4 million and promised not to develop it. For Marining and others, the entire protest was a salutary lesson in the efficacy of direct action, one that helped galvanize various elements of the city's disparate and inchoate alternative scene. David Garrick, a friend of both Watson and Marining and a future Greenpeacer, called the protest "an expression of the collective level of liberation of everyone having anything to do with it." It was another example of the overlap between the city's counterculture and its politically active radicals. In Vancouver, the Trotskyites dropped acid and the hippies went to protests, and people such as Rod Marining could comfortably keep one foot in each camp.[13]

One of the journalists who supported Marining's actions was Ben Metcalfe, a former *Province* columnist who had become a well-known radio personality on the CBC station in Vancouver. Among the local press fraternity, Metcalfe had a reputation for being something of a maverick. He was a steadfast and vocal critic of British Columbia's development-at-all-costs mentality and of Premier Bennett's Socred government. Upon hearing of the All Seasons Park protest, Metcalfe, accompanied by a *Province* reporter, took several gallons of wine down to Marining and his friends as an expression of solidarity and admiration. Despite the fact that he was old enough to be the father of most of the protestors, Metcalfe had more in common with them than the average middle-aged, middle-class Vancouverite. He had tried the various drugs that were popular within the counterculture, including LSD, and had become a strong supporter of Vancouver's growing environmental movement, coming to it from a more traditional conservationist background.[14] Furthermore, Metcalfe was McLuhanesque in his view of the media and was therefore attracted by the Yippie antics of the local alternative scene. Among his various media pursuits, Metcalfe also had a regular spot on the radio as a theater critic, a vocation he claims to have invented for himself. Not surprisingly, the old adage "the world is a stage" was not merely a cliché as far as Metcalfe was concerned but approximated something closer to a philosophy of life—one that blended very well with McLuhan's idea of a "global village."[15]

Born in Winnipeg in 1919, Metcalfe's parents were middle-class English immigrants who instilled in their son a strong affinity for the mother country. At the tender age of fourteen, and in the midst of the Great Depression, Metcalfe left school and traveled to England. Like many teenage boys, he dreamed of flying an airplane. When he was fifteen, he tried to join the Royal Air Force but was rebuffed by the recruiting officer who told him to go home and wait until he was old enough to lie about his age. He was finally recruited in 1936 and served as an RAF officer in Singapore, India, and North Africa during World War II, experiencing, in his own words, "a very simple, unextraordinary war." As soon as it was over, he found work as an information officer for the British foreign office in occupied Germany. After one year in Germany, he set off for Paris, where he became a reporter for London's *Daily Mail*, and for the next decade he had no trouble finding work as a journalist throughout Europe. He returned to Canada in 1953, having spent more than half his life in Europe, and opened a public relations agency, before taking a job with the Vancouver *Province* in 1956. Soon thereafter, he picked up additional work as a theater critic on CBC radio. Metcalfe's comments on theater were often sociological, and he tried to glean insights into modern society from the plays he critiqued. His erudition and willingness to espouse controversial opinions led a producer to ask him if he would be interested in expanding his repertoire into social and political commentary. Metcalfe jumped at the chance, and over the next decade his dry sense of humor, droll satire, and occasionally impassioned commentaries drew a wide following of devoted listeners. His acerbic criticism of the Social Credit government gave most people the impression that he stood firmly on the political left, but Metcalfe always thought of himself as a political maverick who was more interested in fighting for common sense than promoting any particular ideology.[16]

Metcalfe had a reputation as an investigative reporter who was not averse to risking his health in order to pursue a story. At one point, while investigating organized crime in Vancouver, he was beaten to a pulp and left for dead by the side of the road. He created a minor sensation when he discovered a "lost" tribe of Indians in the BC hinterland. Although the tribesmen did not consider themselves to be lost, they certainly had not been heard from in a while. But the most important experience of his journalistic career, and perhaps of his entire life, was his experimentation with LSD in 1959.[17]

Metcalfe did not have to travel very far to experience his first acid trip—a mere ten miles south of Vancouver to the Hollywood Hospital in New Westminster, British Columbia, where the maverick psychedelic analyst and LSD proselytizer, Albert Hubbard, the man who would turn Timothy Leary on to LSD in 1963, had been allowed to set up Canada's first private LSD therapy clinic. At the time, the most powerful psychotropic drug known to science was mostly being used in experimental therapy with alcoholics and mental patients, but Hubbard's experience with LSD had convinced him that it offered an escape route from the destructive path on which humankind was heading. He therefore hoped to administer it to as many people as possible, particularly to those with power and influence. Metcalfe was not quite Walter Cronkite, but he was certainly making a name for himself in western Canada. So Hubbard invited him in for a two-day controlled LSD session. The session took place in Hubbard's custom-designed therapy room, with a copy of Dali's *Last Supper* over the couch, Gauguin's *Buddha* on the far wall, and a statue of the Virgin Mary bathed in a warm, candlelight glow. The trip initially took Metcalfe on a whirlwind tour through time and space. He rode with Genghis Khan, witnessed the fall of Rome, saw flashes of famous historical battles and glimpses of unmistakably Shakespearian figures. He was transported through the Milky Way and played marbles with the angels, before collapsing into a fit of huge heaving sobs which "Dr." Hubbard (the doctorate was from an obscure diploma mill in Tennessee) soothingly told him was all the repressed material being leached out of his psyche: "what we bury to become men." "I must be insane!" Metcalfe screamed. "We're all insane when it comes to confronting ourselves," Hubbard reassured him gently.[18]

The LSD experience altered Metcalfe's perceptions of both himself and the outer world. It showed him how egotistical his worldview had been up until that point. As a result, he began to take an interest in Zen philosophy and to explore various ways of overcoming his ego, which, over the years, had become inflated with the adulation he had received as a hotshot journalist and debonair ladies' man. He also began to question the Cartesian dualism of modern thought and developed a more holistic view of human beings' relationship with the natural world. A keen fly fisherman (he had fished with BC's most famous naturalist, Roderick Haig-Brown) and a lover of wilderness, Metcalfe began to feel a deeper appreciation for the natural world than he had in his pre-LSD days. Suddenly, fishing became more than just the thrill of catching a trout in a beautiful setting. It became suffused with a reverent, almost spiritual quality as Metcalfe developed an ecological awareness that was strongly influenced by the holistic philosophy that underpinned popular ecology. His incipient ecological consciousness was reflected in some of his journalistic work and political commentaries, and he became particularly critical of British Columbia's frontier attitude toward development. In an effort to spread environmental awareness, he organized and paid for a series of billboards throughout Vancouver with cryptic messages such as: "Ecology: Look it up—you're involved." and "Ecology, the last fad."[19]

The notion that the fate of nature should be left to technocrats and experts, a position taken by many prodevelopment scientists and politicians, rankled Metcalfe. When the chief forester at Macmillan Bloedel, Angus MacBean, criticized the "amateur ecologists" among the general public for daring to question the practices of the industry, Metcalfe unleashed a stream of sarcastic invective over the CBC airwaves:

> It's really quite remarkable how many ordinary dumb citizens, trusting only their eyes, ears, nose and throat, claim to know something about the environment—which, as every forestry engineer, chemist, and miner can tell you right away, belongs to the experts, and they'll look after it thank you very much. It's even more remarkable to hear these same ignoramuses complaining about something that hasn't

> even happened yet. The least they could do is wait till the mercury count gets to a decent level, and they can see the fish vanishing from the river and the birds falling from the sky before they set up their portable wailing walls just outside local industry, which is just trying to make an honest buck. . . . It took real experts to kill Lake Erie—not a bunch of amateur fanatics. It took decades of mechanical genius to turn out enough internal combustion engines to get the fumes into the stratosphere—mere faddists from some church group couldn't have done that.[20]

Metcalfe was also furious at the Socred government's plans, in cooperation with an American utility company, to damn the Skagit River on the BC-Washington border. "The Skagit is our river," he proclaimed on his CBC spot, "and we don't have any more reason to flood it than we have to flood Stanley Park. We don't have to explain anything to the Seattle City Light Company. We don't have to tell them why we don't want it destroyed. We don't have to come up with alternatives for them. All we have to do is tell them not to touch our river, that's all."[21]

Metcalfe's feisty "just say no" attitude toward environmentally destructive development struck a chord with urban middle-class liberals, as well as those in the alternative scene. His passion for theater, his McLuhanesque media philosophy, and his interest in environmental issues were emblematic of the social and intellectual currents that were swirling through Vancouver's subculture and occasionally blending with the social and political mainstream. Little wonder, then, that Metcalfe, like Rod Marining, Paul Watson, and Bob Hunter, would become a prominent member of Greenpeace.

Another important figure in Vancouver's activist community, one who bridged the divide between academic science and the counterculture, was a young ecology graduate student from UBC named Patrick Moore. A curious product of the British Columbian frontier, Vancouver's boarding school elite, and the counterculture, Moore was raised on his family's floating logging camp on the isolated and rugged northwest coast of Vancouver Island. Moore had a secular upbringing and was heavily influenced by his mother, a self-styled frontier intellectual who had Moore reading Bertrand Russell by the age of fifteen. With little experience of life outside the logging camp and the elite Vancouver boarding school where he spent his teenage years, Moore arrived at UBC in 1964 as a somewhat naïve freshman but soon became involved in the various causes and movements sweeping through North American universities at the time. By his junior year he had let his curly hair grow out into a frizzy mop and was well and truly ensconced in the campus countercultural scene. Despite his appearance and hippie affectations, however, Moore kept one foot firmly anchored in the university's forest biology department, and by the end of his undergraduate career he became committed to pursuing a PhD in ecology.[22]

In retrospect, Moore feels that his interest in ecology was partly the result of his secular upbringing. In a sense, it compensated for the absence of religion in his childhood. Where the world had previously appeared to him as a fractured and atomized collection of individual biota and landscapes, ecology provided a connecting thread that allowed him to "gain insight into the mysteries of the rainforest as experienced by a child." Viewing the temperate rainforest through the prism of ecology, things that had previously appeared disconnected, such as the giant cedar trees and the salmon caught offshore by the fishing boats near his father's logging camp, were revealed to be part of the same ecosystem, both reliant on climatic conditions and nutrient cycles that had been responsible for the region's unique ecological development over many millennia. "I realized," Moore wrote, "that the feeling of tranquility and wonder I had experienced as a child in the rain forest was a kind of prayer or meditation. Ecology gave me a sort of religion, and with it the passion to take on the world. I became a born-again ecologist."[23]

Moore also saw in ecology a valuable political and legal tool, one that could be used to bring polluting industries and environmentally recalcitrant governments into line. This politicized ecology was widespread among environmentally conscious UBC students, some of whom formed an activist group called the Environmental Crisis Operation (ECO), the goal of which was to "take thoughtful action against specific pollution crimes in B.C. and to inform profs and students about environmental collapse."[24]

Moore's doctoral adviser, C. S. Holling, a renowned ecologist who specialized in modeling predator-prey relationships and who developed important theories and models about resiliency and stability, encouraged this form of activism. By the late 1960s, *Time* referred to Holling, along with Barry Commoner and Eugene Odum, as one of the "new Jeremiahs" of ecology. The scale of anthropogenic environmental destruction had made Holling "stark terrified at what was going on in the world." Suddenly, the idea of studying nature as though humans did not exist began to look absurd. He was struck by "the profound and striking similarities between ecological systems and the activities of man: between predators and land speculators; between animal-population growth and economic growth; between plant dispersal and the diffusion of people, ideas and money." Given this milieu, it is not surprising that Moore's dissertation was as much an exercise in environmental activism as a work of scientific research.[25]

In the late 1960s, the Utah Construction and Mining Corporation had applied for a Pollution Control Permit that would enable it to daily dispose of 9.3 million gallons of mine tailings into Rupert Inlet, near the Moore family's lumber camp on Vancouver Island. Moore's dissertation research constituted a kind of de facto environmental impact statement: He conducted experiments to determine the water circulation in the inlet and the effect of the tailings discharge on water turbidity, as well as analyzing the environmental impact of several other similar mining operations in southern British Columbia. He argued that the results conclusively showed that the inlet was not a suitable environment for mine-waste disposal and organized a campaign against Utah Construction, presenting his information to various government departments and the media. Both he and his department came under pressure from the mining industry and the provincial government, and Moore was warned that he would have difficulty finding a job if he continued to pursue the topic. Undeterred, he persisted with his research—even after the Pollution Control Branch had given Utah Construction its permit—arguing that the structures of decision making and the province's evaluation process were inadequate to deal with resource conflicts that involved major environmental considerations. "The role of economics in resource use decisions," he wrote, "should be a limited one and . . . environmental considerations should be paramount in determining the constraints within which the development should take place." In the end, his dissertation was passed, but only after the dean, the venerable Canadian wildlife ecologist Ian McTaggart Cowan, brought in an adjudicator to break a tie within the PhD committee, which, according to Moore, had been stacked with people sympathetic to the mining industry. For Moore, therefore, ecology, in addition to being an academic discipline, was also Bob Hunter's "secular religion" and Paul Shepherd's "subversive science," and a hippie lifestyle was not necessarily incompatible with any of them so long as it did not descend too deeply into mysticism and esoterica.[26]

The Vietnam War was another factor in Vancouver's transformation from a conservative provincial town into a thriving hub of alternative politics and lifestyles. Apart from providing an issue around which various radical groups could coalesce, the war brought to Vancouver a steady stream of Americans who refused to serve in the U.S. military or to allow their sons to be conscripted. Some of these were older men and their families, such as the Bohlens and the Stowes, but the vast majority were young men evading the draft. Furthermore, they were not just any young men. Canada's immigration regulations replicated the class and race biases of American draft laws. Just as a poor, working-class, black youth would find it difficult to get a draft deferment, the same youth would be hard-pressed to gain immigration status in Canada. On the other hand, those middle-class, college-educated, and mostly white men who, for whatever reason, were unable to obtain a deferment, could quite easily qualify for immigrant status in Canada. Inadvertently, therefore, the system promoted the immigration into Canada of a large number of well-educated, middle-class white men who were firmly opposed to American involvement in the Vietnam War, and who, by extension, tended to align themselves with the more radical elements in the body politic. Not surprisingly, these people frequently came to be influential figures in the Canadian antiwar movement. Furthermore, many Canadian activists encouraged the influx of draft-resisting Americans, whose experiences in the United States tended to make their own struggles seem rather trivial. The Canadian Student Union for Peace Action (SUPA), for example, published *A Manual for Draft-Age Immigrants to Canada,* which sold 65,000 copies in the latter half of the 1960s.[27]

Many of the most radical Americans to cross the border had been associated with the American New Left and found it easy to fit in with its Canadian counterpart, which, both in terms of its demographic makeup and the causes for which it was fighting, was very similar to the movement they had left behind. There was, however, one significant difference between the two groups: since Americans were opposing the actions of their own nation, they tended to abjure nationalism in all its guises and instead embraced the internationalist perspective that was traditionally adhered to by the Old Left. Canadian New Leftists, however, came to espouse certain forms of Canadian nationalism as a defense against what they perceived to be the increasing dominance that the rampantly capitalist and militarist United States was asserting over their nation.[28]

According to historians John Thompson and Stephen Randall, the 1960s "stand out as the decade of greatest Canadian domestic divergence from the United States."[29] While the Johnson administration's efforts to build a Great Society were being sacrificed on the altar of Vietnam, Canadians were building an extensive welfare system, which included universal health care. Also, the rise of the New Democratic Party in the 1960s gave the Canadian Left a viable parliamentary presence for which there was no equivalent in the United States. Little wonder, then, that many Canadians felt that their society was heading in a different direction from that of the United States and that they resented any actions that smacked of American imperialism. Even the election in 1968 of Pierre Trudeau, a man with no fondness for nationalism and distaste for populist rhetoric, did little to counter the growing antipathy that many Canadians were feeling toward their neighbor.[30]

The wave of anti-Americanism in Canada—and particularly in British Columbia—peaked in the late 1960s and early 1970s when the American military began conducting a series of underground nuclear tests on a remote island in the Aleutians, a chain of islands that extends like a disjointed tail from the Alaskan Peninsula to the Kamchatka Peninsula in northeast Russia. In 1964, one of these islands—Amchitka—was chosen by the Department of Defense and the Atomic Energy Commission as a potential site for large underground tests deemed too dangerous for the Nevada proving grounds, given their proximity to the burgeoning casinos and high-rise buildings of Las Vegas. The first blast, Longshot, occurred on October 29, 1965, and was primarily an effort at gauging the ability of the U.S. military to detect Soviet tests in the Far East.

There was virtually no publicity about the 80-kiloton blast, and protest was nonexistent. The next test, Milrow, was a one-megaton "calibration test" designed to determine if the island could withstand an even larger device that the AEC was planning to explode as part of its Spartan anti-ballistic missile development program.[31] Unlike Longshot, however, Milrow, which was scheduled to take place on October 2, 1969, provoked a storm of outrage across Canada and particularly in Vancouver, which was the closest major Canadian city to the blast site. Many feared that the explosion might trigger an earthquake, causing a huge wall of water, as Bob Hunter vividly put it, to "slam the lips of the Pacific Rim like a series of karate chops."[32] Something similar had actually occurred in 1964, when an earthquake centered in the Aleutians unleashed a huge tidal wave that battered the west coast of Vancouver Island, causing over a hundred deaths and millions of dollars worth of destruction.[33] The fear and outrage prompted by the Amchitka tests was shared by the *Vancouver Sun's* editorial writers:

> The AEC is playing with our marbles. How dare it! Who says that Canadians, or anybody else, are prepared to pay this price for an advancement in nuclear overkill? The AEC may not be responsive to consensus, but an alerted North American community undoubtedly will do its utmost to see that it doesn't get away with its gambles as easily in the future as it has in the past.[34]

The detonation of a nuclear bomb on October 2, 1969, managed to bring together, at least for a day, a spontaneous coalition of students, peace activists, environmentalists, hippies, Yippies, Maoists, Trotskyites, anarchists, and various citizens groups. The crowd, which represented Vancouver's diffuse counterculture, as well as older peaceniks and elements of the postmaterialist middle class, converged on the Douglas Border Crossing between British Columbia and Washington state. For the first time since the War of 1812, a section of the U.S.-Canadian border had to be closed. Bob Hunter turned up and gave a "ranting and raging" speech.[35] Rod Marining brought his street theater company along, and Paul Watson arrived with some of his radical friends. Irving and Dorothy Stowe were there, holding up the Quaker banner and representing Irving's various citizens groups. Jim and Marie Bohlen were also present, along with other members of the recently formed BC Sierra Club. Like a flash flood, the protest receded as fast as it had arrived, and the various elements of the coalition flowed back into their respective pools. The editors of the University of British Columbia student newspaper were full of admiration for the way student leaders had organized transport to the border crossing for 6,000 people. Nonetheless, they were critical of the inability of student groups and others to forge a durable coalition that could focus attention on U.S. imperialism and the nuclear arms race in a more enduring and rigorous fashion.[36]

Irving Stowe and Jim Bohlen were convinced that the protest represented an opportunity to form such a coalition. Seeing students as the key ground troops in any such alliance, Bohlen, in his capacity as conservation chairman of the BC Sierra Club, contacted Paul Coté, a twenty-seven-year-old law student at UBC. Coté was the most mature and least flamboyant of the student leaders who had helped organize the border protest. One of nine children from a wealthy, conservative, West Vancouver family, Coté had had little interest in radical politics or social activism until he went to Paris in1968 to spend a year as an exchange student at the Sorbonne. There, while minding his own business at a student bistro, an over-zealous policeman whacked him in the eye with his truncheon while trying to break up one of the many student demonstrations that occurred during that tumultuous summer in Paris. Coté returned to UBC a little less innocent than when he had left. Like Bohlen and Stowe, he saw the potential of harnessing the somewhat chaotic energy of the border protest and concentrating it into a more effective political weapon. The three of them decided to mount a campaign that would draw support from a wide range of groups and to spend the following two years focusing attention on, and building up a strong opposition to, the next U.S. detonation on Amchitka, planned for October 1971. Although not yet sure of the details, they agreed that the campaign would require a combination of direct action, media mobilization, political lobbying, and solid scientific research. It would need to emphasize the political folly of the arms race, as well as the environmental destructiveness of nuclear weapons testing. And it would not hurt, as far as Bohlen and Stowe were concerned, if it played on the latent anti-Americanism that was pervasive throughout Canada.[37]

Initially, Bohlen thought that the protest could be conducted as a Sierra Club campaign. The Club, after all, had the name recognition and the resources to launch a high-profile operation. Furthermore, as a highly respected American organization, it might have been able to provoke a greater degree of concern about the nuclear tests in the United States. The first Amchitka blast, in 1965, had aroused little public interest or media coverage in either country. The second, in October 1969, drew considerably more attention in Canada but, again, went virtually unnoticed in the United States. Perhaps the Sierra Club, which had trumpeted its intentions of turning wilderness preservation into an international crusade, could stir the American public out of its lethargy.[38] But 1969 was a turbulent year for the Club, with its executive director, David Brower, clashing with the board over various managerial and policy issues. After a failed attempt to elect his own slate of candidates to the board, Brower had no option but to resign as the Club's executive director. He immediately went on to found Friends of the Earth, an organization that better expressed his growing concern with international environmental issues and his belief in adopting more activist tactics. For Bohlen and the BC Sierra Club, it was not a good time to try to persuade the Club to back a Canadian-based campaign against nuclear testing in the Aleutians. Therefore, Bohlen, Stowe, and Coté, with the backing of Terry Simmons and others within the BC Sierra Club, decided to form an independent group to organize the protest against Cannikin, as the next bomb blast was called. They remembered the words on one of the picket signs they had seen at the border protest, "Don't Make A Wave," which, unbeknownst to them, had been written by Bob Hunter, and decided to call themselves the Don't Make a Wave Committee (DMWC).[39]

For several weeks thereafter, the DMWC held meetings at the Stowe or Bohlen residences, trying to come up with a plan that would give the next Amchitka blast as much exposure as possible. They all agreed that it was, in Hunter's words, "a potent symbol of war craziness and environmental degradation wrapped up into one," but they struggled to find a method of protest or action that could encapsulate the issue in a powerful and symbolic way. The media was not particularly interested in an event that was almost two years away, and Coté doubted that a significant number of students could once again be mobilized to block the border, an action that would in any case be difficult to repeat now that the authorities were expecting it. At this early stage, the committee was mostly made up of Sierra Club members, Quakers, and some of the students from Coté's circle, and the meetings tended to be dominated by Stowe's endless monologues, many of which degenerated into rants against U.S. imperialism and various other issues that were on his mind. After one such meeting, the Bohlens were sitting in their kitchen, Jim pouring out his frustrations, when Marie came up with an idea that was so obvious, it almost beggars belief that nobody else had thought of it after two months of campaign planning. Why not simply sail a boat up to Amchitka and confront the bomb? Hardly a revelation for experienced peaceniks and Quakers such as the Stowes and Bohlens, who were very familiar with the exploits of the *Golden Rule* and various other vessels that had attempted exactly such an action only a decade before. Nevertheless, it took Jim completely by surprise, and he immediately became excited by the possibilities.[40]

By some strange coincidence, at that very moment, a reporter from the *Sun* rang Bohlen to ask him about various Sierra Club campaigns that were taking place at the time. When he asked him if he knew of any plans to protest the next Amchitka blast, Bohlen took a deep breath, glanced quickly at Marie, and told the reporter that the DMWC planned to sail a protest boat to the Aleutians to bear witness to the blast. The next day, before most of the DMWC members had heard anything about Bohlen's idea, it was reported in the *Sun* and was effectively a fait accompli. Mistakenly reporting the plan as a Sierra Club campaign, the *Sun* wrote that the group intended to sail a boat to the edge of Amchitka's twelve-mile limit (the area under U.S. jurisdiction) before the blast. "If the Americans want to go ahead with the test," Bohlen defiantly proclaimed, "they'll have to tow us out," an action that would effectively constitute an act of international piracy. "Something must be done to stop the Americans from their insane ecological vandalism," continued Bohlen. In addition to the voyage, it was imperative that Canadians be given access to the relevant data on the ecological impact of the first two blasts, something that the United States had so far refused to do. Bohlen promised that his group would mount a scientific campaign that would expose the ecological effects of nuclear testing to public scrutiny: "We will try to mount the most massive campaign ever, against this mad venture, and we'll make sure the American public is aware of how Canadians think about this matter." Fortunately for Bohlen, all the other members of the DMWC agreed that it was an excellent idea.[41]

Once the DMWC had decided to follow in the wake of the *Golden Rule* and *Phoenix* in order to protest the Cannikin blast, they had to formulate a strategy that would allow them to avoid the mistakes of their predecessors. The Stowes and Bohlens immediately realized that one of the main flaws of the earlier campaigns was their lack of well-formulated media strategies. Bigelow and Reynolds had naively assumed that the free and unfettered U.S. media would accurately and fairly report their protest, without fully understanding the structural constraints within which it operated. If it were to have a greater impact than the earlier voyages, the DMWC would need to develop strategies to ensure that the media could not ignore their protest. One possibility was to bring journalists along on the voyage, a practice now referred to as "embedding." Another problem with the earlier voyages was that they involved U.S. citizens protesting against their own government, which made it relatively easy for the AEC to harass and ultimately stop them. The DMWC would have to ensure that its boat would not sail under the American flag and that its crew included a large number of non-Americans.

By February 1970, Bohlen and Coté were spending most of their spare time searching for a boat, while Stowe set about fund-raising and beating up support and publicity. Several brainstorming meetings were devoted to naming the eventual boat. The Don't Make a Wave Committee, though a vivid name that conveyed many people's fears about Cannikin, was rather clunky for a campaign that intended to rely so heavily on the media, as well as being an awkward name for a boat. After several frustrating meetings, it was a young social worker, Bill Darnell, who put together the magic words. As Irving Stowe was leaving a meeting one night, he flashed his usual V sign and said "Peace," to which Darnell responded, "Make it a green peace!" According to Dorothy Stowe, Darnell's words "lit up the room," and there was an almost instant agreement that when they eventually found a boat for their protest, they would name it the *Green*

Peace. Soon thereafter, the words "green peace" became the single and singular term "Greenpeace." Marie Bohlen's son, Paul, a graphic artist, designed a one-inch button that consisted of the ecology symbol above the peace symbol, with the words "green peace" in between. Finding that he was unable to fit the two words in the confined space, he asked his stepfather what he should do, whereupon Jim Bohlen suggested he simply put the two words together as one.[42]

Given the high degree of overlap between the DMWC and the BC Sierra Club, the Sierra Club head office in San Francisco was naturally quite eager to keep up with the exploits of its rather unpredictable and somewhat unconventional little chapter in Vancouver. Although the Club had decided not to allow the BC branch to conduct the Cannikin campaign under the Sierra Club banner, there was nonetheless much sympathy for the group from within various quarters of the U.S. organization. One supporter was Denny Wilcher, a Quaker and long-time Sierra Club employee. Wilcher had been arrested as a conscientious objector during the Second World War and was later involved in the *Golden Rule* voyage in 1958. He congratulated the DMWC on their decision to mount the campaign.[43] Nevertheless, the post-Brower regime at the San Francisco office remained uneasy and was concerned that the BC chapter might embarrass them or cause some kind of trouble. Although they bore the Club's name, the BC chapter had in fact been set up under Canadian law with only an informal endorsement from the Pacific Northwest Sierra Club chapter. In an effort to regain some control over its first non-American branch, the Club sent the BC group a letter requesting them to sign a licensing agreement that would give them the right to operate under the Sierra Club name while also allowing the Club to revoke the license at its discretion.[44]

Replying on behalf of the BC Sierra Club's board, Irving Stowe declined to sign the agreement. His anger stoked by the Club's refusal to lend their name to the campaign, Stowe reminded them of their intentions to expand internationally and was resentful of the licensing request which, he felt, "lends itself to the implication (presumably unintentional) that we are less eager than yourselves to protect the Sierra Club's good name." Becoming an international organization, he pointedly suggested, "imparts more than the rationale that conservation does not stop at the boundaries—it must respect the good faith and dedication of like-minded people on the other side of the boundary. We believe that we are entitled to that trust." Eventually, the head office became reconciled to the BC chapter and its involvement in the DMWC campaign, even donating $6,000 toward the voyage to Amchitka. The uneasiness of many of the old-line conservationists within the Club, however, was an indication of their discomfort with more radical forms of protest. Similarly, the BC Wildlife Federation, which represented hunters and fishermen throughout the province, found the idea too radical for its constituency. Its director, Roger Venables, while admitting that he personally agreed with their principles, refused to endorse the DMWC, noting that it "may give rise to some adverse publicity, which could be harmful to our organization and not in accord with the feelings of our members."[45]

The Vancouver City Council also expressed some hostility toward the DWMC's plans and refused them permission to conduct a "tag day" in Vancouver, where they hoped to hand out fact sheets and sell buttons to raise funds for the journey. Although a considerable minority of councilors supported their petition, others were skeptical of its aims, feeling that the DMWC was merely a front group for anti-American activities. Earle Adams, one of the more conservative members of the council, felt that the DMWC "should start in Russia, where (they) would probably be shot or sent to Siberia." "To give this group approval for a tag day," he continued in the finest McCarthyist tradition, "would stamp the council as communist sympathizers or fellow travelers."[46]

Many other groups, however, were keen to lend both moral and financial backing to the campaign. The Quakers, of course, were enthusiastic supporters, providing the DMWC with vital funding and a network of support throughout North America. Numerous Friends chapters from as far afield as Nova Scotia and San Diego donated sums from fifty to several thousand dollars, with the Palo Alto and Eugene chapters each giving $6,000. Various SPEC branches throughout British Columbia donated small amounts (fifty to a hundred dollars) and support also came from groups such as the Student Christian Movement, the Confederation of the United Church of Canada, the Faculty of Law at UBC, the Law Students Association of BC, a new environmental group in Toronto called Pollution Probe, Zero Population Growth, and the Vancouver District Labour Council. Dave Barrett, the leader of British Columbia's New Democratic Party and the man who would become the province's premier in 1972, thanked Bohlen and the DMWC for their efforts and assured them that the NDP stood firmly behind them. In addition, many individual Canadians and Americans sent small donations and letters of support.[47]

Another organization that ardently supported the DMWC was the Voice of Women. Founded in Toronto in 1960, VOW was initially a collection of very proper upper-class women who opposed the arms race and nuclear weapons testing. Many were the wives of prominent politicians and public figures who tried to distance themselves from the "less respectable" antiwar groups of the time and to pressure their influential husbands to oppose nuclear weapons. Throughout the 1960s, however, the group's membership became more broadly based and less elitist, forging strong links with Quaker peace activists and adopting a more critical feminist and New Leftist stance.[48] Deeno Birmingham, the president of the BC chapter of VOW, devoted much time and effort to aiding the DMWC. She also persuaded her husband, Dave, a ship's engineer, to join the DMWC journey to Amchitka. Another prominent member of VOW was Lille d'Easum, a friend of the Stowes, who prepared the first ever Greenpeace publication, a brief paper entitled *Is Amchitka Our Affair?* with the Greenpeace logo prominently displayed on its cover. The paper summarized the case against nuclear testing on Amchitka, citing the fears held by reputable scientists at MIT and Caltech that a huge underground explosion so close to a fault line could easily trigger a series of earthquakes that could do untold damage to the Pacific Rim. The paper also cited the work of Ernest J. Sternglass, a professor of radiation physics at the University of Pittsburgh whose controversial research on radiation and fetal infant deaths led him to conclude that nuclear testing was directly responsible for half of all deaths among children under twelve months old in the United States and Britain in the mid-1960s.[49]

The largest single financial contribution to the first Greenpeace voyage—$17,000—came from the proceeds of a sold-out concert, organized largely by Irving Stowe, which took place at the Vancouver Coliseum in October 1970. Initially, Stowe had hoped to persuade Joan Baez to perform, but she was unavailable. His second choice, Joni Mitchell, was reluctant, but Irving, with his bulldog tenacity, cajoled her into it by making a series of vague threats. Mitchell was born in Alberta and raised in Saskatchewan but had not performed a concert in Canada since she had become successful. Irving managed to get Mitchell on the phone and suggested that people in Canada would not be too impressed if a Canadian singer, who had just scored a huge hit with her environmentalist anthem "Big Yellow Taxi," was unable to take time out of her busy schedule to come to her home country and support an important antinuclear campaign with critical environmental implications. Mitchell quickly saw that it would be in her interest, both in terms of her reputation and her political convictions, to appear at the concert, even offering to bring along her new friend James Taylor. Having never heard of Taylor, who was on the brink of stardom, Irving asked Mitchell to hold on for a second, put his hand over the mouthpiece, and asked Dorothy and his kids if they had ever heard of James Taylor. None had. Nevertheless, Irving, with some trepidation, told Mitchell that she could bring him along.[50]

The American folksinger and New Left activist, Phil Ochs, leaped at the opportunity to support the DMWC with his customary zeal, even offering to don a parachute and jump into the blast site on Amchitka. The popular local band, Chilliwack, completed the line up. The concert, the first of many fruitful collaborations between Greenpeace and popular musicians, proved extremely successful, garnering the Greenpeace voyage considerable publicity in addition to the valuable funds.[51] The only sour note was that the city, which owned the coliseum, insisted on taking some $3,000 from the proceeds in lieu of rent, an action that provoked a kind of reverse McCarthyist rant from Irving. "By this arrogant, calculated disregard for the disciplined non-violence of the Greenpeace project," he fulminated in his newspaper column, "the power structure has exposed itself as ANTI-CANADIAN—on the side of USA nuclear violence EVEN WHEN IT THREATENS CANADIAN HEALTH AND LIVES!" Clearly, Stowe was very much aware of the nationalism that pervaded the Canadian Left, a situation that complemented his own brand of strident anti-Americanism.[52]

In addition to the concert, Stowe also managed to organize a public lecture by the former *Phoenix* skipper, Earle Reynolds. Reynolds was a legendary figure to the older generation of peaceniks such as the Stowes and Bohlens, but he had also become known to the younger generation of environmental activists when one of his essays on radiation and human evolution was reprinted in a widely read anthology, *The Subversive Science*, a collection of influential essays on ecology. Reynolds's appearance in Vancouver constituted the most direct connection between the Quaker protests of the late 1950s and the Greenpeace movement they inspired.[53]

While Stowe busied himself with the financial aspects of the campaign, Bohlen and Coté searched long and hard to find an appropriate vessel to bear the Greenpeace name. Their initial expectation, which in retrospect seems grandiose indeed, was to find a boat capable of taking 300 to 500 people through the stormy October waters of the North Pacific. Bohlen approached the Swedish government in the hope of chartering such a vessel on good terms from a presumably sympathetic government but had no luck. He and Stowe even looked into the possibility of negotiating a deal with Cuba, which would surely have destroyed all hope of garnering favorable publicity in the United States.[54]

By early 1971, the campaign was about to enter into full swing without a fulcrum. Bohlen and Coté were becoming increasingly desperate and, as a result, increasingly pragmatic about the kind of boats they were willing to consider. Coté, a superb sailor who would go on to win a bronze medal at the 1972 Olympics, began to tap his contacts on Vancouver's wharves. Few captains, however, were willing to risk chartering their boats for such an unorthodox mission. But there was one who was desperate enough. John Cormack, a sixty-year-old fisherman and owner of an 80-foot halibut seiner, the *Phyllis Cormack*, was deeply in debt after several poor halibut seasons and was facing foreclosure. A balding, heavy-set, curmudgeonly old salt with over thirty years experience in navigating the North Pacific, Cormack was willing to charter his somewhat dilapidated, thirty-year-old boat, with its hulking twenty-year-old engine that could barely propel the boat at a speed of nine knots, in a desperate bid to pay off his debts. A depressed Bohlen, who had sailed through the Aleutians after the Second World War, felt that the boat was entirely inadequate. Yet he knew that they had little choice but to accept Cormack's offer. When Bob Hunter went down to the wharf to take a look at the mighty new *Greenpeace*, he passed it half a dozen times, unable to believe that this was the vessel that Bohlen proposed to use. Although Cormack's involvement was largely pragmatic, he would go on to develop a strong attachment to the Greenpeace group, and the *Phyllis Cormack* would remain an important part of the Greenpeace flotilla throughout the 1970s.[55]

The Canadian government, mindful of public opinion, was critical of the American tests on Amchitka and made its protests known within the constraints allowed by the close relationship between the two nations. In a note to the U.S. Department of State, the Canadian ambassador in Washington made "it clear that [the Canadian government] cannot be regarded as acquiescing in the holding of these nuclear tests . . . [and] would have to hold the United States government responsible for any damage or injury to Canadians, to Canadian property, or to Canadian interests resulting from the tests," a sentiment that the prime minister's office repeated in its correspondence with the DMWC. Nevertheless, this opposition to the Amchitka tests did not necessarily endear the DMWC to the Canadian government, which tried on several occasions to put obstacles in the group's path. The Department of National Revenue, for example, expressed skepticism at the DMWC's application for charity status, arguing that it "fail[ed] to see the relationship between the acquiring of a boat and the publicizing of the adverse environmental effects of nuclear explosions on Amchitka Island."[56]

Bohlen, Coté, and Stowe are commonly credited as the founders of Greenpeace. More accurately, they were the founders of the Don't Make a Wave Committee. Reproduced by permission of Robert Keziere/Greenpeace.

Another obstacle for the DMWC was the refusal by fisheries minister Jack Davis to provide the *Phyllis Cormack* with insurance, an act he justified by pointing out that the crew clearly did not intend to go fishing.[57] This would effectively have put a halt to the protest, since purchasing private insurance would have been prohibitively expensive, perhaps $50,000 or more. Instead, the DMWC had planned to obtain insurance from the Industrial Development Bank, a Crown Corporation that generally offered insurance to fishing boats for $150 per month. Fortunately for the DMWC's plans, Davis's electorate was West Vancouver, a wealthy district on the north shore of English Bay where support for the *Greenpeace* voyage was strong. The DMWC set about pressuring Davis to change his mind. Hunter wrote a scathing column, castigating Davis for aiding the AEC by attempting to scuttle "the only determined Canadian opposition to the test."[58] Ben Metcalfe, who by then had joined the campaign, made similar comments in his broadcast, and before long, Davis's office was inundated with letters and phone calls from his constituents, demanding that he reverse his decision. Caving in to pressure, Davis changed his mind, blaming his original decision on one of his underlings, and granted the *Phyllis Cormack* the right to purchase the standard fishing insurance policy. Clearly, the presence of well-known media personalities within the DMWC's ranks was already paying dividends.[59]

A few days before the *Phyllis Cormack* was due to set sail under the Greenpeace flag, customs officials revealed that they would not let the boat depart because it was not registered as a passenger ship. This forced Captain Cormack to put fishing equipment on board and to register everyone as a crewmember of a professional fishing vessel. Hunter once again questioned the motives of the Canadian government, arguing that they must surely have known that this issue was going to arise and their decision to spring it on the Greenpeace crew at the last minute was merely another attempt to halt or delay the voyage.[60] It is difficult to say whether these obstacles were simply the result of bureaucratic intransigence—the inability of a regimented system to deal with unorthodox requests and events—or if they constituted a genuine effort on the part of the Canadian government to disrupt the campaign. One piece of evidence that points to the latter was a U.S. Atomic Energy Commission report prepared after the *Milrow* blast that argued that the commission should attempt to counteract any future opposition to U.S. nuclear testing. The report suggested that the "hysterical anti-test campaign in British Columbia and other parts of Canada" was motivated as much by anti-Americanism as by genuine nuclear fears.[61] One can imagine that the Canadian government would have been sensitive to such charges, whether implicit or explicit, and may well have felt under pressure to at least be perceived to be doing something about them. Whatever the case, the Canadian government's interference, though irritating, was relatively benign compared to the actions that the United States had employed against the *Golden Rule* and the *Phoenix*.

Once the DMWC had resigned itself to the fact that the *Phyllis Cormack* was going to be the campaign boat, it became clear that the number of crewmembers would have to be severely limited—ten people in addition to the captain and engineer. Where the DMWC had initially been concerned with finding enough volunteers to man a large ship, they now found themselves having to select ten people from among the dozens of enthusiastic applicants. Bohlen decided, therefore, to choose a combination that would best serve the needs of the voyage and the campaign overall. Stowe had made it known from the outset that he would not be on board since he suffered from severe seasickness due to an inner ear disorder he had contracted while flying for the US Civil Air Patrol during World War II. Instead, Irving and Dorothy would continue to try to keep the voyage in the media spotlight from the shore. Coté, too, had made it clear that he would not be able to join the voyage if it happened to conflict with the Olympic sailing trials, which indeed proved to be the case. That left Bohlen as the only member of the original DMWC troika. Apart from representing the DMWC, Bohlen was also a trained naval navigator with plenty of sailing experience and therefore a natural choice.[62]

Terry Simmons would be the Sierra Club representative on board and, as a major supporter of the campaign from the beginning, could hardly be denied a place. Simmons had little sailing experience, but he was a very bright and talented young geographer and had considerable knowledge of the native Aleut cultures. Bill Darnell, the young social worker who had coined the term "green peace," was chosen to be the ship's cook, and a thirty-year-old chemistry graduate student and countercultural activist, Bob Keziere, was brought on board as the photographer. In addition, the final crew would include a hippie medical doctor named Lyle Thurston; three journalists—Bob Hunter and Ben Metcalfe from the mainstream media, and Bob Cummings, a writer from the *Georgia Straight*, representing the alternative press—an ecology graduate student, Patrick Moore; and a political scientist from the University of Alaska named Richard Fineberg, who had experience as an investigative journalist delving into nuclear issues. Fineberg, Bohlen, and Simmons were the only Americans on board, while the rest of the crew was made up of Canadian nationals.

Bob Hunter may have been a journalist by training, but he was not one to give undue respect to the journalistic creed of objective reporting. In his application letter to join the Greenpeace voyage, Hunter made it clear that the DMWC "would get as much coverage as could possibly be squeezed out of one journalist" and if that was not enough, he was willing to "offer [his] services as a public relations man." His regular column in the *Vancouver Sun* gave him "certain freedoms . . . that reporters do not generally have," permitting him to "editorialize . . . without having to stick to the so-called 'objective' frame of reference, a frame which normally functions to water reports down to pabulum."[63] Thin, bearded, and with a mane of brown hair swept back over his head, Hunter styled himself as the classic counterculture intellectual. He was verbally swift, charming, witty, and charismatic but also prone to fits of depression and rage. He wrote in a style that was assured and frequently powerful, though the clarity of his thought did not always match the fluidity of his prose. His hippie sensibilities and preparedness to occasionally ditch rational thought in favor of mysticism meant that he was bound to butt heads with "straights" such as Bohlen.

Despite being a regular presence at DMWC meetings and putting considerable effort into the campaign, Rod Marining was not initially chosen to sail on board the *Phyllis Cormack*. This was largely due to Bohlen's fears that people such as Marining and Paul Watson, who had been on the fringes of the DMWC, might undermine the serious, mature, and scientifically responsible image he hoped to convey. This was quite a blow to Marining, who, by virtue of his considerable fame within Vancouver's countercultural scene, had expected to gain an automatic berth. But Marining was nothing if not persistent and would surprise everyone before the campaign was over. Watson would also be involved, and he and Marining would become influential figures in Greenpeace throughout the 1970s.[64]

Whether by chance or sheer force of probability, the crew of the *Greenpeace* reflected, if not in an entirely proportionate fashion, the influences that went into the organization's making. The crewmen (and they were all men) were the products of the various ideas and traditions that shaped the peace and environmental movements in the United States and Canada in the postwar era. To one extent or another, they represented Quakerism, the radical pacifism of the American peace movement, the New Left and the counterculture, Marshall McLuhan's theories of mass communication, and various strands of environmentalism, all united by a commitment to nonviolent direct action and a shared belief in the revolutionary potential of holistic ecology. This is not to say that all these phenomena could only have come together in Vancouver. However, since they did come together there, the organization that emerged was indelibly shaped by the city's protest culture and its activist milieu. And the Vancouver context would continue to shape the organization for many years. For now, however, the core group was ready to board a rickety old halibut seiner and take on the world's most powerful military in one of the most remote and desolate parts of the planet.

4

Don't Make a Wave

The most successful forms of collective action, one would think, are those that inspire high rates of participation and ultimately spawn mass movements. The success and potential influence of a movement, therefore, are often thought to be directly proportional to the number of people taking part in it. In the popular imagination, successful social movements are frequently equated with the spectacle of huge crowds of people marching through the streets or gathering in the town square. One thinks, for example, of the hundreds of thousands who packed the Mall in Washington, D.C., during the civil rights protests of the 1960s or of the similarly sized crowds that participated in the antinuclear demonstrations throughout Germany in the early 1980s. Even in democracies, however, the sheer size of a movement does not necessarily translate into political success, as the massive demonstrations against the American-led war on Iraq made abundantly clear. Among social scientists there is a strong tendency to study such movements as aggregates, analyzing and explaining them in broad, structural, and frequently rather abstract terms. In the process, movements can sometimes be reified into collective responses to broad socioeconomic changes. While such approaches undoubtedly produce their fair share of innovative and insightful studies, they frequently overlook the face-to-face interaction that occurs among the activists at the core of a movement, thereby neglecting not just an important locus of the decision-making process but also the myriad arguments, debates, and personality disputes that help determine the shape and direction that a movement will take.[1]

In an effort to overcome these limitations, some scholars have sought to examine social movements as settings where the participants are continuously engaged in the process of constructing the meaning of their movement. It is here that much of the movement's spirit, character, and dynamics—in short, its culture—are formed. Such an approach, according to sociologists Gary Fine and Randy Stoecker, tends "to see in the social movement the working out of small-group processes": the face-to-face interactions that shape its values and prioritize its goals.[2] Clearly, it is important to understand the broad structural elements that provide the opportunities for the emergence of social movements. However, it is equally vital to examine how the movement's culture is shaped by the relationships between key participants: to see how their struggles for power, their personality clashes, and their ideological disputes play out at important moments in the movement's history. In this sense, the maiden voyage of the *Greenpeace* constitutes an ideal vehicle for such an analysis. For two months, most of the group's key players were corralled together on a small boat on the rough waters of the North Pacific, cementing their bonds and slugging out their disputes.

The story of the voyage to Amchitka is not only one of how a group of people tried to draw attention to the dangers of nuclear testing, it is also the story of how a movement's values and identity were contested and shaped by its key participants. The people who constituted the Don't Make a Wave Committee came from diverse backgrounds and represented a variety of worldviews. Not unexpectedly, when thrown together in such close quarters, the differences between them frequently erupted into vigorous disputes that, on occasion, almost ended in physical violence. Jim Bohlen and Bob Hunter represented the deepest division. Bohlen's primary goal was to run a sober campaign that would garner the respect of elite groups, such as scientists, politicians, and high-level bureaucrats, as well as attract the attention of the popular media. Hunter, on the other hand, was more interested in creating a mythology for the environmental movement, one that would resonate with millions of people and help bring about the change in mass consciousness that was so vital to his vision of the future. Part of the subtext of the voyage, therefore, was a struggle, not just over the organization's leadership and ideology but also over what kind of organization Greenpeace would become, or if it would become one at all. Would it simply be, as people such as Jim Bohlen and Terry Simmons had initially imagined, a one-off campaign designed to draw attention to the perils of nuclear testing? Or would it, as Hunter hoped, be the vanguard of a massive new social movement that fused together the old peace and environmental movements, while simultaneously transcending both of them by embracing the new consciousness of the counterculture?

On September 15, 1971, on a warm late afternoon in Vancouver, about half the crew of the *Greenpeace*, after loading up the last of their supplies for a six-week journey, set sail from their mooring at False Creek. A small but enthusiastic crowd of well-wishers waved goodbye, and cameramen from the local TV station filmed their departure. The boat sailed about five miles around the tip of Stanley Park before docking, some twenty minutes later, at Coal Harbour, on the other side of downtown Vancouver. The decision to "launch" the *Greenpeace* a few hours early was largely taken at the behest of the local media, who wanted to get a shot of the boat departing before the light faded, as well as to ensure that the story made the six o'clock news. Bob Hunter, Bob Cummings, and several other crewmembers who had felt disinclined to involve themselves in the hard labor of last-minute preparations, spent the afternoon at a local pub. With considerable amusement, they watched their own departure on the evening news before they had even set foot on the boat.[3]

It was, of course, entirely appropriate that the first moments of the *Greenpeace*'s maiden voyage should be dictated by the needs of the media. A third of the crew, after all, were journalists of one kind or another, and the success of the campaign rested on its ability to attract media attention to the expedition. The voyage, therefore, and its lead up and immediate aftermath, were important events in the long-term history of Greenpeace: they were the staging grounds on which the various participants battled to stamp their own particular ideology and brand of activism on the nascent organization, even if they were not fully aware of it at the time. In addition to the media strategy, attention to scientific detail was another key element of the campaign. As a result, the DMWC, and Bohlen in particular, went to considerable effort to project an image of serious professionals who had done their homework and whose objections to U.S. nuclear testing were based on solid scientific evidence.

The crew of the first Greenpeace voyage. Clockwise from top left: Bob Hunter, Patrick Moore, Bob Cummings, Ben Metcalf, Dave Birmingham, John Cormack, Bill Darnell, Terry Simmons, Jim Bohlen, Lyle Thurston, and Richard Fineberg. Reproduced by permission of Robert Keziere/Greenpeace.

The DMWC was far from being the only group to oppose Cannikin. In fact, the only vocal proponents of the test appeared to be the Atomic Energy Commission and the Pentagon. Even the U.S. Congress refused to officially sanction the blast, leaving the ultimate decision up to President Nixon.[4] In Washington, D.C. a coalition of antinuclear and environmental groups, including the Sierra Club, David Brower's Friends of the Earth, the Wilderness Society, the Society Against Nuclear Explosions (SANE), the Federation of American Scientists, and Barry Commoner's Committee for Nuclear Responsibility were attempting to sue the AEC, arguing that they had failed to produce an adequate environmental impact assessment of the blast. A group of thirty-three congressmen filed a similar suit, demanding that the AEC release a secret report on the planned test.[5] The *New York Times* was critical of the AEC's stubborn refusal to postpone or abandon the test in the face of mounting fears regarding its safety.[6] The Canadian edition of *Time* magazine also slanted its reports at a critical angle, arguing that the "furor over Cannikin is but the latest expression of citizen discontent with the relatively unchecked freedom with which weapons are commissioned, tested, and deployed." The days when the AEC could, in the interests of "national security," explode weapons without challenge were, according to *Time*, "clearly over."[7] The opposition was particularly fierce in normally conservative and patriotic Alaska, where Democrat Senator Mike Gravel mounted a strong campaign to have the tests cancelled, insisting that to go ahead with the blast was "courting a risk of magnitude we cannot determine." "Our national security," he insisted, "will not be imperiled if the test is cancelled."[8] Alaska's Democratic governor, William Egan, was also concerned about the potential consequences of *Cannikin* and requested that the AEC hold public hearings in Alaska to examine the possible environmental impact of the blast. Although the commission had, in accordance with the 1969 Environmental Protection Act, conducted an environmental impact assessment of the test, most environmentalists, according to America's flagship scientific journal, felt it was "more like a sales pitch than a catalog of possible environmental effects."[9]

Caving in to pressure, the AEC held a series of public hearings in Anchorage and Juneau in the last week of May 1971. Jim Bohlen and Patrick Moore represented the DMWC, while various scientists, academics, journalists, politicians, and environmentalists also joined the mounting chorus of AEC critics. Of all the fears aroused by Cannikin, the most apocalyptic, albeit the least plausible, was that the blast would trigger a major earthquake, which in turn would generate a tsunami that would pound Pacific coasts from Japan to Mexico. California Senator Alfred Alquist recalled that a 1964 earthquake centered in the Aleutians had "generated a Tsunami that Californians will not soon forget." The giant wave had caused considerable damage throughout California, particularly in coastal towns such as Crescent City, where a twelve-foot wave had inundated twenty-seven blocks of the downtown area, destroying 300 buildings and causing five deaths and $11 million worth of damage.[10] In support of this argument, Bohlen quoted James N. Brume, a scientist at the California Institute of Technology, who suggested that there was "no logical reason why a nuclear explosion couldn't be the initiating event in a series of events causing a major earthquake. The larger the blast, the greater the possibility of its triggering such a series." And there had never been an underground blast as big as Cannikin.[11] For its part, the AEC, while not entirely ruling out the possibility of an earthquake and tsunami, argued that the risk was miniscule and that the blast might, in fact, prevent such a catastrophe in the future by releasing the latent energy in the fault.[12]

Although a tsunami was within the realm of the possible, a more realistic probability was that the blast's immense power would fracture the rock between the blast chamber and Amchitka's surface, creating a series of cracks and fissures through which radioactive fallout could escape. Scientists pointed out that of the 230 underground tests conducted in Nevada, at least sixty-seven had leaked varying amounts of radioactivity. Bohlen cited evidence that some of the fallout had already been recorded in Canada, a fact that constituted a violation of the 1963 Nuclear Test Ban Treaty.[13] Senator Gravel was anxious that, at a time when people were deeply concerned about mercury poisoning from seafood, "even the suspicion that radioactive water is leaking to the surface could devastate the market for all fishery species of the North Pacific."[14] Bohlen also questioned the AEC's assumption, as outlined in its environmental impact statement, that there existed "safe" levels of radiation dosage. Citing Berkeley physicist and professor of medicine John Gofman, Bohlen expressed the DMWC's hard line on this issue, insisting "no conclusive evidence exists to suggest that there are any safe levels of radiation dosage beyond that of nature's normal background radiation."[15]

Richard Fineberg, a young political scientist at the University of Alaska, pointed to another potential danger. Although not a member of the DMWC, Fineberg would nonetheless join the *Greenpeace* voyage at the last minute on the strength of the positive impression he made on Bohlen at the hearings. Through careful research, Fineberg had discovered that in 1947, the U.S. Army had dumped hundreds of barrels of deadly mustard gas into the waters around the Aleutians. In 1968, gas from some of these containers had leaked near a U.S. Coast Guard weather station, forcing the evacuation of the site and causing several people to become severely ill. Given the sudden increase in pressure and the violent ground shock that was likely to accompany Cannikin, had the AEC considered, Fineberg asked pointedly, the potential damage that could occur if even a fraction of these barrels were ruptured, thereby releasing the killer gas into the surrounding water and atmosphere? Here, Fineberg insisted, was yet another oversight to add to the inadequacies of the AEC's environmental impact statement.[16]

Part of the DMWC's response to the AEC's arguments was to develop a two-pronged strategy that has become a mainstay of Greenpeace's modus operandi. On the one hand, they employed a rhetoric of righteous moral anger intended largely for the general public's consumption, a common enough strategy among social movement organizations. Appealing to popular ideas of ecology and to broadly held notions of peace, security, and human rights, the DMWC would use evocative slogans and pithy catchphrases that could be picked up by the media and that would resonate with the masses. For example, they characterized the AEC as "ecological vandals" and argued "Amchitka may be the link in the chain of events which will bring human history to an end."[17] Bohlen spoke of the U.S. defense umbrella as a "death canopy for Canada,"[18] while Stowe claimed that the AEC was creating a "pocket of poison" on Amchitka that was "filled with the most lethal and terrible kinds of polluting radiation on the planet." The AEC demonstrated, continued Stowe, "that power pollutes and nuclear power pollutes absolutely."[19] Moore argued that if the United States wished to "indulge itself" and test a device that they claimed was safe, "why not explode it in the geographic center of the United States in central Kansas?"[20] Metcalfe's "Greenpeacing of America" letter, broadcast from the *Phyllis Cormack* on the second day of the voyage and heard on CBC radio, is a classic example of the DMWC's skillful use of the rhetoric of moral outrage to condense an issue into a concentrated emotional form. "We call our ship the Greenpeace," he explained,

> because that's the best name we can think of to join the two great issues of our time—survival of our environment and the peace of the world. Our goal is a very simple, clear, and direct one—to bring about a confrontation between the people of death and the people of life . . . to bring about a confrontation between the great mass of humanity that refuses to risk the future for a crass political advantage and the men who say that's all a bunch of crap.
>
> We do not consider ourselves to be radicals. We are conservatives, who insist upon conserving the environment for our children and future generations of man. If there are radicals in this story it is the fanatical technocrats who believe they have the power to play with this world like an infinitely fascinating toy of their own. We do not believe that they will be content until they have smashed it like a toy.
>
> The message of the Greenpeace is simply this. The world is our place . . . and we insist on our basic human right to occupy it without danger from any power group. This is not a rhetorical presumption

> on our part. It is a sense we share with every ordinary citizen of the world . . . the crew of the Greenpeace know today that they are part of a massive international protest against the insanity of the Amchitka test. They know, too, of course, that they are confronting a power that has a certain experience in ignoring and opposing, and scorning protest. They are under no illusion that the John Wayne syndrome is peculiar only to John Wayne.
>
> But there's a certain feeling aboard, as there is in what we now call the Greenpeace Ashore, that a new and tougher situation has now developed for the nuclear people. That's what we call the Greenpeacing of America. It could work.[21]

The DMWC recognized, however, that if they wanted elite opinion—scientists, academics, policy makers—to take them seriously, they would have to go beyond catchy slogans and the rhetoric of moral outrage. Hence, they also put together a more "objective" report that took the form of a scientific review article. DMWC member, *Greenpeace* photographer and chemistry student Bob Keziere was the lead author. Eschewing apocalyptic rhetoric and excessive moral criticism, the report summarized both the AEC's position and that of its opponents, buttressing its arguments by citing credible research and distinguished scientists.[22] Such reports would become commonplace and increasingly sophisticated in future Greenpeace campaigns.

While the mood among the *Greenpeace* crew was initially buoyant, the atmosphere among the crew would become increasingly tense as they sailed up the BC coast. This was hardly surprising; it would be naïve to imagine that a dozen men of different ages, backgrounds, and lifestyles, and several strong egos, could spend six weeks together on a small fishing boat without getting on each other's nerves. It seems that at some point throughout the voyage, just about every crewmember clashed with every one of his shipmates. Many of these arguments involved tactical and strategic issues and some were simply due to petty disputes. The most fundamental clash was between Bohlen and Hunter. Bohlen had some sympathy for the counterculture, but he could not abide what he felt was the chaos, hedonism, drug-taking, fuzzy-thinking, and downright irrational behavior that characterized hippies. Hunter, from Bohlen's perspective, embodied far too many of these traits. If the voyage were to succeed at any level, he believed, it would be because of a sound media strategy and the mature, dignified, and rational conduct of the crew. Aside from the more extreme members of the counterculture, nobody would care about a story involving a bunch of spaced-out, drugged-up hippies throwing the *I Ching* and proclaiming themselves "warriors of the rainbow" and members of the "Whole Earth Church." The mainstream media would simply make fun of them, and scientists and public officials would have all the more reason to dismiss them. But if people could see that a group of highly educated, high-minded, middle-class professionals were willing to risk their lives to stop Cannikin, then the world might really take notice.[23]

For Hunter, the voyage presented an opportunity to put his "mind bomb" theory into action, and he was not about to let the chance pass him by. Bohlen's ideas, he believed, were simply too traditional and staid to provoke the kind of change in consciousness that he felt was needed if the world was going to make any serious progress toward ending military conflict and environmental destruction. Winning the support of scientists and other elites was in many ways a redundant ploy as far as Hunter was concerned, since they were the people who were least likely to undergo any fundamental change in their worldview. To reach a broader audience and to demonstrate the synergistic potential of a union of the peace and ecology movements, it was necessary to tap into the mindset of the counterculture and to spread that awareness to those who had not yet experienced it. Drug taking, by this logic, was a legitimate means of ensuring that those on board maintained a countercultural perspective. Furthermore, it was vital to create a new mythology that would add depth and a mystical quality to popular ecology. Getting the science right was fine, but it was just as important to develop new metaphors, slogans, and institutions that the youth of the world could identify with and embrace. When Bohlen would scornfully reproach the excessive drug taking on board, Hunter would challenge him to prove that a society of pot smokers and acid droppers could do a worse job of protecting the environment and promoting peace than Bohlen's rational, high-minded scientists and bureaucrats. This debate played out at various times and in various ways throughout the journey.[24]

Bohlen, though nominally the leader of the expedition, did not possess Irving Stowe's charismatic leadership qualities. The lack of a strong leader left a vacuum that Hunter and Metcalfe, each in his own way, attempted to fill, as well as giving Captain Cormack a greater degree of influence than was perhaps in the best interests of the campaign. Hunter in particular was eager to persuade others of the potential of his broader vision for Greenpeace. Moore, Keziere, and Lyall Thurston, the doctor, although all men with advanced scientific training, were also into drugs and countercultural lifestyles. It soon became clear to Bohlen that Doc Thurston was dispensing more than just aspirin from his pharmaceutical stash. Thurston's standard reply when asked how far he thought it was to the next landmark was "about three joints."[25] Not surprisingly, Moore, Keziere, and Thurston, while respecting Bohlen, were also drawn to Hunter's style and vision. Darnell and Simmons tended to lean toward Bohlen, while Metcalfe felt that neither Bohlen nor Hunter was suited to lead the group, with Bohlen lacking the necessary charisma and Hunter being too young and wild to invite the loyalty of the older crewmembers. Furthermore, Simmons's and Fineberg's precocious academic bearings did not sit well with Metcalfe's wily journalistic sensibilities. He even became convinced that Fineberg was a CIA agent and managed to sow considerable suspicion and mistrust among the crew, making Fineberg's life on board miserable.[26]

According to Hunter, there were also differences between the attitudes of the Americans and Canadians on board. The Americans, especially Bohlen and Fineberg, tended to view the voyage as a skirmish in a broader battle against U.S. militarism and political and economic hegemony. Given that their struggles seemed so much greater than those with which the Canadians had to contend, they had a somewhat patronizing attitude toward those who had only recently begun to feel the sting of the U.S. military-industrial complex. For their part, the Canadians, Metcalfe in particular, had a sense of moral superiority; Canada, after all, was neither waging war in Vietnam nor testing nuclear weapons.[27] Such feelings were symptomatic of a widespread ambivalence among Canadian activists toward their American brethren. On the one hand, they admired them, recognizing that they were in the front line of the battle and had suffered a great deal more as a result. On the other hand, they resented their condescending attitude and continuing belief that the United States was, even in all its depravity, the center of the world.[28]

Bob Hunter and Ben Metcalfe share the wheel of the *Phyllis Cormack* en route to Amchitka, September 1971. Reproduced by permission of Robert Keziere/Greenpeace.

Despite the differences among the crew, however, the atmosphere aboard the *Greenpeace* remained upbeat and the mood was convivial as they sailed the placid waters of the straits that divided Vancouver Island from the mainland. Engineer Dave Birmingham was surprised at the copious quantities of alcohol that the crew managed to consume and found it difficult to hide his disappointment that they were not the epitome of Quaker sobriety and earnestness. "I had expected the crew of the *Greenpeace* to be men of religion," he grumbled when he stuck his head in the galley during one of the many happy hours.[29] Cormack, too, appeared to have little respect for his crew, whom he frequently disparaged as "thirty-three pounders" and "mattress lovers." At times they did little to earn his respect. At one point, Cormack discovered that the boat had been sailing around in circles for half a day because Doc Thurston had left his tape recorder on top of the magnetic compass. Curmudgeonly and constantly irritable, Cormack insisted that the crew respect the many sailors' superstitions that he lived by. Mugs had to be hung facing outward and cans opened the right way up or else tossed overboard, and Cormack would explode with fury whenever his rules were transgressed.[30]

On September 23, the AEC announced that it was delaying the test until October 20. The *Greenpeace* may have figured in the AEC's calculations, though it is doubtful that they were the sole, or even the major, reason for the postponement; the imminent Canadian visit of Soviet Premier Kosygin was the more likely cause. Whatever the case, the AEC's decision meant that the *Greenpeace* crew suddenly found themselves with an extra three weeks on their hands. The problem was compounded by the fact that they had entered the stormy waters of the Gulf of Alaska, where most of the "thirty-three pounders" became violently seasick. Finally, Cormack suggested that they dock at the small Aleutian island of Akutan, a few miles northwest of Unalaska, where they could save fuel and possibly pick up some supplies.[31]

It was on Akutan, an enchanting, misty island of treeless sub-Arctic tundra, that the first elements of Greenpeace's mythology were crystallized from the swirling currents of popular ecology, countercultural consciousness, science fiction, and mind-expanding drugs. Among the reading materials that Hunter had brought with him on the journey was a book called *Warriors of the Rainbow*, a compilation of native American prophecies and myths collected and interpreted by Vinson Brown, a prolific nature writer with a strong interest in Native American culture, and William Willoya, an Alaskan Indian who visited dozens of tribes throughout the northwest in order to gather material for the book. Willoya and Brown argued that the precontact Indians had great spiritual and psychic power, that they foresaw much of what was going to happen to their peoples, and that even though much of this power had been lost because white civilization had crushed the tribes, there would come a time when they would regain their strength and force the white man to see the error of his ways.[32] A typical piece of sixties esoterica, the book added to the growing image of the wise "ecological Indian" that became such a powerful symbol of 1970s environmentalism. This construct blended romanticized primitivism with moralistic environmentalism. Another of its most famous manifestations was the "How can you buy or sell the sky?" speech attributed to Chief Seattle but actually written by Ted Perry, a film scholar who had just received his PhD from the University of Iowa, as part of a script for a film on ecology called *Home* being produced by the Southern Baptist Convention's Radio and Television Commission.[33]

Hunter's rather fanciful story was that the book had been given to him by a mysterious dulcimer maker who claimed to be both a Jew and a gypsy and who had suddenly appeared on his property several years before. Hunter could not remember the man's name but described him as "full-bearded, long-haired, hook-nosed, wearing a black frock coat and a skull cap, pants with buttons instead of a zipper, and beaded moccasins." His Chevy pickup, furthermore, "had a cedar-shake Hobbit house mounted on the back, complete with crooked stove pipe and a macramé God's Eye in the window." This curious figure marched up to Hunter, "his manner as brusque and impersonal as a cop delivering a parking ticket," and, with a copy of *Warriors of the Rainbow* in his hands, said to him: "Here, this is for you. It will reveal a path that will affect your life." In return he asked Hunter if he could cut down some old fence posts on his property, the dry wood being excellent material for dulcimers.[34] At the time, Hunter thought little of the incident—Vancouver attracted all sorts of eccentric characters—and placed the book on his bookshelf. It remained unread until he began to search for books to read on the voyage to Amchitka, whereupon it fell off the shelf as though demanding to be taken along.[35] The story is somewhat undermined by the fact that Hunter quotes a passage from Willoya and Brown's book in *Storming of the Mind*, so presumably he had at least given it a cursory read before taking it on the journey.[36]

One of the central prophecies in *Warriors of the Rainbow*, and one that seemed particularly pertinent to Hunter given the circumstances of the journey, was the story of an old Indian grandmother named Eyes of Fire and her grandson. The name, Eyes of Fire, coincidentally or not, evoked one of the most famous passages of modern environmental literature—Leopold's encounter with the wolf with the "fierce fire dying in her eyes."[37] In order to make the tale as widely applicable as possible, Willoya and Brown had not specified which tribe she had belonged to. When Eyes of Fire's grandson asks why the "Grandfather in the sky" had allowed "the White Men to take our lands," Eyes of Fire explains that it was all part of a greater cosmic plan. Initially, the Indians would be defeated and humbled by the white colonizers. In the long term, however, this would prove to be beneficial, for it was only through "the white man's conquest" that the Indians would be "cleansed of all selfish pride" and made ready for the "great awakening."[38] Given Hunter's interest in Gestalt therapy, Freudian psychology, Eastern mysticism, and LSD, it is not surprising that such myths resonated deeply with his own worldview. All of these phenomena, in their own way, urged people to let go of their egos, a critical element of Hunter's hoped-for revolution in human consciousness.

As the story continues, Eyes of Fire sends her grandson on a spiritual quest. On this journey, he learns that in the future, "just when the Indians seemed to be all becoming like the more foolish white men, just when everybody thought they had forgotten about the ancient days, at that time a great light would come from the east" revealing a path for the Indians to follow. This great light, in the form of a rainbow, would lead Eyes of Fire's grandson to become a warrior of the rainbow. In this capacity, he would "[spread] love and joy to others" and work toward "a new, spiritual civilization . . . [which] will create beauty by its very breath, turning the waters of rivers clear, building forests and parks where there are now deserts and slums, and bringing back the flowers to the hillsides." The warriors of the rainbow will

> brighten the understanding of the ignorant destroyers. They will soften the hearts of the would-be killers so the animals will once more replenish the earth, and the trees shall once more rise to hold the precious soil. In that day all peoples will be able to walk in wilderness flowering with life, and the children will see around them the young fawns, the antelope and the wildlife as of old. Conservation of all that is beautiful and good is a cry woven into the very heart of the new age.[39]

Whether Willoya and Brown's rendering of these legends was faithful to their original spirit is perhaps open to question. What is beyond doubt, however, is that they closely paralleled the sentiments of the counterculture and holistic ecology. Much of Greenpeace's association with Indian mythology comes from Hunter's reading of this book and his experiences on the journey. For example, as the *Greenpeace* approached the Kwakiutl village of Alert Bay on the northern coast of British Columbia, they received a radio message inviting them ashore to accept a blessing and a gift of Coho salmon. Once the boat was docked, the daughters of the chief came aboard to wish the crew well and offer their total support for the mission. The Kwakiutl were at the time carving what they hoped would be the world's largest totem pole, and they invited the crew to call in again on their way home and have their names inscribed onto it. Native children seemed particularly drawn to the crew, especially to those who looked the most like archetypal hippies—Hunter, Keziere, Moore, Fineberg, and Doc Thurston. For Indian children, wrote Hunter, trying to explain this affinity, "there can be no figure more beloved, for hippies are whites turned Indian." The children in the tiny Kitasoo village of Klemtu spontaneously composed a song, based on the tune *We Love You Conrad* from the Broadway musical, *Bye Bye Birdie,* which they sang boisterously to the crew:

> We love you Greenpeace
> Oh yes we do
> We love you Greenpeace
> Oh yes we do
> Oh Greenpeace
> We love you![40]

The connection to the *Warriors of the Rainbow* myth was further strengthened by the fact that the famous Indian actor, Chief Dan George, one of Doc Thurston's patients, was a vocal supporter of the *Greenpeace* mission. This was in stark contrast to the world's most famous cowboy, John Wayne, who had pulled into Victoria harbor on his privately owned minesweeper a few days after the *Greenpeace* had departed from Vancouver. When asked what he thought of the protest, Wayne reportedly snapped: "They're a bunch of commies. Canadians should mind their own business." To the hippie crewmembers on board the *Greenpeace* the symbolism was crystal clear: the AEC were the arrogant, violent, gun-slinging cowboys, while *Greenpeace* was on the side of the morally and ecologically superior Indians.[41]

The further they sailed and the closer they came to the awesome power of Cannikin, the more the old Indian prophecy appeared to make sense to Hunter:

> The whole West Coast was a ghostland of wrecked totems and abandoned canneries. The Indian civilization had been smashed by white technology. Then the Indians were forced to adopt the white man's ways—toiling in warehouses and punching clocks. And now the white man's technology had all but ruined the fishing. The herring populations were down so low no one was allowed to fish them. Halibut were on the wane. The northern West Coast was already a lonely deserted land, with the Indians struggling to survive in the ruin of their world. The white men, meanwhile, were pulling out, closing down the canneries behind them and taking away the Indian's fishing licenses, leaving them trapped in scenic concentration camps, forced to live on welfare.[42]

The Indians' spirit, it seemed to Hunter, had almost been broken but was not yet dead. It almost appeared as though the *Greenpeace* was fated to be the messenger between the destructive white man and the ecological Indians; it was, in short, the first tribe of rainbow warriors.

Hunter shared his copy of *Warriors of the Rainbow* with the rest of the crew. Predictably, the older members were less impressed than the younger ones. Yes, they agreed, Native Americans were among the major victims of European civilization's environmental despoliation. True, they were more responsible environmental stewards and there was much they could teach us. But their salvation and everyone else's lay in political struggle, not in soppy appeals to myths and prophecies, the veracity of which was in any case suspect. The fact that rainbows appeared throughout the voyage, however, only added to Hunter's feelings that there was some kind of mystical, enchanted bond between themselves and the Native Americans and that the *Greenpeace* was under the aegis of intangible and imperceptible forces.[43]

On Akutan, according to Hunter's typically vivid description, the crew "entered a perceptual world as unbearably intense and brilliant as the other worlds of schizophrenia and LSD." With its landscape of sub-Arctic tundra and its cloud-covered hilltops, the island seemed strangely surreal, as Hunter's description makes clear:

> The slopes of the hills were snaggled with tundra and scrub willows and moss, a mattress of vegetation. You could take seventeen-foot leaps and bounce. Bunchberries throbbed like tight little brains quivering to be set loose. Rippling, blinding sheets of light. Cow parsnips arrayed in whole alien civilizations at our feet. Moss. Lupen. Horsetails. Blueberries. Daisies like flourishing bits of Holy Nimbus. It was precisely like a Psychedelic Rush.[44]

The *Greenpeace* spent six days anchored in Akutan's harbor. While the island on the whole may have looked enchanted, its one settlement—a dilapidated Aleut village—was depressing. Unlike other native villages that the crew had encountered, the Aleuts on Akutan seemed to despise Caucasians, hippies or not, and spat at some of the *Greenpeace* crew who ventured among them.[45] Apart from passing time on the ship and trying to send out stories and receive news of when the blast was scheduled, the men had little to do other than explore the island. To help relieve the boredom, Bohlen, Darnell, and Simmons, the three Sierra Club members, decided to organize a group hike up the hill that towered over the harbor. Hunter, Moore, Keziere, Thurston, and Fineberg joined them. Along the way they discussed their mission, whom they represented, and what the future may hold for their "movement." It seemed they had fashioned a broad, albeit loose, alliance of groups that, in many instances, were unlikely to agree on much other than the danger posed by nuclear testing. At one extreme were groups such as the radical Vancouver Liberation Front, at the other, those represented by the Vancouver Real Estate Board, and in between were various churches, unions, student groups, and countless unaffiliated ordinary citizens. Terry Simmons felt that the alliance had the potential to become a tremendous future power base if any environmental group could find a way to harness it. It seemed that a transpolitical environmental alliance had come into broad outline around Amchitka.[46]

There were also more whimsical, though no less important, discussions. Hunter and several of the other hippies were Tolkien fans and liked to imagine themselves as Hobbits bearing the Ring toward the volcano at Mordor. They called themselves Greenhawks, the ecological counterparts to the 1950s comic book heroes, the Blackhawks, and practiced their "Hawkaaaaa!" battle cry as they leapt off large rocks onto springy mattresses of Aleutian vegetation. They styled themselves as the Merry Pranksters of ecology and the shock troops of the conservation movement, and formed a peace and an ecology symbol from the volcanic rocks scattered among the vegetation. At one point, Patrick Moore began to gently dig into the moss and soil with his bare hands and others kneeled down and joined him, marveling at the miniature ecosystem that existed just below the island's grassy surface. Moore, the ecology graduate student, began to give a spontaneous lecture on the interconnectedness of life, how all species were, at base, interdependent. Western man's understanding of nature—his taxonomies and his splintering and dissecting approach to the natural world—had served to obscure this holism, which, to the men kneeling in Moore's little circle, was never more apparent than at that moment. A wide grin appeared on Moore's face as he found the perfect hippie metaphor to describe this holistic ecosystem. It means, he exclaimed jubilantly, "That a flower is your brother!" Hunter immediately ordained them all ministers in the Whole Earth Church.

> It was a religious experience of some kind. At least it was connected to the root of what we *thought* of as religion. . . . The emotion we know as awe. . . . We drew closer together at that point. At one level it was all a surrealistic fantasy. Greenhawks, Eco-cairns. Surf-Clouds. Anti-spacemen. Environmental Merry Men. Brotherhood with flowers. With all living things. With the Earth. At this level, it was a joke to ordain everyone. . . . But the theological degree was tacked up on my bunk, and if there was ever going to be a moment in anyone's life when they would feel like part of some mystical universal force—it was at that moment, in the wind on Akutan Mountain, in the shadow of the H-bomb.[47]

Hunter was clearly in his element on Akutan, high on the surreal Aleutian scenery, on the possibilities of a new ecologically inspired religion, and on whatever chemicals Doc Thurston was dispensing. There was a kind of proto deep ecology in ideas such as the Greenhawks and rainbow warriors and simple metaphors such as "a flower is your brother." This sort of rhetoric and spirit, with its deep countercultural influences, would become a strong undercurrent of Greenpeace culture, waxing and waning throughout the 1970s and remaining in constant tension with the views represented by Bohlen, a tension that was captured in the binary phrase, "the Mystics versus the Mechanics," which became a kind of leitmotif of Greenpeace's development throughout the 1970s.

After spending a few days on Akutan, the *Greenpeace* sailed to Sand Point, a small, nondescript town on the Alaskan Peninsula. Upon their arrival, Bohlen and Cormack paid a visit to the harbormaster's office. They were treated politely, though coolly, and eventually cleared customs after the immigration officer soundly admonished them for transgressing several U.S. maritime and customs laws, warning them that any further infractions would lead to the impoundment of the boat. After emerging chastened but relieved from the harbormaster's office, Bohlen located the nearest phone booth and called Irving Stowe in Vancouver. Bohlen explained their situation to him, outlining the difficulties they would encounter should they continue their voyage to Amchitka. It was then that Stowe excitedly told him of his "backup plan," which involved chartering a larger and faster boat in case the test was further delayed and the *Greenpeace* was unable to complete its mission. More funds had been flowing into the DMWC coffers while the *Greenpeace* was at sea, and they now had $25,000 in the bank with all expenses paid. Although heartened by the news, Bohlen was skeptical about the plan's feasibility, especially at such short notice. He decided to share the news only with Metcalfe in case nothing came of it.[48]

The crew's spirits were fortified by the news that over 8,000 high school students had held an anti-Cannikin demonstration in front of the U.S. consulate in Vancouver. Principals throughout the city had refused students' calls to cancel classes for the day, but thousands had nonetheless joined the demonstration, making it the largest antinuclear protest since the Aldermaston marches of the mid-1960s. Irving Stowe's sixteen-year-old son, Bobby, who was chairman of the Student Action Committee Against Nuclear Testing, was a key organizer of the event, and Irving himself gave a lengthy and impassioned speech to the assembled throng.[49] Chief Dan George and former Canadian Prime Minister Lester Pearson took out a full-page advertisement in the *Washington Post* in the form of an open letter to the American people, urging cancellation of the blast.[50] In Washington, D.C., lawyers representing the coalition of antinuclear and environmental groups who had earlier attempted to sue the AEC were continuing to mount a last-minute legal challenge and organizing demonstrations and protests.[51]

For much of the voyage, the crew had felt like they were in another world, as though normal everyday conceptions of time and space had given way to a new temporal regimen where the only date that mattered was the elusive day on which the AEC would choose to detonate its bomb. At Sand Point, however, mundane reality intruded on their trip. It looked increasingly likely that the AEC would postpone the test until November, thus forcing many of the *Greenpeace* crew to think about their responsibilities and duties back home. Bohlen had to get back to his job at the Forest Products laboratory once the six weeks were over. Moore and Keziere were concerned about losing their graduate school places. Metcalfe's freelance business was suffering and Thurston was paying $100 a day to maintain his medical practice. In addition, the North Pacific seas were becoming increasingly rough and dangerous as early winter storms began to roll in across the Bering Sea and the Gulf of Alaska. Several of the crew began to discuss the possibility of giving up on Amchitka and heading back to Vancouver.[52]

The chief protagonists in the ensuing debate were Metcalfe and Hunter. Of the two journalists, Metcalfe was the more experienced, pragmatic, and cynical. He had managed political campaigns, run a public relations firm, and was a talented spin-doctor long before the term had been invented. In his view, the campaign had "peaked." Instead of continuing on what would be a dangerous and most likely futile attempt to reach Amchitka, the *Greenpeace* would be better off returning to Vancouver, where they could continue the media campaign and take advantage of the demonstrations and protests occurring throughout Canada. Before the crew had a chance to discuss the issue, Metcalfe made the following announcement on his CBC broadcast:

> The crunch is coming and [the crew] know it. They know that soon, very soon, they'll have to decide whether they should try to wait out bureaucratic delays of the Amchitka test and take the risk of being ploughed under by the sands of time, or whether to call their own shot, which would mean realizing that Canadian opinion against the bomb is at its peak now and that they can move more usefully by going home to help that opinion there than stay and watch it decline . . . So now the *Greenpeace* must contemplate its nettle. Whether they grasp it will depend to some extent on whether their voyage was planned as a practical protest to help raise public opinion in Canada or whether it was a hero trip for the gratification of a few egos.[53]

Bohlen agreed with Metcalfe. Continuing the journey, he insisted, would only undermine the hard work they had done so far in convincing mainstream opinion that they were not a bunch of "loonies." It would be far wiser, he felt, to head home and drop into as many Alaskan and Canadian towns as possible. En route they could hold meetings with locals, conduct open-ship visitations, and test public opinion to assess how effective their political and media strategies had been.[54]

Hunter, however, continued to steadfastly cling to his own McLuhanesque theories. For him, the *Greenpeace* remained "an icon, a symbol . . . a kind of wobbling control tower from which we might affect the attitudes of millions of people toward Amchitka specifically, and their environment as a whole." Nothing had yet happened to convince him "that the boat was not, after all, a mind bomb sailing across an electronic sea into the minds of the masses."[55] Giving up on the voyage's primary goal, therefore, would immediately undermine the group's credibility and break whatever spell they had cast on the mass mind of public opinion. Terry Simmons, though hardly a fan of Hunter and his theories, also insisted that the voyage would be a failure if they did not do all that was in their power to make it to Amchitka. Fineberg was of the same opinion.[56]

The issue flared into a furious table-pounding, fist-shaking argument that threatened to erupt into violence. The crew attempted to restore some semblance of order by passing around a hat procured during a visit to an Alaskan crab-processing factory. The hat, all agreed, would bestow upon its wearer a divine right to uninterrupted speech. Metcalfe began referring to those who insisted on continuing the voyage as the "psychedelic kamikaze squad" and accused them of putting the voyage, and the crew's lives, at risk in order to gratify their own egos. Hunter recognized that there was some truth in this but nonetheless insisted that they go on. He tried to manipulate Cormack by appealing to his stubborn pride in his ship and his ability as a captain, asking him repeatedly if he felt the ship could make it to Amchitka. Initially, the ploy seemed to work, with Cormack angrily insisting that the *Phyllis Cormack* was the finest boat on the West Coast and could handle any kind of weather. He ended his speech, however, on a desperate, pleading note that none of the others had heard from him before and which clearly influenced their decision. The boat, Cormack insisted, could make it to Amchitka, "*but you'd be crazy to try it.*"[57] Cormack's influence was decisive; his obvious trepidation, combined with Metcalfe and Bohlen's arguments, swayed six of the crew to vote against the four "kamikaze nuts," as Metcalfe referred to Hunter, Simmons, Darnell, and Fineberg. These four, not coincidentally, were also the crew members who had the least to lose by continuing, with Hunter and Darnell continuing to draw pay for as long as they were on the voyage, and Simmons and Fineberg having no pressing engagements to return to.[58]

Hunter was furious with the decision, smashing the door on his way out and sobbing with rage and frustration, feeling completely and utterly defeated. His violent outburst prompted Doc Thurston to prescribe some medication to calm him down. Fineberg was also bitterly disappointed and decided to buy a plane ticket to Anchorage. Hunter thought about joining him and flying

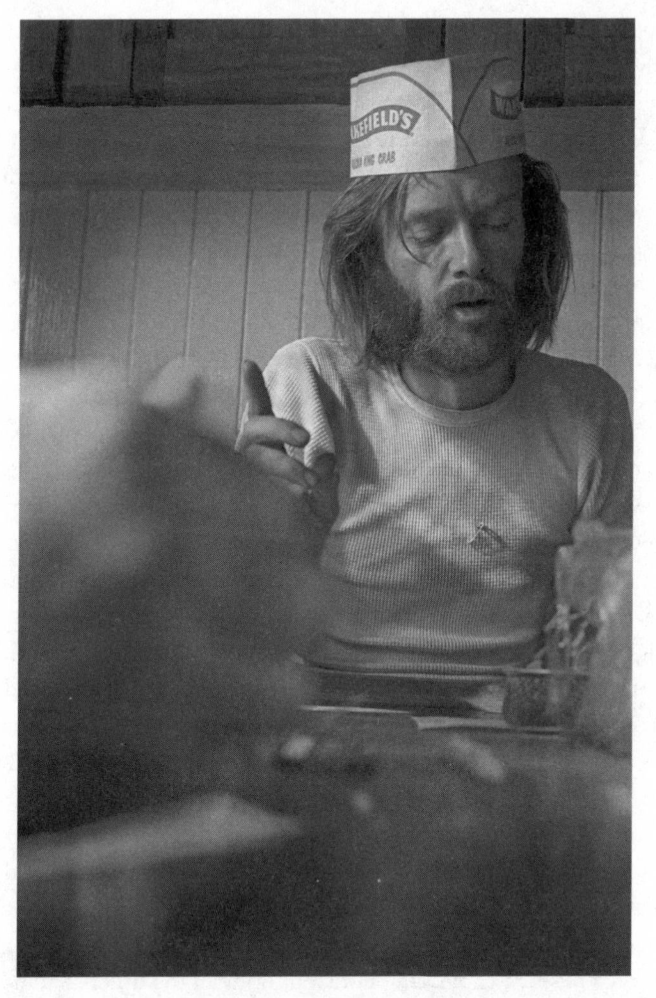

Bob Hunter makes a case for continuing to Amchitka. Sand Point, Alaska, October 1971. Reproduced by permission of Robert Keziere/Greenpeace.

back to Vancouver but was persuaded to remain on the *Greenpeace* by a combination of Thurston's sedatives and a conversation with Keziere, who convinced him that it was important to see the whole thing through, as much for Hunter's karma as for the campaign itself. Hunter turned thirty that night, and Metcalfe used his birthday as an opportunity to write a light-hearted report on how a *Greenpeace* crewman was turning thirty during a protest against the establishment, thereby deflecting attention from the decision to give up on the goal of reaching Amchitka.[59] Nevertheless, Hunter's bitterness and disappointment were reflected in his newspaper column that week. "The *Greenpeace* voyage to Amchitka was over," he wrote, "and we had failed and had been defeated. We did not make it to the bomb. The only hope left was that the voyage of *Greenpeace*—the idea, not the boat, had just begun." Such sentiments once again illustrate Hunter's hope that the voyage would spawn a larger and more permanent movement.[60]

Once the crew decided to adopt Bohlen's plan of heading for home and making ports of call along the way, the *Greenpeace* suddenly became less like the *Golden Rule* and the *Phoenix* and more like the pacifist sloop, *Satyagraha,* which had sailed along the New England coast during the early 1960s, dropping into seaside towns to raise awareness about nuclear issues.[61] As far as Bohlen was concerned, the boat was always supposed to be a vehicle for both direct action and education. Even though they had failed in their ultimate aim of bearing

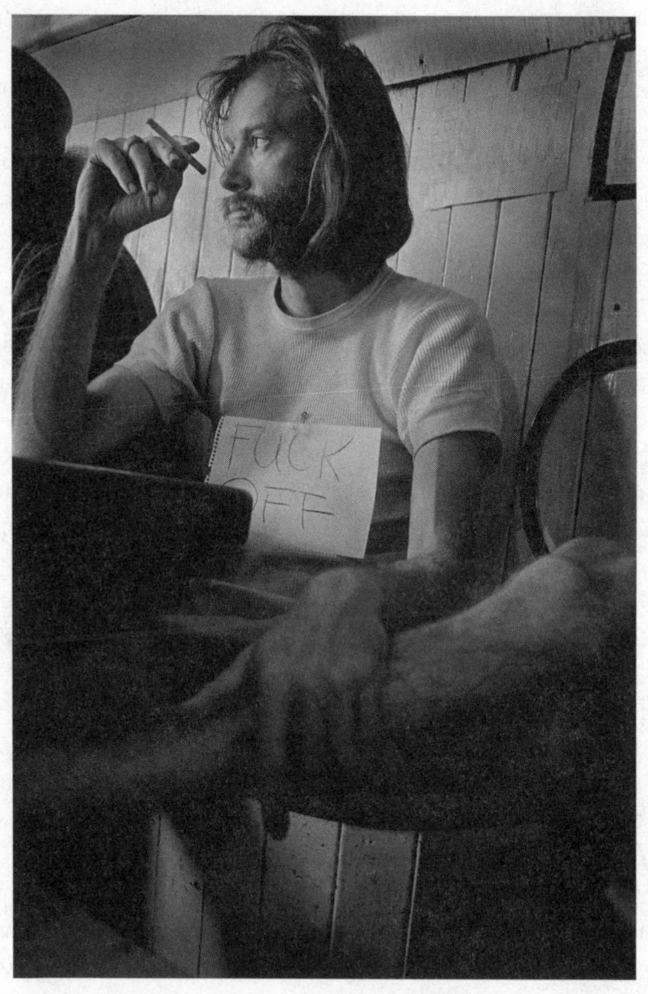

The majority of the crew decide not to continue to Amchitka. Hunter makes his feelings known. Sand Point, Alaska, October 1971. Reproduced by permission of Robert Keziere/Greenpeace.

witness to the bomb at Amchitka, they could still perform the important, if less glamorous, task of stumping from town to town like eco-evangelists and raising people's awareness of nuclear and environmental issues.[62]

Their next port of call was Kodiak. With a population of 20,000, it was the largest town they had visited since leaving Vancouver. The mayor had invited the *Greenpeace* to stop over as a gesture of the town's support for their cause. As they chugged into the harbor, the crew were surprised and delighted to find a large crowd welcoming them and bearing banners reading: "Thank You Greenpeace." Hunter and several other crewmembers, though not Metcalfe, had become so caught up in the drama that was taking place on the boat that they had temporarily forgotten that the outside world did not know of their "failure" and had formed their opinions largely from the positive messages that Metcalfe and Hunter had projected throughout the journey. A local group of antiwar activists had helped organize the reception and a number of young Coast Guard sailors were also among the crowd. They presented the crew with a three-foot-long dope-smoking pipe with a surgical mask at one end, a gift that some appreciated more than others. Much to everyone's amazement, Rod Marining was also waiting on the dock, having flown and hitchhiked his way up the coast, determined to take part in at least a small portion of the voyage.[63]

The mayor of Kodiak was somewhat taken aback by the sight of several of the Greenpeacers, who looked as though they had just walked out of Golden Gate Park rather than the well-groomed Quakers and scientists he was apparently expecting. Nevertheless, he organized a banquet in the crew's honor and was doubtless relieved when Bohlen, rather than Hunter or Marining, stood up to deliver a speech that did not call for an immediate revolution or an attempt to levitate the Pentagon. Bohlen, in fact, had prepared an "alphabet speech" of crimes against the environment—A is for "arms race," and so forth—and proceeded to steadfastly deliver all twenty-six points. Meanwhile, Hunter, Marining, and some of the other crew members went searching for the local alternative scene and wound up smoking dope at the house of a resident Gestalt therapist whose wife threw the *I Ching* for them.[64]

Overall, the three-day Kodiak visit was a positive experience. They met with members from an organization called the Concerned Servicemen's Movement, a group of young military personnel who were skeptical of Pentagon and AEC propaganda. They thanked the *Greenpeace* crew for exposing the AEC's "whitewash" of Amchitka, and the Greenpeacers, in turn, were once again forced to question their image of the military as a monolithic brainwashing machine.[65] The crew held several other meetings with various concerned organizations. Their conversations with locals demonstrated that their voyage was having an impact and that even Alaskans, traditionally among the most conservative and patriotic of Americans, could be opposed to U.S. nuclear testing and open to learning more about environmental issues. The local newspaper also gave them positive coverage, asking Alaskans to put aside their prejudices and embrace the *Greenpeace* message:

> Because their speech was spiced with the jargon of a generation yet to be recognized here and because beards and long hair were in abundant evidence, many locals found it easy to ignore the implications of the visit... [But] if the *Greenpeace* visit helped even a few people see Kodiak's own stake in the tests more clearly, then the delicate confrontation of ideologies will have been worth the negativism that the visit also spawned.[66]

After Kodiak, the *Greenpeace* continued down the west coast, stopping at the southern Alaskan port of Ketchikan and at Prince Rupert in British Columbia, before returning, as promised, to the Kwakiutl fishing village of Alert Bay. The Kwakiutl, along with the Haida, were the major native tribe of the Queen Charlotte Islands and the northern BC coastline. They had developed a cooperative fishing industry that had brought them a measure of economic stability, enabling them to embark on a miniature cultural renaissance. Much to the crew's surprise, the Kwakiutl invited them to participate in a ritual ceremony that was normally reserved for weddings, funerals, and the election of chiefs. All twelve were led into the village longhouse and asked to stand in front of a great cedar wood fire surrounded by giant totems. As old Indian men "with hands like roots pulled from wet clay" banged away on huge wooden drums and women in red-beaded robes danced to the pulsating rhythm, the Greenpeacers were anointed with water and eagle down feathers and made into blood brothers of the Kwakiutl people. They were then urged to abandon their egos while observing, and eventually participating in, a series of native dances. "It was as though the Kwakiutl were somehow able to read our minds," Hunter later recalled, "or had somehow succeeded in understanding perfectly the experiences we had just been through."[67] All of the crewmembers were moved by the experience, but for Hunter in particular, it was further confirmation that the *Greenpeace* was in some profound way connected to the myth of the warriors of the rainbow.

> Here I was, just having been *adopted* as an Indian. And fresh from having seen abandoned whaling stations on the bone-strewn shores of North Pacific islands, the ocean and sky almost empty, nuclear bombs being tested whose shock waves killed sea mammals en masse... and now here were these Indians, who spoke of animals as their brothers, if I understood this stuff properly, welcoming us into their tribal embrace, teaching us to let go of our egos. [Afterward], I wandered

> around outside by myself, staring through the blazing new eyes of somebody who has been through a transformative experience.[68]
>
> Vividly, I recalled the prophecy of the old Cree grandmother, Eyes of Fire, who had said a time would come when the Indians would teach the White Man how to have reverence for the earth, that we would all go forth as Warriors of the Rainbow. . . . Until a decade before, the Kwakiutl religion had been outlawed by the Canadian government. It had had to go underground, so few white men had witnessed these rituals, let alone taken part in them. Now her prophecy began to come true.[69]

While the *Greenpeace* crewmen were giving speeches, mingling with "groupies," and taking part in tribal initiation ceremonies, Irving Stowe and other DMWC members back on shore were doing their utmost to keep the campaign alive. Irving, Dorothy, and several other demonstrators protested Soviet Premier Kosygin's visit to Vancouver, condemning the Soviet nuclear program and undermining the criticism of those who accused them of being mere communist stooges.[70] The Stowes gave speeches and held press conferences throughout Canada, and Irving was determined to raise Canadian anger toward the test at every opportunity.[71] During one particular Air Canada flight across the country, Stowe became incensed by the fact that the airline was screening a John Wayne movie. He demanded to speak to the passengers over the PA system and alert people that Wayne had characterized Canadian concern over Amchitka as "a bunch of crap." The film was screened despite Stowe's plea to the contrary, but the airline did write back to assure him that they did not plan to screen any more of Wayne's films in the foreseeable future.[72]

Apart from venting his spleen at Air Canada, Irving Stowe also spent the weeks leading up to the blast trying to organize another voyage to Amchitka. Now that the *Greenpeace* had become a household name across Canada, there was no shortage of willing shipowners. Stowe decided to charter a converted minesweeper, the *Edgewater Fortune,* owned and captained by a forty-one-year-old navy veteran whose political views and lifestyle were diametrically opposed to most of those involved with the DMWC. A Rotarian, Shriner, Mason, and member of the New Westminster Chamber of Commerce, Hank Johansen had never been involved in any kind of protest. However, like many conservatives, he chafed at the perceived arrogance of the U.S. military and liked the idea of putting his new boat to the test by sailing her to the North Pacific in the early winter. Furthermore, he imagined, the media exposure would be a boon for his new charter boat business.[73]

Over fifty volunteers turned out on the morning that the *Edgewater Fortune,* dubbed the *Greenpeace Too* for the duration of the voyage, was due to set sail. Johansen chose those he thought would be most useful and put them to work.

Among the new crew were Kurt Horn, the SPEC representative from Victoria, DMWC members Paul Watson and Chris Bergthorson, and an experienced U.S. Navy veteran and trained navigator named Will Jones. Jones's life story was similar to those of Bohlen and Stowe. A well-educated professional who had worked for IBM, he became disillusioned with America's involvement in Vietnam and its overall military agenda. He and his family left when his sons were approaching draft age, in part because Jones had become "disgusted at the arrogance of the country." A Quaker, Jones had seen a television report about the *Greenpeace Too* and had immediately packed his duffel bag and made his way to the wharf, leaving his teenage sons to take care of the house and to inform their mother, who was visiting San Francisco, of his decision.[74] In addition, the *Greenpeace Too* had a substantial media contingent on board, including a six-man CBC television crew, a *Time* magazine photographer, a reporter from the Vancouver *Sun*, and two radio broadcasters.[75]

The *Greenpeace Too* left Vancouver on October 27 and that afternoon it pulled up beside its smaller sister vessel, the *Greenpeace*, at the Union Bay wharf, within sight of the town of Courtenay on Vancouver Island. After crewmembers exchanged revolutionary handshakes and embraces, Bohlen passed the *Greenpeace* flag to Captain Johansen, who had it raised on the *Greenpeace Too*'s masthead.[76] Places on board the larger vessel had been reserved for any of the original crew who wished to continue the journey, but only Simmons, Cummings, Birmingham, and Marining took up the offer. By this stage, Hunter felt he had pushed his sanity to its limits, and he agonized about whether he should go on or not. Given his almost hysterical insistence that the *Greenpeace* should have tried to reach Amchitka from Sand Point, he felt that it would be grossly hypocritical of him not to continue. Furthermore, the fact that Cummings, who was also pushing the bounds of his sanity, had chosen to persevere with the voyage only made Hunter feel greater pressure to swap ships. With barely concealed *schadenfreude*, Bohlen and Metcalfe watched Hunter, the lead "kamikaze nut," phone his wife and meekly ask her whether he should continue or not, hoping that she would urge him to come home. But Zoe Hunter was herself committed to the cause and was not about to dissuade her husband from continuing, instead telling him that she would support whatever decision he made. In the end, he decided that he simply could not face the prospect of continuing the journey on a different boat, with dozens of new people, and go through the decisions and arguments all over again. He took solace in the advice of the Kwakiutl and Fritz Perls, let go of his ego, and sailed back to Vancouver with the seven remaining *Greenpeace* crewmen.[77]

During the last stage of the *Greenpeace* voyage, several members of the crew began to discuss how the campaign had gone and what the future might be for the DMWC. Hunter and Simmons had been especially aggrieved at some of

Irving Stowe's rhetoric during the journey, feeling that he was undermining their efforts to build a transpolitical alliance by constantly demonizing the United States in his speeches and press statements. Hunter in particular was beginning to form the notion that Stowe was a hardcore Maoist and that his strident ultra-left-wing rhetoric would taint the rest of the DMWC, destroying the hard-won credibility they had achieved through the *Greenpeace* voyage. Coté had come to a similar conclusion while working with Irving back on shore, and even Bohlen was beginning to wonder if Stowe's inflammatory rhetoric was doing the campaign more harm than good. Bohlen had phoned Coté during one of the *Greenpeace's* earlier ports of call and the two discussed what to do with the DMWC once the AEC had completed the test. With some reluctance, they decided that it would probably be best just to dissolve the committee. Hunter, however, felt it would be foolish to close down an organization that, in his opinion, was just beginning to create a transpolitical alliance of citizens concerned with peace and the environment. Instead, he suggested that they should change the organization's name to the Greenpeace Foundation and register it as "an all-purpose ecological 'strike force,' rather than a one issue add-hoc [*sic*] committee." For once Bohlen agreed with Hunter and told him that he would recommend the idea to Stowe and Coté as soon as they were ashore.[78]

On November 4, a coalition of antinuclear groups in Washington lodged an appeal to stop the test with the Supreme Court, which surprised many people by taking the unprecedented step of agreeing to hear the case within twenty-four hours.[79] At this point, the *Greenpeace Too* was some thousand miles from Amchitka and making good headway through calm seas. If the Supreme Court had postponed the test for another three days, they would almost certainly have reached the island. However, just as it had agreed to hear the case in record time, the court also gave a quick and, according to some, hasty ruling, voting four to three against any further delay of the test. Dissenting Justices William J. Brennan, William O. Douglas, and Thurgood Marshall argued that the blast should be delayed until a commission could determine if the AEC had violated the National Environmental Protection Act by not filing an adequate environmental impact statement, while the other four justices gave no reason for their decision.[80]

The AEC wasted no time. Within minutes of the Supreme Court's decision, an arming party drove the twenty-odd miles from the Amchitka camp to the test site to make the final connections to the detonating mechanism. At 11:00 a.m. local time, AEC Chairman James R. Schlessinger, who had brought his wife and two daughters with him to Amchitka as a demonstration of his faith in the commission's safety procedures, gave the command to trigger the bomb. With the *Greenpeace Too* still 700 miles away, Cannikin exploded with a force that was 240 times greater than the bomb that had been dropped on Hiroshima. The intense heat vaporized solid rock, creating an underground cavern some 400 yards in diameter and unleashing a shock wave that registered 7.2 on the Richter scale.[81] The whole island wobbled like a giant pudding, and entire cliff faces slid off their edges and into the roiling sea. Birds standing on rocks had their legs driven through their bodies, and hundreds of otters, their ear drums perforated by the sudden change in water pressure, died slow and agonizing deaths, their bodies washing up on the shore for the next several weeks.[82] Nine-year-old Emily Schlessinger, protected in a thick spring-mounted, steel-reinforced concrete bunker, described the tremor as being "kind of like a train ride."[83] Her father was "elated" with the test, and General Edward Giller, the director of the Division of Military Application at the AEC, was pleased that everything had "gone to plan," pronouncing the blast a great success.[84] Ron Zeigler, the White House press secretary, was pleased to point out that the fears of environmentalists and peace protestors had been completely unwarranted. "The government," he calmly assured everyone, "would not have continued [with the test] if careful study had not been given to its environmental impact and other negative aspects."[85] AEC monitors reassured the world that "not a trace" of radiation had leaked into the atmosphere. Meanwhile, the crew of the *Greenpeace Too* could only meekly report that they had "felt nothing" of the blast.[86]

Upon the *Greenpeace Too*'s return to Vancouver, the first order of business for the DMWC was to make the appropriate arrangements for the future of their organization. Prior to this, however, the group had to deal with a number of disputes and issues that primarily revolved around the ideological and personal clashes that had occurred between the campaign participants. The first of these presented itself to Terry Simmons literally the moment he stepped onto the dock, where a jovial Irving Stowe was waiting to greet the returning crew. As Simmons remembered somewhat bitterly, before Stowe had "the courtesy to say hello," he began to hand out copies of the latest issue of the *Georgia Straight*, which bore his picture on the cover.[87] Inside was a two-page, center spread interview with Stowe in which he gave free reign to his strong anti-American sentiments and, from Simmons's perspective, grossly misrepresented the views of the DMWC. For regular *Straight* readers, Stowe's comments were merely a particularly cogent summary of ideas and issues that were discussed in the paper virtually every week: the relationship between U.S. militarism and international capitalism, Canada's complicity in the nuclear industry, and the role of organized labor in the environmental movement. For Simmons, however, whose vision for the organization was predicated on building an international, transpolitical alliance, the interview contained a number of gratuitous and inflammatory anti-American statements that could only damage the organization's reputation in mainstream society. The AEC, according to Stowe, was filled with "psychopaths" who were "educated beyond their intelligence . . . and beyond their humanity." More generally, the "so-called American Way of Life" was, in fact, "inflict[ing] illness and death and extinction" on Americans, whose much vaunted standard of living was not leading to improvements in the quality of life but, rather, was becoming a "death trip." The United States "has polluted and polluted and destroyed its own natural resources and now is coming up to Canada to take our resources." Assisting it in obtaining such resources for its weapons and its unsustainable standard of living would be like offering "aid and support to [Canada's] executioner."[88]

As a disillusioned former American, Irving Stowe had his own reasons for deeply resenting the United States, and he was certainly not going to forgo the opportunity to incite such sentiments among his newly adopted countrymen. His argument with Simmons, however, illustrates not just their different feelings toward the United States but also their different conceptions of the DMWC's and, ultimately, Greenpeace's strategies and goals. For Simmons, the DMWC was founded in the spirit of the *Golden Rule* and the *Phoenix* with the sole purpose of sending a protest vessel to Amchitka. They were concerned with nuclear weapons testing and its immediate environmental impact, not with broader issues such as Canada's participation in NORAD, NATO, or any other military agreement, or with nickel and uranium exports or any other specific U.S.-Canadian trade pacts. Most importantly, Simmons emphasized, "The Don't Make a Wave Committee is not anti-American." In addition to the Amchitka protest, they had formally objected to Soviet and French testing, though it had not been practical to protest against them in the same way. "The nuclear tests at Amchitka don't make Americans Nazis," he insisted, and "the AEC scientists are not psychopaths; they are bureaucrats. The American government, like all the rest of the nuclear nations, is merely responding to age-old military problems with the best of twentieth century technology." Stowe, according to Simmons, found it difficult to distinguish between his own opinions and the official views of the DMWC, an error he had made on several occasions and which had made it necessary for the committee to publicly denounce him.[89]

In his reply to Simmons, Stowe, at base still a Yale-educated lawyer, quoted directly from the DMWC's constitution to counter the claim that the committee was a narrowly conceived, single-issue protest group. According to its certificate of incorporation under the BC Societies Act, the DWMC's chief objective was "to foster public awareness of the possible environmental effects of the detonation of nuclear explosives" and "to support or conduct research in the area of environmental preservation."[90] Whether this definition was flexible enough to encompass some of Stowe's more extreme public comments depends on how broadly or narrowly it is interpreted. Perhaps a clever lawyer such as Stowe could claim that the definition was vague enough to cover all of his statements: that informing the public that the AEC was run by psychopaths, for example, "fostered public awareness" of the environmental impact of nuclear weapons testing.

Whatever the case, the dispute indicated that even after the Amchitka campaign was over, the founding members of the DMWC still had divergent views about the exact nature of the organization's objectives.

Just as Stowe's public anti-American comments had proved troublesome to Simmons, Bohlen's similar attitude was problematic for Hunter. Although generally less strident than Stowe, Bohlen would still air views that did not sound altogether appropriate for a leader of a nominally pacifist organization. For example, he told one journalist that he had left the United States because he could no longer stand the country's militaristic and imperialistic posture. If he had stayed there much longer, he confessed, he "might have blown up a draft board or something."[91] On board the *Greenpeace*, he became so enraged when the AEC finally announced that they were proceeding with the test that he went storming into the galley and announced: "That's it. When we get back, I'm going straight down to the consulate's office and I'm going to burn an American flag and renounce my American citizenship."[92] Like Stowe, Bohlen had become, in Hunter's words, "a Thoreau-like character, an absolute foe of everything of which he had once been a part." Hunter, like Simmons, felt compelled to publicly reiterate that the *Greenpeace* crew, "whatever else [they] might have been . . . were not as a group or movement, anti-American."[93]

A few days after the return of the *Greenpeace*, Bohlen was asked to give a speech at a rally in front of the U.S. consulate in Vancouver. Hunter was concerned that Bohlen was going to exploit the anti-American mood of the crowd and make good on his flag-burning threat. He was, therefore, relieved to find that Bohlen "restrained his anger and stuck to the initial ground rules of the Greenpeace mission—no violence, no blind anti-Americanism, no racism." For Hunter, Bohlen's self-control represented a "personal triumph over (his) own angry impulses." The moment he overcome his anti-American feelings, wrote Hunter, he became "something much closer to a Canadian," a somewhat ironic statement given the wave of anti-Americanism sweeping across much of Canada at the time.[94] According to Hunter, the fact that the DMWC had targeted the United States was, after all, "an accident of history." If Alaska had still belonged to the Russians, Hunter insisted, then the *Greenpeace* would have been protesting against the USSR. Furthermore, according to Hunter, the United States was "the most likely crucible of rapid social change." Thus the DMWC's "assault on Amchitka [was] intended to assist the Americans who are trying to reshape their society."[95]

Despite his deep disappointment in his country of birth, Bohlen by and large recognized that Simmons's and Hunter's arguments made sense. So, apart from the odd angry outburst, he generally tried to maintain, at least in public, a calm and rational attitude toward the United States. When a UBC professor commented that the DMWC had lost credibility by protesting against the United States in Amchitka while the French were still conducting above-ground tests in the Pacific,[96] Bohlen corrected him by pointing to the notes and telegrams that they had sent to both the French and Chinese authorities: "The fact that these activities did not capture the headlines the way the Amchitka protest did was not of our choosing. We, professor, do not control the media."[97]

That the DMWC did not control the media was certainly not for want of trying. Throughout the voyage, Hunter, Metcalfe, and Cummings had "hammered out propaganda broadsides" with great zeal and diligence, and the Canadian media coverage, particularly in British Columbia, had been favorable.[98] It would not be an exaggeration to suggest that the DMWC helped shape the Canadian media's general discourse regarding Amchitka, and Bohlen, Hunter, and Metcalfe were pleased with the amount of coverage they received in Canada.[99] The U.S. media, however, proved a tougher nut to crack, and the *Greenpeace* crew was disappointed with the tepid interest shown by the American press. "Apart from a few informed and dedicated Alaskans," Bohlen told the crowd in front of the American consulate in Vancouver, "we found the American media Great Wall of Silence was still as large as ever."[100] Hunter, too, despaired of the American mainstream media's "conspiracy of silence."[101] Subsequent examination, however, indicates that the media resonance in the United States was greater than the DMWC had thought, with one analyst arguing that the campaign demonstrated "the power of a well-conducted pseudo event."[102]

Somewhat ironically, given their disappointment with the U.S. media coverage, the DMWC was quick to take credit for the fact that the AEC abandoned Amchitka as a test site after Cannikin, a view that subsequently entered the official version of Greenpeace history.[103] According to Hunter, they had helped create "too much of a stink for the AEC to dare to try to use Amchitka as a test site again." Bohlen also believed "that the Greenpeace campaign had greatly influenced" the AEC's decision to abandon Amchitka.[104] Metcalfe, too, felt that the voyage had been successful in these terms, since it had "created a focal point for Canadian protest," a view he did not subsequently alter.[105] While there is some evidence to support Metcalfe's contention, to argue that the *Greenpeace* played a significant role in the AEC's decision to abandon Amchitka is drawing a very long bow. Six months prior to the Cannikin blast, the AEC had already announced that it would be the final test in the Amchitka series.[106] Although the commission had broken such promises in the past, the trend toward building more maneuverable missiles with smaller warheads meant that once Cannikin was complete, Amchitka would have served its purpose and could be abandoned.[107] The historical importance of the *Greenpeace* voyage, it would appear, lay more in the organization it spawned than in the dubious proposition that it gave the U.S. military a black eye.

How, then, can the *Greenpeace* voyage most accurately be characterized? Was it a brave and idealistic protest against American militarism and environmental destruction? A quest to ignite an eco-consciousness revolution? A cynical propaganda mission designed to manipulate public opinion against the United States? Or merely a hapless and poorly executed protest? To some extent, of course, it was all of these. Ben Metcalfe managed to capture the essence of the voyage as well as anyone: "The *Greenpeace* crew," he told his CBC audience upon the boat's return to Vancouver,

> constituted the perfect metaphor of modern protest. We were the classically absurd motley of our time; powerless nobodies representing all of the modern ages of frustration . . . young cynics and old romantics, Quixotes and Panzas, Ahabs and Ishmaels, synonyms and antonyms of each other, made compatible by a kind of paranoid grandiosity that was consummated in our over-weening belief that we could, as a matter of fact, sail in under the guns of the mega-machine and destroy it.[108]

Although neither the *Greenpeace* nor the *Greenpeace Too* had come within 700 miles of the enemy's guns, the experience nonetheless provided those on board with a glimpse of the potential power of their little movement. A group of "powerless nobodies" could, if properly organized and committed, attract the support of thousands of ordinary people in a crusade against militarism and "ecological vandalism," thereby forcing the "mega-machine" to take notice of the waves of protest rippling outward from their wake. Those on board the *Greenpeace* felt their campaign had generated the embryonic stirrings of a broad international transpolitical alliance. All agreed that such a possibility was too important to abandon, regardless of how amorphous the alliance or how difficult the task of mobilizing it may prove. And such feelings were not without justification. Despite their failure to reach their destination and the flakiness that characterized some aspects of the campaign, it was nonetheless a substantial achievement. Unlike similar voyages of the past, such as that of the *Golden Rule*, the *Greenpeace* protest managed to attract considerable media attention. Furthermore, as well as employing the direct-action tactics of its predecessors, the campaign, which was almost two years in the making, made a genuine effort to unite two of the major social movements of the twentieth century—environmentalism and the peace movement. The DMWC managed to lay the groundwork for such an alliance in a deliberate and thoroughgoing way. Where previously the two movements had merely overlapped, now, at least among a certain segment of the Canadian population, the values and tactics of the peace and environmental movements, as well as their respective critiques of militarism and environmental destruction, were on the way to being integrated.

The question for the DMWC, then, was what shape should the organization take to provide such an alliance with a more concrete form? According to Hunter, who from the beginning had possessed the grandest vision for the DMWC, the new organization needed to abandon the traditional political goals of pressuring governments or trying to replace one political party with another; this would only result in illusory change. Instead, it would have to focus all its energy on bringing about a consciousness revolution on a world scale, using cameras, rather than guns, to fight a McLuhanesque war for the hearts and minds of the masses. The ultimate goal should be nothing less than the creation of a green version of the United Nations.[109] How exactly such an organization should be structured and managed was not precisely clear, but then again, organizational matters were never Hunter's strong suit.

Metcalfe also had an essentially McLuhanesque vision for any new organization that might emerge from the DMWC, but it was one that was unencumbered by the kind of utopianism that characterized Hunter's thinking. Instead, the more cynical, elitist, and conspiratorial Metcalfe felt that the most useful thing that they could accomplish would be to create an organization that would do for ecology what Madison Avenue had done for corporate America. If brainwashing was the only way to save the earth from humanity, then so be it.[110] Bohlen and Simmons, who had never given much thought to the creation of an ongoing organization, were essentially satisfied with the DMWC as it stood, feeling that with some minor structural tinkering, it could be set up to run multiple campaigns based on direct action, scientific research, educational outreach, and solid media work.[111] For Irving Stowe, the DMWC had the potential to empower various disenfranchised social groups by acting as an organizer, facilitator, and funder of progressive social and political movements. The committee, he told the *Georgia Straight*, could use "its funds and influence, and speaking and organizing abilities [to help] those groups in the community which have a base for action to actually translate that concern into action." Students and women, Stowe felt, were particularly aware of the systemic problems of modern industrial societies since they were among its victims. "My feeling is that the best expenditure that the people in the Don't Make a Wave Committee can do is to help these groups in whatever way they call upon us to become politically active, politically motivated, and take action."[112]

In the weeks following the voyage to Amchitka, the DMWC held a series of meetings to determine the organization's future. Their ranks swollen by new members, many of them crewmen from the *Greenpeace Too*, the meetings frequently erupted into disputes and quarrels as various participants tried to influence the form the new organization would take. Bohlen, Stowe, and Coté, the three original signatories on the DMWC's charter, decided that they would no longer lead the outfit. After two years of almost full-time work they were exhausted and wanted to devote their energies to other projects for a time. Nevertheless, Bohlen and Stowe were determined to maintain an active interest in the group and to ensure that it would remain in the hands of somebody they felt they could trust. Naturally, this meant that Hunter, who was eager to take on the job, was out of contention. Bohlen in particular had developed a strong antipathy toward Hunter and regretted that he had been able to play such a major role in the campaign. He was certainly not about to let him turn the sober and professional DMWC into some wacky hippie-esque league of rainbow warriors. For the time being, at least, Hunter was forced to take a back seat to Ben Metcalfe, whom Bohlen and Stowe appointed as the first chairman of the DMWC's successor, which, as Hunter had suggested, would be called the Greenpeace Foundation.[113]

The name "Greenpeace Foundation" was itself emblematic of the antagonism between Hunter and Bohlen and the worldviews they represented. For Bohlen, the word "foundation" described a nonprofit organization interested in promoting research and funding campaigns; it was synonymous with professionalism and respectability. Hunter, however, had specifically chosen the term in reference to one of his all-time favorite works of science fiction, Isaac Asimov's *Foundation* trilogy. Asimov's novels described a galactic empire that, though corrupt and in decline, still clung to power at the expense of the other creatures in the galaxy. Dissidents within the galaxy organized an oppositional force, called the Foundation, whose task was to hasten the collapse of the Empire so that its brutal and destructive reign would only last another 1,000 years instead of the expected 30,000. In more than one sense, then, Hunter and Bohlen's conceptions of the new organization were worlds apart.[114]

The new Greenpeace Foundation, simply referred to as "Greenpeace" by most of those associated with it, came into being in early 1972. Its broad parameters were well established: like the DMWC, it would continued to employ direct action and a savvy media strategy to protest nuclear testing and environmental destruction and to educate people about such issues in an effort to raise ecological awareness. Beyond that, however, the organization was for the most part a blank slate. What campaigns should it concentrate on next? How should they be run? Along what lines should it be structured and administered and how would the decision-making process function? Should the organization be allowed to grow and spread in an organic, unstructured fashion like a social movement, or should it be centralized and tightly controlled? Clearly, throughout the Amchitka campaign, key figures such as Hunter, Bohlen, Simmons, and Stowe had begun to formulate their own ideas about what any future organization should look like, and they would continue to try to shape the new Greenpeace Foundation in ways that reflected their own styles and ideological proclivities. For the moment, however, Greenpeace was driving down a wide-open road, and for the first vital year Ben Metcalfe would be behind the wheel.

5

Not a Protester in the Usual Sense

Organizations that are part of broad social movements—what social scientists refer to as "movement organizations"—are seldom well structured or coherently organized in their early stages. Instead, they are usually in an emergent phase during which they rely on protest activities as a fundamental resource with which to mobilize supporters, rather than on more conventional internal and external means, such as direct mail.[1] In the first few years of its existence, Greenpeace was no exception. Key members were uncertain about what they wanted Greenpeace to become or how it should best function. Should it try to be a broad social movement in itself—one that would benefit from the synergistic unification of various segments of the peace and environmental movements? Or should it be more circumscribed in its aims and endeavors, content to mount campaigns that drew people's attention to the political and environmental repercussions of nuclear weapons testing? For Jim Bohlen, the latter scenario was the most realistic, since it made the best use of scant resources. Bob Hunter, on the other hand, was determined that the term "Greenpeace" should be more than merely the name of an organization. Rather, it should become an all-purpose label for a new ecological sensibility.[2]

During the first two years of Greenpeace's existence, however, neither Bohlen nor Hunter was in charge of the organization. Despite the vital role he had played in founding Greenpeace, Bohlen's influence was already on the wane, though this was largely his own decision. Hunter's time, on the other hand, was still to come. Instead, the conspiratorial scheming and "paranoid grandiosity" of Ben Metcalfe dominated the next Greenpeace campaign. Much to his subsequent regret, Metcalfe drafted a boom-and-bust Canadian businessman named David McTaggart into the organization. McTaggart, who had never so much as contemplated political activism in his forty years on the planet, was an unlikely candidate to lead a Greenpeace protest. Yet, he would go on to exert greater influence over the organization than any other individual in its four-decade history.

As the first president of the newly created Greenpeace Foundation, it was Metcalfe who determined that the group's next campaign would be against

French nuclear weapons testing in the South Pacific. His plan was conceived as a virtual one-man show: he would pull all the strings and be the only one with full knowledge of the entire strategy. In McTaggart, however, Metcalfe found that he had an obstreperous puppet.

In 1972 and 1973, McTaggart led two Greenpeace protests to the French testing zone in the South Pacific. Unlike the first Greenpeace voyages to Amchitka, McTaggart reached his goal on both occasions and proved to be a major political and military headache for the French. Despite this success, the relationship between McTaggart and Greenpeace was fraught with difficulties. It led to Metcalfe's withdrawal from Greenpeace and brought the young organization to the brink of collapse. Ironically, however, it would be McTaggart who, somewhat unintentionally, would rescue the group despite developing a deep suspicion of Greenpeace's political motives and managerial competence. In the short-term, the campaign likely played a significant role in ending French atmospheric testing, although underground blasts would continue for many years. It also inspired several similar protests against nuclear testing, as well as promoting the establishment of Greenpeace's first non-Canadian branches. In the long-term, it set the stage for McTaggart's eventual control of the organization in the 1980s, as well as creating a lasting tension with the French government—one that would culminate in the sinking of one of Greenpeace's vessels, the *Rainbow Warrior*, in Auckland in 1985, a deed that can accurately be described as an act of state terrorism.[3]

The first post-Amchitka "action" for the DMWC members occurred in January 1972 and was organized by Ben Metcalfe. Metcalfe and his wife Dorothy, along with the Stowes, the Bohlens, Hunter, Moore, Doc Thurston, and several others disrupted a Liberal Party function held in the plush suburb of West Vancouver, where they heckled the speaker, Canadian External Affairs Minister Mitchell Sharp, for what they felt was the Canadian government's timid stand on nuclear weapons.[4] Afterward, they went back to Metcalfe's house and discussed the future of the organization and what actions, if any, they should plan next. It was at this meeting that the group agreed to dissolve the Don't Make a Wave Committee and to adopt the name, "Greenpeace Foundation," with Metcalfe as chairman.[5] During the discussion, the group debated a wide variety of ideas about what the new organization's goals should be and how it should be organized. Although there were still considerable differences among them, they agreed on certain broad principles. The Greenpeace Foundation, they decided, should be international in scope; it would try to coordinate the activities of various protest groups—many of which had come together for the first time to oppose the Amchitka blast—in order to develop the organizational muscle to oppose any further AEC tests in the North Pacific; it would try to publish a magazine that would cover a broad array of antinuclear and environmental issues; and it would set its sights, in Hunter's words, "on a wide variety of targets, from oil tankers to unnecessary hydro-electric dams, pollution in all its forms, and anti-war activities." The foundation's broad objective would be "to sustain the new ecology and anti-war alliances which have come into being and to be ready to give active support to further protests along those lines."[6]

By early 1972, the name "Greenpeace" would have been familiar to many Canadians, particularly in British Columbia, and to Americans who read the newspapers closely and were interested in such issues. Beyond North America, however, few people had ever heard of it. A noteworthy exception, and one which serves as an example of Greenpeace's nebulous status somewhere between a one-off campaign, an organization, and a social movement, was a group of antinuclear anarchists in the United Kingdom who were associated with the long-time pacifist publication *Peace News*. The group had subscribed to the *Georgia Straight* and was impressed with Stowe's "Greenpeace is Beautiful" columns. Borrowing from Stowe, they published a broadsheet supplement to *Peace News*, which they called *Greenpeace*. They used the term not to describe an organization but rather, as a general philosophy of life denoting a marriage of peace and ecology. By 1972, the group, which, in true anarchist fashion, contained no "members" but only "people," began to refer to itself as Greenpeace London and in effect became the first Greenpeace group outside Canada.[7]

In some ways, the new Greenpeace Foundation was similar to David Brower's Friends of the Earth (FOE). Established in San Francisco in1969, FOE had by the early 1970s become a loosely knit international environmental organization. FOE published the very popular *Environmental Handbook* just in time to take advantage of the publicity generated by the inaugural Earth Day, providing Brower with enough funds to travel to Europe in an effort to build up a network of FOE offices. With the aid of two young Americans, a physics student named Amory Lovins and a Paris-based lawyer Edwin Matthews, Brower set up Les Amis de la Terre in France, choosing a radical young economist, Brice Lalonde, to lead the new organization.[8] Similar events occurred in the United Kingdom and West Germany, so that by 1971, FOE was already providing a model for the establishment of more activist-oriented and internationally focused environmental organizations. Both Greenpeace and FOE concentrated primarily on antinuclear campaigns, with FOE focusing on the civilian uses of nuclear power and Greenpeace on nuclear weapons testing. Each of them, furthermore, had a desire to internationalize the environmental movement and to encourage a greater degree of activism. Unlike Greenpeace, however, FOE was largely the brainchild of one man and started life with a clearer vision of where it was heading and how it would be organized. Therefore, although Greenpeace quickly took direct action and "mind bombing" to new levels, it would be several more years before they would develop the kind of coherent organizational structure that FOE already possessed by 1971. Nevertheless, the early histories of the two organizations frequently intertwine; throughout the 1970s, for example, FOE activists were often responsible for setting up new Greenpeace offices, particularly in Europe.

It was Joe Breton, a member of the *Greenpeace Too* crew, who suggested that the Greenpeace Foundation turn its attention to French nuclear testing in the South Pacific.[9] The French, who were not signatories to the 1963 Test Ban Treaty, were still detonating atmospheric nuclear devices on coral atolls in their Polynesian colonies. The idea drew mixed reactions: some were immediately enthusiastic about it, insisting that what the French were doing was far worse than what the Americans had done on Amchitka and that an organization such as Greenpeace was one of the few in a position to contemplate sending a boat into the South Pacific. Others, however, were skeptical. Irving Stowe, according to Metcalfe, was not altogether keen to shift the focus of his wrath away from the United States, while others felt that the voyage would be extremely difficult from a logistical point of view, since the French were bound to prevent a boat from sailing from Tahiti or any of their other colonies, thereby necessitating a long voyage from either New Zealand or South America. In the end, however, it was Metcalfe who made a unilateral decision to embark on the campaign, a decision that was prompted by a provocative newspaper article in the *Vancouver Province*, in which a columnist attacked Greenpeace for being anti-American and challenged them to prove him wrong by protesting against French nuclear testing. The next day, Metcalfe phoned Bohlen and informed him of his decision, and Bohlen promised that neither he nor Stowe would stand in his way.[10]

France's Pacific empire was not particularly large when measured in land and coral acres, but it ranged over a vast area of ocean approximately the size of China. From New Caledonia in the western Pacific, French colonial outposts dotted the seascape for the next 5,000 miles to the Tuomotu and Marquesas Islands in the east. To the Western imagination, the region conjured up romantic, Gauguin-inspired images of deserted beaches, palm-fringed tropical lagoons, and lovely, sexually permissive Polynesian women. By the early 1970s, however, the societies on which such myths were based had long been transformed by the dictates of French colonial policy and capitalism's global reach. French fries, pizza, and Coca-Cola had largely replaced the traditional coconut milk, yam, and fresh fish diet in all but the most remote of the islands, while many a palm-fringed lagoon had become a concrete harbor for French war ships.[11]

In 1958, French president Charles de Gaulle made a proclamation that would have far-reaching consequences: to ensure "that others should not become masters of our destiny," the French government had decided to embark on a program to "endow ourselves with nuclear weapons."[12] Apart from ensuring that they remained "masters of their destiny," the French government and military believed that the United States was an unreliable ally that was prepared to sacrifice Europe to save itself. Therefore, a nuclear-armed France would be the best guarantor of West European freedom. It was also felt that, despite the enormous costs of the nuclear program, it would still be cheaper than maintaining the huge conventional force that would be needed to deter a Soviet invasion. Perhaps the most obvious, if unstated, reason lay in France's difficulty in accepting the fact that it had declined from a great to a middling world power. Joining the nuclear club undoubtedly compensated for this loss of prestige and influence.[13]

In February 1960, the French military began testing nuclear weapons in the Sahara Desert. However, as France became mired in an increasingly bloody war against the Algerian independence movement, it soon became clear that they would have to search further afield for a more secure testing facility. Without consulting the Polynesian population, the military decided to use an atoll some 800 miles east of Tahiti as a long-term testing facility for the *force de frappe*.[14] It was de Gaulle himself who announced this to the Polynesians in 1963, brazenly declaring that it was a gift to repay Polynesian loyalty during the war: "I have not forgotten all that you have done, and this is one of the reasons I have chosen to install this base in Polynesia. . . . Not only will this test centre promote French military research, but it will also be highly beneficial from the point of view of the economy of the inhabitants of Polynesia."[15]

In preparation for the blasts, the Centre d'Experimentation du Pacifique (CEP), the French agency in charge of nuclear testing, annexed a large atoll just north of Gambier Island. Known as Moruroa to the locals, the French misspelled the name as "Mururoa," which has remained the most common spelling ever since.[16] The first test on Mururoa was symptomatic of the French military's haste to develop its nuclear program and of de Gaulle's imperative to put French military interests before all else, regardless of the cost. De Gaulle had traveled to Polynesia to observe the blast, but strong easterly winds had forced the CEP to postpone the test for several days. The CEP had promised to test only on days when the wind was from the west so that atmospheric radiation would blow across the largely uninhabited waters of the southeast Pacific, rather than back toward Tahiti and other populated islands. Nature, however, remained uncooperative, and the strong easterlies continued to blow while the CEP admirals grew nervous at de Gaulle's growing impatience. Finally, with his work piling up in Paris, de Gaulle let it be known that he could not continue to cruise around the South Pacific until the conditions were just right, leaving his officers with the firm impression that if the bomb did not explode soon, the president would. Deciding that the latter possibility was more frightening, the CEP detonated the bomb on September 11, 1966, sending a cloud of radiation blowback across the South Pacific. Over the next few weeks, while French authorities continued to deny the test's environmental impact, the National Radiation Laboratory of New Zealand measured significant increases in radiation levels in both the water and the atmosphere as far west as Fiji.[17]

Between 1966 and 1972, the CEP tested twenty-nine nuclear devices on Mururoa and on nearby Fangataufa Atoll. In 1971, as the AEC was preparing to detonate Cannikin deep below the Aleutian bedrock, the CEP exploded five atomic bombs. None were even remotely close to the American bomb in their explosive force, but all were detonated above ground. Throughout this period, the governments of Australia and New Zealand, as well as those of smaller Pacific nations, had registered their protest with the French and all were met with the same response: the tests were environmentally innocuous and were in the broader strategic interest of those who condemned them.[18]

In France itself, the force de frappe enjoyed widespread popular support, and Mitterrand and other socialists realized that they would never attain power by opposing it. Unlike its British counterpart, the French Left did not have a strong history of antimilitarism, and in postwar France it was quite possible to be on the left and not be opposed to the army, colonialism, or nuclear weapons. No CND-type antinuclear umbrella organization existed in France, while the major peace organization, the Mouvement de la Paix, had close links with the French communists, thereby restricting its potential support base.[19] French politicians, the mainstream media, and the military denigrated antinuclear protestors by referring to them as "pacifists," which, in the postwar French context, connoted the appeasement and defeatism of the Second World War. Furthermore, unlike in Britain, Germany, and Canada, there was never any question of American missiles being stationed on French soil, which meant that the anti-Americanism that was an important part of the peace movements in other countries was not a major factor in France. Even the relatively weak French environmental movement, which had been a strong foe of nuclear power, largely caved in to the consensus view on nuclear weapons. While many activists may have opposed the force de frappe in private, the movement as a whole did not make the issue a focal point of its campaigns for fear of alienating the majority of the French people.[20]

After the war, Ben Metcalfe had spent several years in France and had been briefly married to a French woman, with whom he had had a daughter. Although Metcalfe had enjoyed life in the south of France and had admired many aspects of French culture, he nonetheless retained the anglophile's mistrust of French foreign policy. He was particularly suspicious of the French military and contemptuous of the Foreign Legion, which, he believed, hired the world's worst criminals and psychopaths to put down revolts in the colonies. This ambivalent attitude toward the French gave the Mururoa campaign certain piquancy for Metcalfe, and he threw himself into the work with gusto.[21]

Metcalfe considered himself McLuhanesque in his media philosophy, though this was due more to the fact that McLuhan appeared to agree with *him* rather than a case of Metcalfe being persuaded by the force of McLuhan's arguments. That is, Metcalfe's own extensive experience with the media in the post–World War II era had led him in the same intellectual direction as McLuhan, who had been the first to articulate the concept of the "global village" in a systematic manner, creating a theory of media that resonated deeply with Metcalfe's experience. The idea of a world temporally and spatially compressed by a global media, combined with his Machiavellian view of society, led Metcalfe to adopt a condescending and cynical attitude. The media, he argued, was "fundamentally stupid." A hard-working, well-read reporter with common sense and a good nose for a story could easily manipulate the mass media and create pseudo-events virtually out of thin air. The secret was in packaging the stories as much as their content. So long as the clever journalist was able to manufacture a compelling narrative with the appropriate element of conflict, particularly of the David versus Goliath variety, the mass media would rise to it, regardless of the event's actual significance.[22]

According the Metcalfe, the Amchitka campaign could be characterized as "naïve bourgeois" because its organizers had announced its schedule and its limitations—the fact that they could only afford to stay on the boat for six weeks, for example—thereby providing the "enemy" with a huge tactical advantage. Furthermore, they had been very distant from the center of power in Washington, D.C., which greatly reduced their visibility in the U.S. media and their commensurate influence on American public opinion.[23] To avoid a similar fate, the Mururoa campaign would need to be more cunning in order to keep the French guessing. It would also have to bring the protest directly to the French by conducting an act of civil disobedience in Paris to alert the French population to the impact the nuclear tests were having in the South Pacific and to demonstrate the strength of international opinion against the force de frappe.[24]

Unlike the more open, consensus-oriented approach that had characterized the DMWC, this Greenpeace campaign was planned and run as a virtual one-man show. Metcalfe would sit up late at night in his upstairs home office, which he self-mockingly referred to as the Ego's Nest, developing ideas and strategies. To maintain an element of secrecy, he never informed anyone of more than part of his overall plan so that only he was aware of the big picture. When he needed something ratified by other members of the group he would "call meetings backwards": that is, he would reach a decision unilaterally and then run the meeting in such a way that the majority would agree with him. Many of the meetings were held in Gastown, the funky, dilapidated countercultural quarter in inner-city Vancouver, and were attended by dozens of hippies, street kids, and various social outcasts. This, Metcalfe knew, would alienate some of the older, straighter activists, such as the Stowes and the Bohlens, who would have been in a better position to challenge Metcalfe's authority.[25] Hunter, who had hoped to be the first Greenpeace leader, resented Metcalfe. Nevertheless, he had to admit a certain degree of grudging admiration for Metcalfe's "beautiful one man, McLuhanesque show." Others, however, were alienated by his aloof and conspiratorial style, and Stowe in particular grew suspicious of Metcalfe and backed away from the campaign.[26]

Metcalfe was keen to ensure that the campaign was truly international, and he planned accordingly. He would try to organize at least two boats to sail into the test zone—one from New Zealand or Australia and one from South America. He also planned to organize protests in Paris and at the United Nations Environmental Forum in Stockholm, which he hoped would lead to spontaneous antinuclear protests in other European capitals. In order to influence Catholic opinion throughout the world, he would seek an audience with the pope and try to persuade him to publicly endorse the Greenpeace campaign. He would also send activists to New York to lobby the United Nations to condemn French nuclear testing. The whole campaign, he hoped, would be distilled into a simple message for the mass media: The arrogant French were ignoring world opinion and endangering people's lives and the environment to develop weapons of mass destruction.[27]

Metcalfe took a "great man" approach to history, believing that "all great movements, like Christianity and Islam, started with a committee of one." As far as Greenpeace was concerned, he envisioned it as movement that would spread throughout the world in much the same way as Christianity, with himself in the role of the grand strategist and Hunter, Marining, Moore, and others as his apostles. While certainly aware of the hubris involved in such a grandiose analogy, Metcalfe nevertheless felt that the apostolic model was a sound one for the propagation of the movement, particularly when combined with a clever media strategy.[28] However, Greenpeace's growth throughout the world during the 1970s was more haphazard than Metcalfe could have predicted.

Like the Amchitka campaign, Metcalfe's plans for protesting against the French involved direct-action protest at sea. Thus he would need to find a boat and a willing crew. Logistically, the country that offered the most favorable point for embarking on the journey was New Zealand, a nation that prided itself on its sailing prowess. Rather than simply placing advertisements in New Zealand newspapers, Metcalfe, remembering how difficult it had been for the DMWC to find a boat, adopted a different strategy. He typed up a cable in the form of a correspondent's report, which he felt would be more impressive than a mere press release, and sent it off on the wire service of Reuters, one of the world's largest press agencies and his former employer. Just as he had hoped, the story soon reached New Zealand, and instead of a mere advertisement in the classifieds, Greenpeace became front page news in the *New Zealand Herald*, the nation's major newspaper. Metcalfe's story conjured up the image of a large and successful organization: "Our people are in France, New Zealand, Australia, Japan, Peru, the United States and several other countries," he wrote, and they would "make themselves known at the appropriate time." Furthermore, he continued, they were planning a huge demonstration in Paris and intended to seek an audience with President Pompidou in an effort to "arouse . . . a sense of horror and disgust" among the French people at their government's activities, of which most of them were clearly unaware.[29] Metcalfe's portrait of Greenpeace was fanciful at best, but to many New Zealanders, who had been resentful of French testing for the previous six years, the organization appeared to offer the prospect of more serious opposition than the tepid protests that had so far been tendered by their conservative government.[30] Moreover, for New Zealanders, who were always sensitive about their relative insignificance on the world stage, Metcalfe's story gave the impression that they were somehow honored to have been chosen to host the campaign that was being organized by this rather impressive-sounding international organization.

One of the many people who read the *New Zealand Herald* article with interest was Gene Horne, a man in his mid-forties who had lost his right arm in a logging camp accident two decades before and who was living with his wife in the town of Hamilton, seventy miles south of Auckland. The article particularly caught his attention because it mentioned that the mysterious Greenpeace Foundation was from Vancouver. His nineteen-year-old daughter, Ann-Marie, was at the time dating a thirty-nine-year-old sailor who also hailed from Vancouver. His name was David McTaggart, and Horne made a note to ask him if he knew anything about Greenpeace. McTaggart, however, had not lived in Vancouver for over a decade and had never heard of the outfit. Given his background as an aggressive businessman from a conservative family, an organization like Greenpeace would in any case have been of little interest to him. It was therefore a surprising turn of events, to say the least, when barely a fortnight later, McTaggart found himself skippering the *Greenpeace III* as it sailed toward the French testing zone in Mururoa.[31]

Throughout the 1980s and 1990s, when he was the most influential figure at Greenpeace International, McTaggart's pre-Greenpeace days became fertile ground for investigative journalists with the time and resources to pursue his restless and peripatetic life. Those that attempted this discovered some revealing details, but their efforts were frequently compromised by their attempts to use McTaggart's colorful past as an instrument to discredit Greenpeace.[32] Like his fellow eco-warrior, Paul Watson, McTaggart had a reputation for exaggeration and historical revisionism and tended to place himself at the center of important events, regardless of whether he was involved. He was also stubborn, aggressively competitive, cunning, and emanated a rugged charm that many found irresistible. He was clearly no saint. Nevertheless, he possessed a combination of traits that proved useful to a fledgling antinuclear organization with global ambitions.[33]

McTaggart was born in Vancouver in 1932 and raised in a comfortable, upper-middle-class suburb near the University of British Columbia campus. The youngest of three children, David was a precocious child who was heavily indulged by his devoted parents. Reflecting on his childhood, McTaggart recalled that he was always made to feel special by family members, a fact which, he believed, contributed greatly to the development of his strong, often overweening self confidence and his refusal to accept rejection. Even as a child, McTaggart recalled, he was extremely competitive and would often provoke fights with his older brother Drew.[34]

As a teenager, McTaggart was far less interested in schoolwork than he was in sports, chasing girls, and drinking. Confident, handsome, and extremely competitive, McTaggart developed a win-at-all-costs attitude to life and became particularly fond of cuckolding older men.[35] His poor academic performance and hedonistic lifestyle were a constant disappointment to his parents, who belonged to the fundamentalist Plymouth Brethren church, a faith that McTaggart never embraced. At sixteen, he was expelled from his school for getting into a vicious argument with one of the teachers during a school rugby match (the teacher was refereeing and McTaggart simply refused to accept one of his decisions). His long-suffering parents then sent him to St. George's, the same elite boarding school that Patrick Moore would attend a decade later.[36]

While St. George's did not improve McTaggart's academic performance or alter his lifestyle, it did introduce him to the sport in which he was to have his greatest success—badminton. The game suited McTaggart's strengths, requiring constant quick movement and lightening-fast reflexes. Within a year, he was the best player in Vancouver. Throughout his entire life, McTaggart always had a strong desire to try to control whatever situation he was in, a tendency that was reflected in his preference for individual sports—badminton, golf, tennis, squash—over team sports. In the 1950s, badminton was still perceived as something of a gentleman's sport. McTaggart, however, played with ruthless determination, diving all over the court to save points and constantly swearing and disputing umpire's calls. Seemingly unaffected by his heavy drinking and smoking, McTaggart won tournament after tournament, and his appearance frequently drew large crowds (by badminton standards), with many people hoping to see the arrogant, young upstart lose. In 1956, he won the first of three Canadian national championships.[37]

In his late teens, McTaggart began working in the construction industry, where he quickly climbed his way up into various high-level executive positions. By the time he was thirty-two years old, he was already twice divorced and living in San Francisco, where he worked for a firm that specialized in constructing hotels.[38] Through one of his colleagues, he heard about a man who owned a large ranch in the Sierras about 130 miles east of San Francisco and was looking to develop it into a ski resort. The land had been in his family for five generations, and McTaggart smelled an opportunity. He packed a case of beer and drove up to Bear Valley to meet the rancher, Bruce Orvis, and see if they could strike a business deal. Orvis was about the same age as McTaggart, had played football in college, and was the kind of no-nonsense, do-it-yourself man that McTaggart respected. The two hit it off immediately. Within a week, in early 1965, they had struck a deal and formed the Bear Valley Development Corporation, with Orvis as president and McTaggart as vice president and CEO.[39]

Although McTaggart received a sizable income at Bear Valley, he was still a mere salary earner and therefore lacked the controlling influence that ownership affords. In 1969, he purchased a small, rundown hotel a few miles from Bear Valley with the aim of turning it into a restaurant and nightclub for young singles. With Orvis's approval, he scaled back his involvement in the ski resort and took all of his own capital, in addition to borrowing money from a bank and from his third wife's mother, and invested it in the new project.[40] A few days before its official opening, disaster struck. According to McTaggart, he had arranged for one of the members of the U.S. ski team to appear at the opening of the new club, which he called MegaBear, and he and his workers were busy putting the final touches on the complex. One of the employees told him that he had smelled gas in the building, but McTaggart was in too much of a hurry to investigate the matter closely; an inspector had, in any case, already checked the gas fittings and found everything in order. The next day, a huge explosion ripped through the complex, destroying much of the main building. McTaggart and one of his workers raced to the building and found a screaming maintenance man, half-buried in collapsed timbers. Surrounded by flames and fearing more explosions, McTaggart and his powerfully built worker pulled the maintenance man out with such force, that one of his legs was left behind in the rubble. A woman who served as a cook at the lodge was also severely injured and lost an arm as a result of the explosion.[41] On top of all of this, McTaggart had not adequately insured the enterprise, claiming that he had sought insurance but had been told that he could not obtain it because the building was more than forty miles from the nearest fire station. The explosion bankrupted McTaggart and left him in considerable debt. In desperation, he tried to sue the gas company for negligence.[42]

McTaggart was now thirty-eight years old, broke, and unemployed. His third marriage was in tatters; he was heavily in debt and had become deeply depressed. He gathered what little money he had left, made his way to Los Angeles and, without telling a soul, boarded a plane for Tahiti, where he hoped that Gauguin's sun-drenched beaches and palm-fringed lagoons would soothe his ravaged nerves.[43]

It may have been a new start, but it was very much the old McTaggart who, before the plane even landed, had managed to chat up the Pan Am stewardess who had served him during the flight. Although his graying hair had receded, McTaggart still had his rugged good looks. When he was in the mood, few people, men or women, could resist his piercing blue eyes and easygoing charm. No surprise, then, that the young stewardess was more than happy, in McTaggart's words, "to help me out during my first days in Tahiti." He then moved on to the island of Moorea, where he hoped to find work at the Club Med resort. Instead he found François Ravalo, a local artist who specialized in Gauguin-like oil paintings of the Polynesian landscape and people. In exchange for his company, Ravalo fed McTaggart and took care of him for a few weeks. During the evenings, McTaggart would make marauding runs on the resort and its abundant population of young women. He had no solid idea about what to do with his life. He only knew that he was sick and tired of striving for success and wealth in conventional terms and that he wanted to step out of the stressful, high-risk life of property development. He began toying with the idea of purchasing a small yacht with his remaining funds and cruising the high seas in the hope that, by giving full reign to his restlessness, he could somehow overcome it. He could make what little money he would need by occasionally taking tourists on short cruises, perhaps spicing up the prospect with a bikini-clad all-female crew. His mind filled with such vague thoughts, he left Tahiti a few weeks later and flew to Auckland, New Zealand.[44]

Clearly, McTaggart was a man at loose ends. The world of business, like his personal life, had been a roller coaster ride of exuberant highs and depressing lows. A successful and comfortable bourgeois life was his for the taking, yet somehow, his ambition always caused him to overreach: to risk everything he had for a little more. Introspection, however, was not one of McTaggart's strong points. Unlike Jim Bohlen, who had been in a not dissimilar situation, McTaggart possessed neither a reflective predisposition nor the intellectual inclination that might have led him to explore other philosophies of life or formulate a coherent critique of the culture that had shaped his frenetic lifestyle. Instead, his inchoate yearning for something more meaningful in his life was expressed in a headlong rush to flee from his problems and to sail the ocean, free from the pressures and responsibilities of the everyday world.

Soon after arriving in Auckland, McTaggart bought a 38-foot ketch, the *Vega*, a boat that had been entirely hand-built from local Kauri pine by one of New Zealand's leading boat constructors, Alan Orams, in the 1940s.[45] The *Vega* was destined to become one of the most famous boats in the Greenpeace flotilla. For the next year, McTaggart sailed her along the coast of New Zealand and around the South Pacific. He spent several relaxing months in Fiji, chartering himself and the *Vega* to tourists and earning as much in a week as he used to make in an hour back in California, but he was content nonetheless. Slowly, he began to feel better about himself and finally, after more than a year away from North America, he wrote to his family in Vancouver and let them know his whereabouts.[46] Certainly, he had not experienced any kind of radical religious or political conversion; the changes he was undergoing were occurring around the edges of his consciousness, rather than at its core. He was becoming a little mellower and a little less competitive. A worldview that had been a fortress for the crudest brand of bourgeois materialism was starting to open just a little bit.

By January 1972, he was back in New Zealand and was anchored off the town of Russell, in the Bay of Islands. While having lunch in a local café, he was served by a petit, blond, nineteen-year-old student who was waitressing during her summer vacation. Before she had taken the job, Ann-Marie Horne had promised herself that she would not accompany strangers onto their yachts, but McTaggart's charm and persistence eventually persuaded her to abandon the resolution. McTaggart viewed his meeting with Ann-Marie as a key moment in his life; without it, he insisted, his involvement with Greenpeace would never have eventuated.[47] Thus, he started a relationship with Ann-Marie that would last for several years, and it was because of her that he ended up in Gene Horne's living room on April 12, 1972, wondering what this mysterious group called Greenpeace was all about.[48]

Gene Horne, Ann-Marie's father, was unable to find the copy of the *New Zealand Herald* with the article about Greenpeace, but he remembered that the organization was looking for a volunteer crew and a boat to sail into the French nuclear-testing area around Mururoa Atoll. McTaggart looked Mururoa up in an atlas and, out of idle curiosity, began to consider the logistics of such a voyage. He estimated that it would be a round trip of about 7,000 miles and would involve being at sea for three to four months with few, if any, stops on dry land. Furthermore, whoever took up the challenge would only have two weeks to prepare for the journey. It would require someone with considerable ocean-going experience and a reliable "ocean whalloper," as sailors referred to hardy cruising yachts. With a slight sense of unease, it began to dawn on McTaggart that the ideal person for such a voyage, from a logistical standpoint, at least, would be someone such as himself.[49]

Wandering through the streets of Hamilton the next day, McTaggart noticed a handbill on a store window, the kind of thing he normally ignored. It urged people to get involved in the protest against French nuclear testing and was signed by Mabel Hetherington, the honorary secretary of the New Zealand Campaign for Nuclear Disarmament. He took down her number and called her from the nearest phone booth. She was not in, so he left a message asking her to call him at the Hornes' number. That evening, he called various local radio stations and newspapers, trying to get more information about the Greenpeace plan and asking people to pass his details on to anyone who could put him in touch with the organization. "With each unfruitful call," he later recalled, "my own involvement with the idea began to intensify, and over the period of a day, I had almost talked myself into it."[50]

For the rest of his life, McTaggart would claim that, as with many of his actions, he never fully understood his own motivation for getting involved in the Greenpeace campaign. After all, he admitted, "French imperialism was not a subject which interested me."[51] He had never protested against anything or been involved in a volunteer society or charity, and the closest he had ever come to developing an environmental consciousness was the recognition that ski resorts looked nicer if a few trees were left standing. For McTaggart, "ecology had meant aesthetics, and that was tied to continuing profits. Just plain, hard-driving good business sense."[52] In principle, he opposed nuclear testing, but that was scarcely enough motivation to read about the issue in the newspapers, let alone to risk his life protesting against it. In the final analysis, the best explanation he could come up with was the challenge presented by the journey and France's insistence, illegal under international law, that 150,000 square miles of sea around Mururoa remained off-limits to all vessels during the testing periods. Given McTaggart's obsession with personal freedom and his hatred, verging on phobia, of being geographically restricted, there seems little reason to doubt his word.[53] However, McTaggart's actions during and after the voyage have led some of the other participants to conclude that his motives were more venal.

The day after learning about the Greenpeace campaign, McTaggart received a call back from Mabel Hetherington, a seventy-year-old English woman who had come to New Zealand after the Second World War and who had been active in the peace movement for over thirty years.[54] Greenpeace certainly had not entered an antinuclear vacuum in New Zealand and probably would not have gotten very far had there been one. Hetherington and other activists had formed the New Zealand CND, the local version of its famous British namesake, in Christchurch in 1960 by amalgamating a number of smaller pacifist and antinuclear organizations. Many antinuclear activists in New Zealand traced their involvement with the cause back to July 9, 1962, when the night sky was lit up by the afterglow of a U.S. high-altitude test on Christmas Island, some 3,000 miles away. David Lange, a future New Zealand prime minister who, during the 1980s, caused an international furor when his government banned American nuclear ships from New Zealand ports, recalled seeing

> an eerie lightening of the sky. Beams of light radiated from the northern horizon and intersected with each other through the blackness of the

> night. . . . I was arrested with awe as the sky pulsated with these brilliant shafts of light. . . . They extended across the night like ribs of a fan. They were spinning, they were intermingling. . . . The sky was diffused with a ghastly brush of red. . . . I couldn't reason away the chill sweat of dread.[55]

In 1964, New Zealand CND launched a "No Bombs South of the Line" campaign to create a nuclear-free southern hemisphere, a move aimed specifically at the French plans to test nuclear weapons in Polynesia. They collected over 80,000 signatures (in a nation of less than three million people) on a petition against French testing and repeatedly protested outside the French embassy and alongside visiting French warships. They also urged the New Zealand government to send a frigate into the test area to register the nation's protest.[56] One of their members, an expatriate American named Bob Stowell, had proposed the kind of campaign that Greenpeace would conduct in the 1970s. Writing in the *CND Bulletin*, he argued that "a well planned protest against the tests, including sail boats, rafts, or even small aircraft placed in the testing area by private organizations and manned by crews from several countries," would be a vivid way to draw attention to the issue.[57] In 1965, CND supported an Australian protest vessel that had attempted to sail from Sydney to Mururoa but which had ultimately been abandoned by an inexperienced crew in Raratonga.[58]

From 1966, many in the antinuclear movement turned their attention to New Zealand's entry into the Vietnam War, which dominated antiwar and pacifist activities for the next five years. Nevertheless, committed CND activists continued to keep the issue of French testing on the political agenda. In 1965, Richard Northey, a University of Auckland student and a member of CND's youth branch, formed the Committee for Resolute Action against French Tests, and throughout the late 1960s, CND forged contacts with leaders of the Polynesian independence movements. CND also helped to organize a lecture, attended by a wide array of protest organizations and Maori groups, by the veteran Tahitian independence leader, Pouvanna a Oopa, who was in Auckland for medical treatment in 1969. In 1971 in Wellington, a group of Pacific nations, including New Zealand and Australia, founded the South Pacific Forum in order to better coordinate regional political and economic matters. This event, combined with the French announcement the same year that they were embarking on an expanded program of testing that would continue into the foreseeable future, helped once again focus attention on nuclear testing in the South Pacific. Symptomatic of the shift in attention from Vietnam to Mururoa was the formation, in 1971, of the Peace Media Research Project by Barry Mitcalfe, a well-known poet who lectured in Polynesian Studies at Wellington's Victoria University and was chairman of the Wellington Committee on Vietnam.[59]

The morale and the prominence of local pacifists was boosted in 1971 with the visit to New Zealand, sponsored by the New Zealand Society of Friends, of the English Quaker activist, Donald Groom. Groom had worked with Gandhi, and his lectures and weekend workshops on nonviolent resistance aroused considerable interest among antiwar protest groups and the media. This was followed, in 1972, by the visit of George Lakey from the New Life Center in Philadelphia, a Quaker foundation that promoted social and political change through nonviolent action. Lakey had been chairman of the Quaker committee that had organized the voyage of Earle Reynolds's boat, the *Phoenix*, to Vietnam to deliver medical supplies. He, too, gave a series of lectures and held workshops on nonviolent direct action, as well as making an appearance on New Zealand's most popular TV current affairs program, *Gallery*.[60] Like its popular culture, it seemed, America's protest culture was a readily exportable commodity.

Although antinuclear groups found the New Zealand government's protests against French testing feeble, successive governments since 1957 had, in fact, consistently opposed nuclear testing. The Labour government of 1957–60 declared itself against all nuclear testing and had deserted traditional allies, such as the United States and Great Britain, by voting against the French tests in the Sahara at the United Nations. The conservative National Party government of the early 1960s had expressed "profound dismay" at U.S. and Soviet tests of the early 1960s and called the 1962 test on Christmas Island "surprising and regrettable."[61] By the early 1970s, however, the National Party government of Prime Minister John Marshall began to temper New Zealand's opposition because of fears that the French would block New Zealand lamb and dairy sales to the United Kingdom, which would have been a massive blow to the domestic economy. The French Minister for Overseas Territories, Pierre Messmer, underlined this threat, telling reporters in New Caledonia that although "we have condescended to grant New Zealand guaranteed exports to Great Britain for a period of five years . . . this could be reconsidered or reduced and we could look at the application with disfavour."[62]

In 1972, French testing became an election issue with the Labour opposition vowing to toughen New Zealand's stance. They even promised, in the spirit of the times, to send a naval frigate to Mururoa to "bear witness" to the tests.[63] J. B. O'Brien, the leader of the Social Credit Political League, New Zealand's third-largest political party, also criticized "the weak policy of the New Zealand government," which, he believed, "did not reflect the attitude of New Zealanders." He further declared that he would not "stand idly by and watch the French abuse our environment in order to satisfy their own nuclear ambitions," before joining the ever-growing queue of those proposing to send a protest vessel to Mururoa.[64]

When the CND heard about Greenpeace's plans, they offered to act as their agent in New Zealand and to help find a boat and a crew to sail to Mururoa.

Thus, it was Mabel Hetherington who put McTaggart in touch with Metcalfe. The outcome, for both McTaggart and Greenpeace, could well have been quite different if Hetherington's first choice had been available. On April 10, 1972, a week after the Greenpeace article first appeared in the New Zealand newspapers and two days before McTaggart became aware of the organization, an old Baltic trader called the *Fri* sailed into Auckland Harbour.[65] A solid, 105-foot oak schooner, the *Fri* had been constructed in Denmark in 1912. Sixty years later, it belonged to David Moodie, a twenty-six-year-old American from a wealthy New England family who had turned the boat into a floating hippie commune. With his partner, Emma Young, Moodie had taken a well-worn, almost hackneyed, countercultural path on his way to purchasing the *Fri*. From college at Syracuse University, he went on to manage a health food cooperative in Vermont, and then traveled to San Francisco, where, among other things, he wrote for an underground newspaper. He purchased the sturdy old trader in March 1970, just after she had been used to deliver water supplies to a group of Native Americans who had seized Alcatraz Island from the U.S. government, apparently intent on making a symbolic stand against the American dispossession of native lands.[66]

In late 1971, when the original *Greenpeace* was sailing toward Amchitka, Moodie and his crew heard reports of the protest on the radio from their base in Hawaii. "Those two evocative words instantly struck a chord with me," Moodie later recalled. He also took the *Greenpeace* campaign more seriously than most, sailing the *Fri* out to sea before the Cannikin blast, just in case it caused a tsunami to roll across the Pacific and submerge Honolulu's harbor. After several months of island hopping around the Pacific, the *Fri* arrived in Auckland, where Hetherington asked Moodie if he would be willing to sail his schooner, under the Greenpeace flag, to Mururoa. Clearly the *Fri*, with its idealistic skipper, pacifist crew, and ample storage space for long-haul cruises, would have been the perfect protest vessel. However, though Moodie was more than willing, his crew was exhausted after several months at sea and the boat was in need of repairs. Instead, the *Fri* spent the rest of the year in various ports in northern New Zealand, where it constantly attracted members of the local counterculture and helped mobilize the grassroots base that would go on to set up the first significant Greenpeace office outside Canada.[67]

With the *Fri* out of the running, and none of the other candidates looking very promising, Metcalfe was still searching for a boat when McTaggart called him at his West Vancouver home. Although Metcalfe had been a household name in Vancouver for over a decade, he had only risen to prominence after McTaggart had left his hometown in the late 1950s. McTaggart was somewhat suspicious of the organization and its supporters, fearing that they may have been militant radicals or communists. Metcalfe assuaged his fears, assuring him that they were not involved with any political party or movement; they were just an ordinary group of people committed to opposing nuclear testing. To further impress McTaggart, Metcalfe told him that Greenpeace was supported by international luminaries such as Jacques Cousteau, Linus Pauling, Jean Paul Sartre, Simone de Beauvoir, and Buckminster Fuller and that they had the endorsement of the World Council of Churches, the Sierra Club, Les Amis de la Terre, and numerous other organizations all over the world. All this was technically true, although it gave the impression that the organization was far more established and widely known than was actually the case. He also told McTaggart that Greenpeace would cover all of his costs, including any legal and medical expenses, as well as purchasing any extra gear that he and his crew would need for the voyage. McTaggart replied that he would need to think about it for another night and that he would call back the following day with his answer.[68]

Initially, Metcalfe was impressed with McTaggart. He was clearly not a flaky hippie and appeared to possess the requisite degree of sailing experience to undertake the voyage. Furthermore, his lack of idealism and his experience in the cutthroat world of the construction business would stand him in good stead should he encounter the French military.[69] For his part, McTaggart retained his conservative businessman's skepticism of protestors and other do-gooders and decided to discuss the matter with Gene Horne, whom he considered to be "a clear-thinking and conservative individual, not given to flights of fancy." Horne was convinced that there was something deeply wrong about the way the French were treating the Pacific and supported any sensible and legal protest. McTaggart never had any time for religion or mysticism in any form, yet the more he thought about the voyage, the more it seemed that some kind of cosmic finger was pointing at him. On the next day, he gave Metcalfe the answer he had been hoping to hear. Furthermore, he had also calculated the costs of supplies and gear for the journey and had resolved to set up a joint Greenpeace-CND bank account for donations and expenses.[70]

For the first time in his life, McTaggart found himself working among a group of idealistic and committed individuals, and the experience was a revelation. Former youth CND leader, Richard Northey, now a law professor and the leader of Auckland CND, introduced McTaggart to the students and CND members who would help organize the campaign. "I was impressed," McTaggart later recalled.

> Student volunteers showed up by the score, each of them seizing on any task and applying themselves vigorously. I had not worked with any volunteer groups before and I was astonished at their discipline and the

> amount of work they had got done—for nothing but the honour of helping. . . . In the midst of the steady hum of activity, I paused often during those days, astonished by how much *love* I felt around me. I was used to groups that were competitive, raucous, defiant, proud. But here was something different. Here was something that I didn't want to even try to put into words.[71]

McTaggart's first task was to assemble a crew for the lengthy voyage. From the outset, Metcalfe made it clear that he expected to go along, though McTaggart was skeptical. The *Vega* could only carry five people, and McTaggart wanted to reserve places for a good navigator, a mechanic, and somebody who could operate a ham radio. He was also concerned that Metcalfe, at fifty-two, might not be in optimum shape to undertake such an arduous journey. Metcalfe, however, reassured him that he was as fit as a thirty-five-year-old and that he could operate a radio.[72] The man whom McTaggart most wanted to bring along with him was Nigel Ingram, a twenty-five-year-old Englishman who had studied at the Royal Naval College and who, despite his relative youth, was an experienced navigator and ocean sailor. Ingram had lived in London throughout the late 1960s and early 1970s, and although he was not politically active, he was generally sympathetic to the protest movements of the era. He left England in 1971 and went to Australia, where he sailed in the prestigious Sydney to Hobart yacht race, before moving on to New Zealand in early 1972 and finding work at the Westhaven Boat Harbour in Auckland. It was there that he met McTaggart, and the two of them quickly became good friends. McTaggart's offer appealed to Ingram. He had never been involved in a serious protest before, but here was a chance to use his sailing skills in the service of a just cause.[73] The fourth member of the crew was to be Gene Horne, who, despite having only one arm, was an excellent all-round mechanic and someone whom McTaggart felt he could trust in a difficult situation. The final crewman chosen was Roger Haddleton, another former Royal Navy sailor with a solid knowledge of diesel engines and plenty of sailing experience.[74]

As chance would have it, McTaggart's brother-in-law's brother, David Exel, was a journalist on New Zealand's popular *Gallery* TV program, and he invited McTaggart on for an interview. Initially, he was nervous about the idea, since he knew very little about nuclear and environmental issues or South Pacific politics, but Exel promised he would avoid such questions and concentrate on McTaggart's personal motivation for undertaking the journey. McTaggart felt that the interview went well and that he had managed to convey the fact that he was not a run-of-the-mill antinuclear activist: "I'm not a protester in the usual sense," he insisted. "My reasons are personal. I'm not going to make a spectacle of myself."[75] Apart from demonstrating a rather poor understanding of the motivation of more committed antinuclear protestors, it was clear that McTaggart still felt it necessary to convince himself that he was not like the activists and do-gooders he had ridiculed for most of his life.

Over the next several days, the New Zealand government made numerous ham-fisted attempts to stop the protest voyage. A police search of the *Vega* yielded several Seiko watches and earned McTaggart a smuggling charge and a short stint in an Auckland jail before his lawyer paid the fine.[76] Having failed with the smuggling charge, authorities then insisted that the *Vega* undergo a survey to prove its seaworthiness, a process that usually took several weeks to organize. Fortunately for McTaggart, a sympathetic friend was able to organize an immediate survey in less than a day.[77] Another act that looked suspiciously like government interference was the Department of Immigration's failure to issue Gene Horne's passport on time, thus forcing McTaggart to ask Grant Davidson, a young man from Sydney, to join the crew at the last minute. Davidson, an itinerant laborer at the Westhaven Boat Harbor, had been keen to go on the voyage right from the start, but McTaggart felt he lacked experience and the necessary skills. However, Davidson was a good cook with a lively sense of humor and a gregarious personality, so McTaggart decided to ask him along literally a few hours before they were due to depart. Despite the lack of preparation time, Davidson accepted the offer enthusiastically.[78]

Another problem was that McTaggart and Ingram were not getting along with Metcalfe. When they had picked him up at the airport, they were taken aback by his large physique, and both felt that he did not look like a particularly good candidate for a long journey on a small, crowded yacht. They also found him a little too boisterous for their liking and were rankled by what they felt were his bourgeois affectations, particularly his habit of always being impeccably dressed and well-groomed. His eloquence and erudition smacked of intellectual arrogance, and the fact that he planned to go fishing during the frantic few days before their departure only increased their antipathy.[79]

It was not, on the whole, the most promising start to a grandly ambitious campaign. The major protagonists—Metcalfe and McTaggart—were two of the least representative Greenpeace activists one could imagine. Metcalfe was imperious, secretive, elitist, and manipulative. McTaggart shared at least three of those qualities, and one could add ruthlessly opportunistic and unashamedly exploitative to the list. This is not to vilify either: they also had their fair share of positive traits. Nor is it to suggest that Stowe, Bohlen, Hunter, and the others were saints. Nevertheless, neither Metcalfe nor McTaggart possessed the stalwart political conviction of Irving Stowe, the tempered idealism of Jim Bohlen, or the countercultural utopianism of Bob Hunter. Grassroots democracy, consensus decision making, and egalitarianism were of little interest to either. Nor did they belong, in historian Michael Bess's lovely description, to "that heterogeneous portion of humanity attuned to the sensibility of transcendence."[80] On the other hand, given what they were about to embark on, none of this was particularly important. The more pertinent qualities were self-confidence, hubris, mistrustfulness, and courage. In this respect, both were well equipped to deal with the French, if not with each other.

6

Mururoa, Mon Amour

McTaggart and Ingram were relieved to finally depart from Auckland and escape the legal and bureaucratic attentions of the New Zealand government. However, it was not long before new problems emerged. It soon became clear that Metcalfe could not operate the ham radio on board the yacht. A quite complex piece of equipment, the radio required a license, and it seems that Metcalfe, who had been busy dealing with media issues in his few days in Auckland, had merely assumed that he would know how to use it. He had certainly used similar radios before, most recently on the *Phyllis Cormack*, but this one appeared beyond him. An angry McTaggart was forced to turn back toward the coast, anchor offshore so as to avoid customs, and paddle ashore in a dinghy to find a radio technician. The technician showed them how to operate the radio but discovered that it required far too much power for the *Vega*'s batteries, rendering it essentially useless.[1]

For his part, Metcalfe, whose mistrust of McTaggart was growing by the day, could not fathom why the radio had not worked and began to suspect that McTaggart had somehow sabotaged it to undermine his credibility, a charge that seems highly improbable. After all, the technician was able to operate it without any problems and any advantages McTaggart may have gained from undermining Metcalfe's authority were far outweighed by the drawback of being without a ham radio.[2] McTaggart was further angered when he discovered that Metcalfe's eyesight was too poor to navigate the boat at night, thereby forcing others to take up the slack during the demanding night watch, which was particularly unpleasant while sailing through the stormy seas of the "roaring forties." The tension aboard was exacerbated when Metcalfe tried to challenge McTaggart's authority, insisting that, as campaign leader, he was entitled to make some decisions regarding their course. He was not convinced by McTaggart's decision to remain in the southern latitudes, thus taking an indirect course to Mururoa, and was not assuaged by McTaggart and Ingram's insistence that, although less direct, such a route would be faster because of the favorable winds. On the odd day when the wind died down, Metcalfe urged McTaggart to run the diesel engine, but McTaggart refused, claiming that it was necessary to conserve fuel to recharge the batteries.[3]

Clearly, Metcalfe and McTaggart were not the type of personalities suited to spending long periods of time together in confined spaces. Both were used to being in charge and neither was willing to subordinate himself to the other. From McTaggart's point of view, Metcalfe was endangering the voyage: he was challenging McTaggart's authority as the skipper, he could not navigate, he did not know how to use the radio, which was the main reason for bringing him along in the first place, and he was generally not pulling his considerable weight on the boat. Metcalfe, on the other hand, found McTaggart crude and boastful. He seemed more interested in talking about how much money they could make from the publishing rights of their story than he was in the protest itself, and he was prone to wild, foaming-at-the-mouth temper tantrums. He also bragged incessantly about his sexual conquests and kept asking Metcalfe if he could borrow his athlete's foot cream for a rash that developed on his penis, occasionally displaying the irritated organ in case Metcalfe was curious about the condition (he was not).[4]

The upshot of the deteriorating relationship was that McTaggart decided to change course and head for Raratonga, where he planned to kick Metcalfe off the boat and give the crew a break. Upon their arrival, Ingram contacted his girlfriend in Auckland and received some stunning news. The *Greenpeace III*, apparently, was merely a decoy vessel to divert attention from the real protest boat, which was heading to Mururoa from Peru. The aim was to have *Greenpeace III* arrested, thus distracting the French and allowing the "mystery boat" in through the "back door." The story was reported in a New Zealand newspaper and the source, according to the article, had been Metcalfe himself.[5] McTaggart and Ingram were furious with Metcalfe and confronted him in his hotel room, demanding to know the truth. It was not so much that they were against the idea of acting as a decoy; rather, they resented not being privy to Metcalfe's plan and felt they had been unconscionably manipulated.[6]

Metcalfe, however, was in no mood to assuage their anger. All they needed to worry about, he reminded them somewhat patronizingly, was sailing the boat to Mururoa. The overall campaign strategy was in his hands and it was imperative, he insisted, that it remain hush-hush, thereby implying that they could not be trusted to keep a secret. McTaggart also discovered from Davidson that Metcalfe, with whom Davidson had gotten along well, had asked him to forward all the tapes and photos from the journey to Vancouver rather than leaving them with McTaggart.[7] McTaggart was incensed at the thought that he had simply been Metcalfe's puppet. Though McTaggart had frequently operated in a similar manner—indeed, such cloak-and-dagger tactics were to become a hallmark of his campaigning style—he could not bear to be a mere pawn in such a strategy.

Furthermore, CND and the other volunteers who had worked so diligently on the campaign were also angered by the revelation that *Greenpeace III* was a mere decoy. McTaggart was now more determined than ever to proceed without Metcalfe and began to contemplate disassociating himself from Greenpeace altogether. He called his brother, Drew, now living back in Vancouver, and asked for advice. Drew told him that Greenpeace was mixed up with communists and that McTaggart should have nothing whatsoever to do with them.[8] Ingram was also upset, but nevertheless felt that it would be imprudent to burn their bridges with the organization that was leading the campaign. He convinced McTaggart to reconsider his position on Greenpeace and to ask Metcalfe to guarantee them that whatever happened from that point on, Greenpeace would stand firmly behind them and take care of any legal and medical expenses they may incur. Metcalfe assured them that they could rely on Greenpeace's total support for the rest of the campaign, despite the problems they had experienced thus far.[9]

Just as McTaggart was preparing to tell Metcalfe that he was kicking him off the boat, Metcalfe saved him the trouble. He told McTaggart that he needed to fly to Peru to continue the campaign and that he was heading out the next day. McTaggart and Ingram confronted him one last time, wanting information about the boat from Peru, including its radio frequency so that they could make contact with her should they both reach the testing area. Metcalfe, however, refused even this request. Instead, he added to their consternation by asking them to maintain radio silence in order to keep the French guessing about their whereabouts. Given Metcalfe's behavior, it was an act of abject humiliation for McTaggart to then have to ask him for $200 to make repairs and purchase supplies. Metcalfe cheerfully gave them $100 and told them to wire his wife, Dorothy, for the rest. She replied that it would take seven days to cable the funds due to problems with the bank linkup. Frustrated, they called Richard Northey back in Auckland, who sent them $100 the next day.[10]

Metcalfe admitted that what might have seemed like a grand strategy from the outside was actually a series of snap decisions made on the fly. The story about the *Greenpeace III* being a decoy for a boat from Peru was itself a diversion designed to confuse the French and force them to expect protestors from all sides. Metcalfe had a wealthy and influential Peruvian friend, Roberto Lett, and the two had discussed the idea of sending a boat, or at least giving the appearance of sending one, from Peru.[11] The Mururoa tests had given rise to strong anti-French sentiments in Peru, particularly when traces of radioactivity began to show up in the fish caught along the South American coastline. Many of the fish in the area migrated through French Polynesia and formed a vital part of the Peruvian diet and economy.[12] However, in mid-April 1972, the French sought to mute Peruvian protests by making available $60 million worth of monetary credits for civil and technological projects. The French made it clear that the money was conditional on Peruvian acquiescence in the Mururoa program, and conservative elements within the Peruvian military and government agreed to the conditions. Almost overnight, the climate of tolerance for nuclear protestors disappeared. Metcalfe's friend, it appears, became a victim of the crackdown and was briefly detained, thus scuttling any plans the two had made.[13]

So, instead of flying to Peru, Metcalfe, for reasons that are still not entirely clear, flew to Mexico City to "lay low" for a while before making his next move. The strategic advantage of Mexico City is not at all obvious and one cannot help but feel that for Metcalfe, the venture was something of a working holiday, an impression that is strengthened by his desire to go fishing in New Zealand while McTaggart and his crew were working feverishly to ready the boat. Even laying low in Mexico City did not work out particularly well; he was tracked down by a reporter from the *Vancouver Sun*. Nonetheless, he stuck to his story, claiming that his decision to leave the *Greenpeace III* in Rarotonga "was always inherent in the plan" and that his main function had merely been to ensure that the boat set sail on time and that the crew was in the right frame of mind for the protest. Clearly, Metcalfe had failed with the latter goal: by the time he left them, McTaggart and Ingram were angrier with Greenpeace than they were with the force de frappe.[14]

While Metcalfe was in the South Pacific, his wife, Dorothy, along with Pat Moore, Bob Hunter, and a few other Greenpeace stalwarts, continued to run the campaign from Metcalfe's West Vancouver house. They organized a benefit screening of the antinuclear documentary, *Hiroshima Mon Amour*, from which they developed the idea of producing buttons and T-shirts with the slogan, *Mururoa Mon Amour*, a phrase that was widely used by antinuclear protestors for the rest of the decade.[15] They also continued trying to raise money, though with only moderate success, and Hunter tried to keep the issue alive in the local media, discussing it regularly in his *Sun* column. Jim and Marie Bohlen, along with Pat Moore, flew to New York, where they knocked on over 160 consular doors at the United Nations headquarters in an effort to convince as many nations as possible to sign a declaration condemning atmospheric testing at the upcoming UN Conference on the Human Environment in Stockholm.[16] Lyle Thurston had gone to London to help start a Greenpeace antinuclear protest and was surprised to find that a group calling itself Greenpeace already existed there (Greenpeace London). He bumped into the small group of anarchists during a street demonstration and was arrested when they got a little too rowdy for the local constabulary.[17]

Rod Marining, meanwhile, had flown to Paris, where he was trying to bring an assortment of antinuclear and environmental groups together for a Greenpeace protest, which, they hoped, would finally prompt the French media to stop ignoring the issue of nuclear testing. Along with Moore, Thurston, and various French supporters, such as Les Amis de la Terre, he organized a protest march that culminated with an occupation of Notre Dame Cathedral. The local gendarmes were unimpressed. Several of them grabbed Marining, dragged him outside, and punched him repeatedly in the stomach, demanding to know if he was a "red." "No," he gasped in between blows, "I'm a green," a statement that apparently made little sense to them.[18]

Ben and Dorothy Metcalfe flew into Paris at about the same time and were met at the airport by immigration officials, who arrested them and told them they would be deported back to Canada immediately.[19] They changed their minds, however, when Dorothy produced a very authentic looking invitation from the pope. So, instead of putting them on a plane back to Vancouver, the police assigned an armed guard to escort them to the Italian border. Moore, Thurston, and several of their entourage met them in Rome, where they did, indeed, receive an audience with the pope, who blessed the Greenpeace flag and told them he approved of their actions.[20]

After Rome, the group traveled to Stockholm for the UN Conference on the Human Environment. The Stockholm conference was without doubt one of the groundbreaking events in the history of international environmentalism. According to John McCormick, the author of a useful work on the subject, it represented, "the first occasion on which the political, social, and economic problems of the international environment were discussed at an inter-governmental forum with a view to actually taking corrective action."[21] Representatives of 113 countries, 19 international agencies, and over 400 NGOs attended the conference, which resulted in the creation of the United Nations Environmental Programme. Accredited NGOs, including Friends of the Earth, were provided with a separate facility in which to hold their own forum, a move that many saw as an effort to remove them from the official conference. As a recently established organization, Greenpeace was not part of this forum and was instead part of the alternative People's Forum set up by various groups who were considered, or considered themselves, too radical for the NGO forum.[22]

Marining and Thurston had managed to mobilize several dozen French and British environmental activists to attend the conference as Greenpeace protestors and, according to their own version of events, they played a leading role in organizing a parade of several thousand people—mostly radicals, hippies, and antiwar activists—who marched through the streets of Stockholm protesting nuclear weapons testing. Never one to join a plebian march, Metcalfe spent his time hobnobbing with the powerful and the famous, and managed to have several lengthy conversations with Margaret Mead, who became a strong supporter of Greenpeace. Metcalfe and the other Greenpeacers were ecstatic when a New Zealand-sponsored resolution condemning nuclear testing was passed by an overwhelming majority of government representatives. After all their international campaigning—the *Vega* in the South Pacific, the door knocking at the UN headquarters in New York, meeting the pope, and the protests in London and Paris—they could not help but believe that they had played a major role in the resolution.[23]

The Greenpeace campaign in Europe did not directly result in the creation of new Greenpeace groups there, but it did widen its name recognition, especially among environmentalists and the media, thereby creating the base for the establishment of more formal Greenpeace chapters in the near future. Metcalfe would frequently receive calls from international journalists wanting to know exactly who was behind Greenpeace and who was funding them. Most refused to believe his story that they were just a handful of activists in Vancouver who knew how to stir up trouble. In mid-1972, however, such a description of Greenpeace was perfectly apt.[24]

While Greenpeace's core activists were in Europe, McTaggart, Ingram, and Davidson, having left Haddleton in Raratonga, continued their voyage to Mururoa.[25] As per Metcalfe's suggestion, they continuously broadcast false positions in the hope of confusing the French military. However, as they would eventually learn, it was to little avail. Later conversations with French military officers revealed that powerful tracking stations had been monitoring their progress since Raratonga. Furthermore, the French Navy had sent several ships from the French fleet in New Caledonia to deal with any protest vessels that managed to make it all the way to the atoll.[26] The *Vega* sailed into strong headwinds along the dangerous Tuamotu Archipelagos, a huge, crescent-shaped chain of atolls where razor-sharp coral reefs lurked just beneath the ocean surface, ready to slice open unwary boats. On the evening of June 1, they arrived in the test zone, exactly on the day that the testing period was due to begin.[27]

For the next month, the crew of *Greenpeace III* played cat and mouse with a small French Navy flotilla. The French frequently harassed and intimidated them with dangerous maneuvers. They would steam toward the *Greenpeace III* at full speed as though they were going to ram her, before veering away at the last second. Planes flew overhead regularly. These were clear signs that the French were taking them seriously. Technically, so long as they remained outside the twelve-mile limit, they were perfectly within their right to sail around the area as much as they wished. The trick, therefore, was to stay close enough to Mururoa to remain within the danger area, but far enough away so that they did not accidentally stray into the twelve-mile zone, where the French could legally arrest them and seize the vessel. They tried to get messages to the outside world on their marine radio but could only reach a Belgian vessel called the *Astrid*, which, they were later to learn, was actually a French naval vessel masquerading as a commercial ship. They also spotted American and British naval ships patrolling the area. The French admiral in charge of operations at Mururoa admitted that these vessels were cooperating with the French, helping them with preparations and sharing data, a fact that was later corroborated by a U.S. congressman who investigated the matter. The presence of the U.S. and British navies violated the spirit, if not the letter, of the 1963 Partial Test Ban Treaty.[28]

Meanwhile, back in New Zealand, interest in various antinuclear protests reached a peak. The media followed *Greenpeace III*'s progress as well as they could, announcing any news of the vessel, while CND helped maintain the media's interest in the campaign, organizing demonstrations and writing to newspapers. The New Zealand Federation of Labour, the umbrella organization for the country's unions, adopted a resolution to ban its members from working on French ships and aircraft, and refused to service the regular French commercial flights through Auckland.[29] The strike provoked retaliation from the French, who withdrew two of their cargo ships that were supplying New Zealand's Pacific island territories, as well as threatening to blockade New Zealand's agricultural exports to Europe. The business community and the National Party government strongly opposed these union actions. As one businessman put it: "Because of the favorable financial balance of trade that we enjoy with the French . . . we need them as trading partners infinitely more than they need us." Nevertheless, the Federation of Labour stuck to its policy and appeared to have widespread support outside business and government circles.[30]

In Australia, one of the leading figures of the Labor Party opposition, Dr. Jim Cairns, requested the socialist Allende government in Chile to allow a Chilean vessel, at the time in Sydney, to take him and an Australian delegation to Mururoa in a joint Australian-Chilean protest. Allende was reported to have seriously considered the request before deciding against it (presumably after being pressured by military leaders in Chile).[31] Also in Sydney, a group of daredevil antinuclear activists called the Save the Earth Society, were planning to parachute into the test zone and rendezvous with *Greenpeace III*. McTaggart and his crew found the whole notion patently absurd when they heard about it on the Australian news.[32] Nevertheless, the group was quite serious and even went to Fiji to start training for the exercise, which, in the end, never amounted to anything. In the not-too-distant future, such apparently outlandish schemes would come to be part and parcel of Greenpeace's modus operandi, but in 1972, sailing thousands of miles into nuclear testing grounds was still a powerful enough mind bomb. The fact that the Australian group referred to themselves as the *Greenpeace IV* project gives us some idea of how Greenpeace was perceived. It also indicates a general sense of confusion about whether Greenpeace was an organization, a movement, or simply a brand name that could be adopted by any antinuclear group that wished to use it. It was an issue with which Greenpeace itself had not yet come to terms.

By the end of June, more boats began to depart New Zealand for Mururoa. Barry Mitcalfe, a Polynesian studies professor and peace activist, and his organization, Peace Media, managed to organize four protest boats on very short notice. Mitcalfe was determined to follow Greenpeace's example and "keep [the boats] going until the French stop testing." Then, he proclaimed, "We'll start on China."[33] Along with Mitcalfe (who was also a poet), another well-known New Zealand writer, Maurice Shadbolt, joined the crew of one of the vessels, the *Tamure*, which got as far as Tahiti before the French finished with their tests. Summing up his motivation for participating in the protest, Shadbolt wrote:

> If governments appear powerless, individuals feel doubly so. And it was this feeling of powerlessness which I wished to escape at the age of forty and as a father of five young children—the feeling that as an individual I was able to do nothing to halt the slow slide of the human race toward extinction, and that as a writer I was doing no more than stand a passive doomwatch.[34]

In Canada, various members of Parliament were challenging the ruling Liberal government to commit itself to supporting the *Greenpeace III* voyage. External Affairs Minister Mitchell Sharp, however, was evasive whenever the subject was broached, even suggesting that the ship was flying a Peruvian flag and, therefore, was not Canada's responsibility.[35] Metcalfe angrily rebuked the government, accusing it of "chickening out of accepting responsibility for one of its own peaceful citizens." He appealed to the New Zealand government to offer the *Greenpeace III* help and protection, arguing that it would be in character with New Zealand's position as a leading antinuclear nation. The conservative New Zealand government, however, was also unwilling to claim responsibility for the yacht, leaving McTaggart and his crew diplomatically isolated.[36]

On June 29, the bewildered crew of the *Greenpeace III* heard that the French had exploded the first bomb on June 26. Despite being only forty or so miles away, none of the crew had even noticed the relatively small explosion, apparently a triggering device for a much larger H-bomb the French were planning to detonate.[37] Rubbing more salt into their wounds, the French had issued a report that the *Greenpeace III* had "sailed peacefully out of the area on June 21st and had not been seen since." The media picked up the story and New Zealand Prime Minister Marshall commended McTaggart for realizing that "discretion was the better part of valour."[38] McTaggart and his crew may have been doing some phenomenal sailing, but as far as the propaganda war was concerned, the French were winning hands down.

Despite their difficulties contacting the outside world, however, McTaggart, Ingram, and Davidson were causing a considerable headache for the French. The cost in dealing with them, both in terms of delays and in devoting ships to the task, was running to millions of francs. By July 1, they had had enough. The minesweeper, *La Paimpolaise*, was assigned the task of removing the *Vega* from the area so that the testing of larger bombs could finally proceed. In a series of breathtakingly dangerous moves that contravened countless maritime laws, *La Paimpolaise* charged, circled, and cut off the *Vega* in an effort to force her to sail out of the danger zone. But McTaggart and Ingram continued to circle around in the same general vicinity, refusing to be driven from the area. Then, the inevitable happened. *La Paimpolaise* approached *Vega* from the port side, swung in front of her, and then looped back around until she was motoring some twenty yards behind the yacht, slightly to its starboard side. As *La Paimpolaise* approached to within ten yards, thereby blocking the wind from *Vega's* mainsail, her bow wave knocked *Vega* sideways, pushing her in front of the giant minesweeper, which then crashed into her starboard side. According to McTaggart's description (though the words are clearly Hunter's), the "impact of [*La Paimpolaise*'s] four thousand tons crunching into *Vega*'s flank brought a cry of tortured wood, splinters zinged through the air, the hardwood rub rail exploded, and there was a concert of groaning and cracking from *Vega*'s twisted joints."[39]

Despite the huge size difference between the ketch and the minesweeper, the *Vega* was not as badly damaged as she might have been and none of the crew was injured. Nevertheless, they were in no condition to continue sailing and would have to rely on assistance. Despite *Vega*'s difficulties, the French did not come to their aid for over two hours, ignoring the emergency flag McTaggart had raised and only responding after McTaggart had lit two disaster flares. Eventually, the captain of *La Paimpolaise* zipped across to the *Vega* in a rubber zodiac. He appeared almost as shocked by the incident as the crew of the Greenpeace and apologized profusely for the collision. McTaggart believed him when he said that he had not deliberately rammed the *Vega*. Nevertheless, he was certainly the responsible party and had been sailing with reckless disregard for the crew's welfare. McTaggart would later learn that, although several French officers had been in favor of simply ramming the boat, the admiral in charge of operations at Mururoa had prohibited the action not so much because he feared for the crew's safety, but because he was concerned about sparking an international incident.[40]

Too damaged to sail on her own, McTaggart was left with no choice but to allow the French to tow the *Vega* into Mururoa for repairs. He tried to insist that the French contact the crewmembers' respective governments and alert them to what had occurred, but they refused. On Mururoa, the crew met with Admiral Claviere, the high-ranking naval official in charge of military operations on the atoll. The admiral proved to be an agreeable, polite, and shrewd host, insisting that McTaggart, Ingram, and Davidson join him and his officers for lunch. None of the *Greenpeace III* crew spoke French very well and the admiral's English was limited, but several of his officers were able to act as interpreters. Claviere congratulated McTaggart on a remarkable feat of seamanship and told him that, despite the fact that they had been a thorn in his side, he could not help but admire their bravery and skill. As the admiral dominated the conversation, McTaggart, Ingram, and Davidson devoured their first decent meal in over a month, washed down with generous quantities of excellent French wine. Several young women strolled past the diners—an event that was clearly staged—and the admiral used this as another opportunity to point out how safe the atoll was and how little danger there was of contamination from fallout. Several photographers with telephoto lenses lurked around the periphery of the compound, documenting the proceedings. The *Greenpeace III* crew was vaguely aware of their presence, but short of storming off in protest—hardly the polite thing to do at a civilized luncheon—they felt helpless to prevent the photographers from continuing to shoot what would be a series of compromising photos. Meanwhile, the admiral continued to expound on the French justification for the tests, while also admitting that the *Greenpeace III* was the most impressive protest they had yet encountered.[41]

For the next few days, the French workers helped repair the *Vega*. During this time, McTaggart was able to learn just how much the French had been monitoring their communications. At one point, one of the officers asked him how Ann-Marie was, recited her phone number, and, with a broad grin, said he would give her a call next time he was in New Zealand.[42] When the *Vega* had been repaired, McTaggart insisted they would not voluntarily sail from Mururoa until the admiral sent a message to the outside world reporting the collision and the crew's whereabouts. Claviere gave his word that he would do so. Then, with *La Paimpolaise* as its escort, the *Greenpeace III* departed from Mururoa. The crew soon discovered that, despite the repairs, the boat was leaking quite badly. They asked the French to allow them to enter Tahiti to repair it properly, but they were refused permission.

Later that evening, they heard the first news reports based on the message the admiral had released to the media. McTaggart fumed as Radio Australia announced that the collision had been the result of a faulty maneuver on McTaggart's part, that the *Greenpeace III* skipper had requested that the French tow them to Mururoa for repairs, and that the French had generously obliged, repairing the ketch at their own expense and ensuring its safe departure. The NZBC version of the story was further embellished with details of the delightful lunch under the swaying palm trees, a civilized encounter in which both parties had come to respect each other's point of view while literally breaking bread together.[43] Greenpeace and CND supporters were also upset by the reports and felt that McTaggart had let them down. Even those sections of the media that had previously supported them began to report on the voyage in a gently mocking tone. It was, in short, a public relations disaster. As McTaggart later wrote:

> I did not quite realize the press is called the "establishment press" precisely because it listens to the establishment first, and it listens to "kooks" hardly at all. In two months of isolation with Nigel and Grant, both of whom I knew to be intelligent, sensitive human beings, I had completely lost track of the reality that there were legions of cynical news editors sitting *out there* waiting for the slightest opportunity to write the voyage of *Greenpeace III* off as the work of kooks, radicals and frauds.[44]

On July 15, the battered and leaky *Greenpeace III* limped back into Raratonga. Finally, McTaggart had the chance to contact the outside world and report his version of events. However, his experience with Metcalfe and with the media in general had left him feeling suspicious toward the press. The only significant reporter on Raratonga was a New Zealander whom McTaggart described as "an alcoholic," who kept pestering McTaggart to release their photos and tapes to him so that he could write the "big story." McTaggart refused, which prompted the irate reporter to write an article implying that McTaggart was holding out on the media in order to try to make some money from his story—hardly the kind of action that was likely to endear him to the average reader, let alone the antinuclear and environmental activists who had supported the voyage. Penniless, exhausted, and fed up with the media, McTaggart phoned Metcalfe in Vancouver asking for $1,500 to pay for repairs to the *Vega*. After promising several times that he would send the money, Metcalfe finally admitted that Greenpeace was cash-strapped and could not help him. Thoroughly disillusioned with the organization, McTaggart sold the damaged ham radio set for $700 and bought a ticket for Vancouver. Returning to New Zealand seemed pointless. Their credibility had been badly damaged by erroneous press reports and their own naiveté in lunching with the admiral. Furthermore, the media's attention had now turned to the launch of several other boats by Barry Mitcalfe's Peace Media, and it seemed unlikely that the press would be interested in the ordeal of the *Greenpeace III*.[45] Leaving Ingram to sail the *Vega* back to Auckland, McTaggart returned to Canada where he hoped to persuade his government to support him in a legal case against the French.[46]

If one views the situation from McTaggart's perspective, his concern with money seems quite reasonable. After all, *Vega* was his only significant asset and she was badly damaged. Furthermore, Metcalfe had failed to come through with the aid that Greenpeace had promised him, and if he were going to take on the French Navy in court, he would doubtless need considerable money for legal expenses. To people such as Grant Davidson, however, McTaggart's financial concerns detracted from the purity of the protest and made him appear selfish and venal. In his memoirs and autobiography, McTaggart had nothing but praise for Davidson. However, at least in his private correspondence with Metcalfe, Davidson did not return the compliment. He wrote that he could not wait "to put a good 6,000 miles" between himself and McTaggart and described the *Vega*'s skipper as being "as slippery as an eel and as cunning and devious as Nixon." According to Davidson, after Metcalfe had left the boat in Raratonga, McTaggart tried to distance himself from Greenpeace, telling people both there and on Mururoa that Greenpeace was not supporting him and that others had financed the voyage. Davidson also felt that McTaggart was deliberately withholding their story from the "alcoholic" reporter in Raratonga in order to see how much money he could make from interviews and publishing rights, "an obvious indication of a mind plotting and scheming for financial gain." McTaggart also made false statements to reporters, claiming that they had never stepped off the *Vega* while in Mururoa and that the French had not offered them food. The French immediately countered these assertions by releasing photographs of the luncheon. According to Davidson, "these inconsistent reports and the refusal to talk to the press sure questioned the credibility of the voyage and the crew."[47]

There is no doubt that McTaggart was trying to find money wherever he could, but to portray him as self-serving and out for financial gain, at least in this instance, seems unfair. There were plenty of other ways that a man of McTaggart's experience and abilities could make money other than sailing to Mururoa to protest French nuclear testing. Certainly, given McTaggart's sometimes shifty and inconsistent behavior, combined with his unfamiliarity with the rhetoric and culture of activism, one can see why people such as Metcalfe and Davidson, as well as others within Greenpeace, were suspicious of his motives. In light of Greenpeace's failure to support him as promised, however, it is understandable that McTaggart should try to take financial matters into his own hands, even if he did so somewhat clumsily. Years later, when McTaggart was in charge of Greenpeace International, reporters would come to Metcalfe to dig up some dirt about McTaggart, and he would provide them with copies of Davidson's letter, as well as giving them his own rather bitter recollections of McTaggart's behavior during the voyage. It was easy to fit such stories into the broader outline of McTaggart's life and use them to vilify his character and question his commitment to the Greenpeace cause. Such criticism, however, demonstrates little effort to view McTaggart's actions in context. He may well have been a "slippery" character, but his desire to raise funds in the wake of the *Greenpeace III* voyage was clearly motivated by his desire to repair the *Vega* and his growing obsession to take the French to court, rather than for his personal material gain.[48]

On his return to Vancouver, McTaggart received anything but a hero's welcome. Only his parents and a lone reporter were there to greet him. He soon learned that, apart from Hunter's support for the mission in the *Vancouver Sun*, most of the local press had been more interested in the smuggling charges and in the lunch with the admiral than in the protest itself. Metcalfe was at a small college in the BC interior, where he was a visiting professor for a semester, and Greenpeace itself, as far as McTaggart could tell, had ceased to exist. Indeed, McTaggart began to wonder if it had ever existed in the first place or if it was merely another of Metcalfe's ruses. Nonetheless, despite his disillusionment with Greenpeace and with how the campaign had gone, the voyage had been a life-changing experience. He could no longer identify with his old friends, who now only served to remind him of his previous life as a narrow-minded conservative. At the same time, however, he could not overcome the deep cynicism that had been a part of his worldview for most of his life, and therefore did not feel he belonged in the activist milieu. Lonely and depressed, he nonetheless forced himself to begin exploring the various legal avenues that were open to him, and he found a law professor at UBC who was willing to help him free of charge.[49]

McTaggart did not hear from the local Greenpeace crowd until several weeks later, and even then it was purely by coincidence. He was in a supermarket when he bumped into Dorothy Stowe, who asked why he had not gotten in contact with them. She invited him to the organization's next meeting, and McTaggart looked forward to sharing his experience with a roomful of sympathetic listeners and to finally learning more about exactly who these Greenpeace people were. His disappointment only deepened, however, when he arrived at the meeting to find that the only attendees were three middle-aged American couples—the Stowes, the Bohlens, and Will Jones and his wife. Far from being sympathetic, they were skeptical of his motives. Apart from having heard Metcalfe's disparaging description of McTaggart, they had also found out that he had been a rapacious developer in California who had abandoned three wives and daughters. Clearly, he was not one of "them."[50] The way they saw things, Greenpeace was primarily an activist organization. They did not think it would be very productive to use what little money they had to support McTaggart in a long, drawn-out court case against the French Navy. From McTaggart's perspective, they had the barefaced temerity to criticize him for selling the ham radio in Raratonga without handing the proceeds to Greenpeace.[51] His disenchantment was compounded when he received a phone call from Metcalfe, who wanted him to sign away his rights to the *Greenpeace III* story for $3,000 so that Metcalfe could write a book about the voyage. Metcalfe argued that the story was really about Greenpeace, not McTaggart, and that any proceeds from it should go to the organization. McTaggart hung up in disgust and never spoke to Metcalfe again.[52]

McTaggart's legal advisers soon made it clear to him that it would be virtually impossible to sue the French government or its navy and that his best chance was to get the Canadian government to represent him. As one lawyer wrote, "You can face your Government with the dilemma of either claiming your damages from the French Government or, if they do not wish to do so, of paying you themselves because they deprived you of your only international law remedy."[53] The government, however, while expressing sympathy with McTaggart's circumstances and reiterating their opposition to nuclear testing, seemed bent on stalling the case in the hope that McTaggart would give up in frustration. Minister of External Affairs Mitchell Sharp told McTaggart that it was "necessary for an individual claimant to exhaust the available legal remedies and to show a denial of justice before a government can consider any espousal of his claim."[54] By sheer coincidence, McTaggart managed to appeal his case directly to Prime Minister Pierre Trudeau, who happened to be holidaying in Buccaneer Bay, BC, the same time McTaggart was there. Feeling nervous and deeply self-conscious, McTaggart rowed out to Trudeau's yacht. Gripping the star rail and craning his neck at the prime minister, who did not invite him on board, McTaggart told Trudeau the story of his ordeal. Trudeau informed him in no uncertain terms that he was "not very sympathetic to the Greenpeace Foundation." Given the bad press his government had received from the likes of Hunter and Metcalfe, this was not surprising. McTaggart once again distanced himself from Greenpeace. "I am discussing my case, not theirs," he assured the prime minister. After a forty-minute conversation, Trudeau finally shrugged and made an open gesture with his hands. "Mr. McTaggart, do you expect us to go to war over your boat?"[55] Things did not look good: his own prime minister had told him, face-to-face, that his case was doomed. All avenues, it seemed, had been exhausted.

McTaggart refused to give in. He was flown to Toronto to appear on one of Canada's more popular news programs, *W5*, and his spirits were boosted by the sympathetic treatment he received there. While in Toronto, a journalist put him in touch with Jonathan Bingham, a Democratic U.S. congressman from New York who had been investigating the extent of America's involvement in the French atmospheric testing program. He confirmed McTaggart's finding that the U.S. military was heavily involved in the French program, in clear violation of the 1963 Partial Test Ban Treaty. When McTaggart asked if he intended to pursue the matter further, the congressman told him that he had made his statement, which was now part of the *Congressional Record*, and that he had been told to leave the matter there. He also met Canada's most famous nature writer, Farley Mowat, who told him that he greatly admired his protest and advised him to write a book about the entire affair. McTaggart figured that if he were to somehow bring the case to trial, he would need to organize his arguments and materials. Writing a book might be a good incentive to accomplish this task, as well as generating some publicity and much-needed cash. He convinced Ann-Marie to join him in British Columbia, and the two of them spent a damp and gray winter in a dilapidated shack on Vancouver Island, with Ann-Marie taking care of domestic duties and McTaggart spending all his time writing letters, working on his manuscript, and reading about nuclear issues.[56]

In early 1973, McTaggart was visited by an Australian lawyer who wished to get an affidavit that could be used as evidence against the French. In the months since McTaggart's departure from the South Pacific, leftist governments had been elected in both Australia and New Zealand. Each promised to take a tougher stand against French nuclear testing. Prime Minister Norman Kirk of New Zealand and his Australian counterpart, Gough Whitlam, had decided to launch an appeal against the French at the International Court of Justice in The Hague. A high-level delegation, including New Zealand's attorney general, traveled to Europe to present the case. The people of the South Pacific, they argued, "actively resent the contamination of the air they breathe and of the waters from which their food supplies are drawn. . . . It is an obvious imposition that it should be necessary to maintain in New Zealand itself, and in the Pacific Islands, outposts to keep watch on the fluctuating levels of an unnatural and unsought hazard."[57] The delegation also made it clear that they were responding to the "intense activity by private individuals and groups to impress upon the New Zealand Government their anxiety about the tests." Such activities had been "supported by the churches, by local bodies and community organizations, by trade unions, by student and other youth organizations, and by virtually every other grouping of public opinion in a vigilant democratic society."[58] In June 1973, the Court announced an interim order urging France not to further aggravate the dispute by detonating more bombs. Although the French were strong supporters of the International Court, they refused to recognize its jurisdiction in matters of national defense. Nevertheless, the trial attracted widespread publicity and contributed to the mounting international pressure being brought to bear against the force de frappe.[59]

Protestors, too, had become more ambitious in their activities. Barry Mitcalfe's Peace Media had spread throughout New Zealand over the preceding year, while also forging alliances with groups in France, Fiji, and Peru. Mitcalfe convinced David Moodie and the crew of the *Fri* to lead a flotilla of yachts to Mururoa during the 1973 testing season. Combined with the Labour government's promise to send a New Zealand frigate to the test zone, it appeared the French would be facing a more organized and concerted protest effort than ever before. In late 1972, Kurt Horn, one of the crewmen on the *Greenpeace Too*, had immigrated to New Zealand, adding another link to the activist chain that stretched from Vancouver to Auckland. Horn's arrival in New Zealand coincided with Mitcalfe's preparations for the 1973 campaign, and he soon joined the Peace Media effort. Like Moodie and other activists in New Zealand, Horn considered himself to be working for Greenpeace as well as Peace Media and the CND. The organization mattered less than the cause, and "membership" in any particular group was determined by participation rather than by legal affiliation.[60] In the end, Horn, a sailor who believed in a strong captain and maritime discipline, was unable to come to terms with Moodie's mellow attitude and consensus-seeking leadership, and he decided not to sail on board the *Fri*. Ironically, however, it would be the *Fri*, rather than the *Greenpeace III* or anyone with direct connections to Vancouver, that would provide the core group of activists that set up Greenpeace's New Zealand branch.[61]

McTaggart, meanwhile, was growing restless. His manuscript was nearly complete, and he was dubious about whether the *Fri* and the other protest boats would be able to make it all the way to Mururoa, a prospect that probably got his competitive juices flowing. His effort to enlist his government to support him in his case against the French was not going anywhere fast, and he was angered by the French decision to ignore the verdict of the International Court of Justice. These factors all played a role in his decision to sail to Mururoa once more. Clearly, however, the overriding motive was vengeance:

> I had hoped to have had the chance to meet the Admiral in court, to be able to confront him with the truth, so that the ordinary man in the street could finally see to what lengths governments will go to hide the truth from their people. A return voyage to Mururoa might not get us into court, but if I could spend just a few minutes face-to-face with the Admiral to tell him what I thought of his smooth-talking, lying ways, I would feel vindicated.[62]

Whether in badminton or in business deals, McTaggart had always felt most comfortable confronting problems mano a mano. Trying to get at the French through the courts and through his own intransigent government was a source of endless frustration. Instead, he would compress the complex political and military forces swirling around him into one compact, manageable figure he could confront face-to-face. He would sail back to Mururoa and slug it out with the admiral.

By early 1973, less than a year after it was officially registered as a legal non-profit group, the Greenpeace Foundation was in disarray. Fragmented, disorganized, and effectively leaderless, it was in danger of collapsing altogether. Metcalfe, despite being the group's official leader, was barely involved any more, and the group's meetings were poorly attended. The divide between the older Quakers and peace activists on the one hand and the countercultural ecology freaks on the other was wider than ever, with each faction sometimes unaware of what the other was doing in the name of Greenpeace. In February, for example, a group led by Hunter, Marining, and Watson staged a protest against a pair of visiting French warships, an action that turned into something of a fiasco. They had once again hired the *Edgewater Fortune* (a.k.a. the *Greenpeace Too*) as a protest vessel, but at the last minute, Captain Johansen raised the gangplank when he could not settle on a fee. Hunter and Watson rushed off to Hunter's little yacht and sailed her toward the approaching warship, while Marining stood atop a bridge dropping mushrooms and marshmallows on the bemused sailors, before getting arrested for his troubles.[63] Marining's description of how the protest was conceived reflects the group's fragmentation and haphazard planning style: "There were six of us in a living room trying to figure out what to do about these French warships. That was two days before. It was just a little Greenpeace meeting. I had called everybody together but only six came."[64]

Despite the confusion and lack of planning, or perhaps because of it, the event still managed to attract plenty of local media attention. Even at this early stage, however, Marining was somewhat ambivalent about what press coverage alone could achieve: "The press picks up on all the sensational things. They say Greenpeace did this, Greenpeace did that. They make it look like there's thousands of people caring and bringing on the revolution, when there was really only about six of us. The rest is all myth.... All that Greenpeace Power is illusory. It looks like there's a lot of people worried about what's happening thousands of miles away in the South Pacific but they would really only be worried if it were happening in Squamish."[65]

Understandably, McTaggart remained unimpressed with Greenpeace, but felt he had nowhere else to turn to find support for his second voyage to Mururoa. While the old guard of the Bohlens and the Stowes remained cool toward him, McTaggart struck up an instant rapport with Hunter. Despite coming from different worlds, the two rapidly became good friends. United by the grudges they bore against Metcalfe, they met regularly in Hunter's kitchen, drinking beer while discussing politics, environmentalism, and life in general. Hunter was impressed by the older man's life experiences and intrigued by his exploits as a cunning and ruthless businessman. For his part, McTaggart was able to learn much from Hunter's knowledge of environmental issues, the media, and the counterculture.[66] By this stage, Metcalfe had begun pursuing other interests and was no longer actively involved with Greenpeace. As a result, Hunter and Marining became the de facto interim leaders. Working closely with McTaggart, they managed to raise $13,000 for McTaggart's next protest voyage, including a $900 donation from the recently elected New Democratic Party government in British Columbia.[67] They also convinced the organization's more skeptical members, who felt that McTaggart was planning to use the money to repair the *Vega* rather than to sail to Mururoa, to back the campaign, albeit somewhat grudgingly.[68]

Ironically, therefore, despite his antipathy toward the organization, McTaggart was in large part responsible for rescuing it.

McTaggart's announcement that he was planning to sail back to Mururoa had an immediate impact, indicating that both his own government and the French were taking him seriously. The flow of letters and telegrams between McTaggart's lawyer and the Canadian Department of External Affairs began to increase markedly. Prime Minister Trudeau assured McTaggart that "the Canadian Government is continuing to exert its good offices with the French Government on your behalf" and was "in the process of formally notifying the French Government of the interest of the Canadian Government in the earliest possible settlement" of his claim. The government was also "seeking legal advice on the question of what local remedies may be available in France."[69] Just before he left Vancouver, McTaggart received a message from the French government that they were willing to pay $5,000 for damages to the *Vega*, provided he called off the voyage.[70] At this stage, however, his battle with the French had reached a point where such compensation was of little consequence. In fact, it only strengthened his resolve to once more confront the force de frappe.

McTaggart convinced Nigel Ingram to once again join him on the voyage, and they then flew to Auckland, where the two of them set about working on the *Vega* to get her in shape for the journey. Ingram would have been quite happy to paint over the *Greenpeace III* emblem on the side of the boat, but McTaggart decided against the idea. Although he remained somewhat skeptical of Greenpeace's competence, his meetings with Hunter had convinced him that the outfit, for all its organizational inadequacies and internal squabbling, was nonetheless sincere in its opposition to nuclear testing and in its support for his mission.[71] Due to the extensive repairs that were necessary to prepare the ketch, the *Vega* was not able to be part of the broader Peace Media flotilla that had set sail at the end of March 1973. This actually suited both McTaggart and Ingram, who were quite happy to keep their distance from the "bizarre peace flotilla" and the hippies and peaceniks they felt were largely associated with it, a decision that was to some extent vindicated by the fact that only two of the twenty-five or so vessels actually made it to the testing zone.[72]

One of these was the *Spirit of Peace*, a yacht of similar size to the *Vega* and made up of a small crew of older pacifists, including Kurt Horn. The *Spirit of Peace* spent more time in the restricted zone near Mururoa than any other vessel before or since, but because no one was arrested and, no doubt, because its crew was decidedly unflamboyant, they never received much media coverage.[73] Not so the *Fri*. A crowd of over 3,000 people cheered as it departed from the dock, and its large and colorful international crew helped ensure a healthy media interest. Along with David Moodie and his wife, Emma, the crew comprised several members of the French antinuclear organization Bataillon de la Paix. The most famous of these was Jacques Paris de Bollardiere, a fifty-five-year-old former general who had left the military in protest at France's brutal treatment of prisoners during the Algerian war. Also aboard was Jean Toulat, a fifty-year-old Catholic priest who had written a book entitled *La Bombe ou La Vie?* and Jean-Marie Muller, a thirty-two-year-old philosophy professor and author of several books on nonviolent action. Brice Lalonde, the twenty-eight-year-old head of Les Amis de la Terre was another high-profile participant. Martin Gotje, a young navigator from Rotterdam, who would go on to have a long and distinguished career with Greenpeace, was also on board, as were a Maori seaman, a pregnant university student, and several photogenic young female crew members.[74]

The *Fri* made it all the way to the test zone, where it was boarded by French commandos and arrested. The crew used nonviolent resistance to make their captors' task as difficult as possible. Moodie and several others dived overboard, forcing the French to rescue them on several occasions, while others lay limp on the *Fri*'s deck and refused to cooperate with efforts to tow the ship. When they were finally under tow toward Mururoa, Gotje managed to cut the towropes and the French-speaking crew members gave speeches and sermons to the sailors and commandos. Although the commandos had made an effort to cut the *Fri*'s communication lines, Moodie was able to relay messages to the nearby New Zealand frigate *Otago*, which had been sent as the New Zealand government's official protest vessel, and its captain passed their messages on to the international press.[75] The crew of the *Fri* was flown to the French military base on Hao Atoll, where several went on a hunger strike, forcing the French to take them to the hospital in Papeete. There they managed to join forces with local antinuclear activists and, overall, proved a considerable pain in the neck to the French military. In terms of the amount of media coverage they received and the disruption they caused to the tests, the *Fri* was probably the most successful protest vessel to that point.[76] Although officially part of the Peace Media flotilla, Moodie and his crew also felt that they were part of Greenpeace's antinuclear campaign. In fact, Moodie frequently wore a Greenpeace T-shirt given to him by Kurt Horn.[77] Nobody in Vancouver objected to this fact, if they were even aware of it, thus further illustrating Greenpeace's status as something of a hybrid between a formal organization and a broad social movement.

Greenpeace protests were also underway in Europe. For example, the London outfit associated with *Peace News* organized the Greenpeace London to Paris walk. Some seventy people managed to sneak into Notre Dame Cathedral and chain themselves to the pillars. Unlike the previous year, however, the archbishop allowed them to stay, and the police did not force them out, a fact that Hunter attributed to the pope's support for Greenpeace's cause.[78] Following the church sit-in, a group of about 200 English and French Greenpeace supporters marched toward the Elysée Palace, leafleting along the way, before being rounded up by police.[79] In Bonn, a small group of West German peace activists and environmentalists gathered under a Greenpeace banner and marched through the capital's streets to the French Embassy.[80] Another group of people using the Greenpeace label presented an antinuclear petition to the French government. Several Australians and New Zealanders among them demanded sanctuary in France, claiming that their own countries were being poisoned by radiation from the French tests.[81] Clearly, at this stage the term "Greenpeace" could be used by anyone who supported the cause, without needing to ask the Vancouver Greenpeace Foundation for permission. While such a laissez-faire position had the advantage of encouraging widespread protest among like-minded activists, its results could sometimes be less than professional. For example, the efforts of a group of London activists to protest at the French tourist office in Piccadilly did not go quite according to plan. "Sadly," the *Guardian* reported, "the demonstrators chose the wrong office for their demonstration, and invaded and leafleted the Ceylon Airlines and Air Afrique offices by mistake. The French tourist office was next door. The man from Ceylon Airlines said: 'I quite agree with them.' The policemen outside the embassy applauded after the performance and said they had enjoyed the show. 'It gets chilly out here and this sort of thing passes the time.'"[82] It was a harmless enough farce, but too many such incidents would not do much for Greenpeace's credibility.

Throughout June and July, while the peace flotilla continued to draw all the media attention, McTaggart and Ingram spent the cool and damp New Zealand winter in a boat yard, working frantically to prepare the *Vega* before the French ended their tests for the season. According to McTaggart, he was unable to track Grant Davidson down, but even if he had found him, it is unlikely that Davidson would have agreed to spend another two months cooped up in a small boat with McTaggart.[83] After the previous year's experience, McTaggart and Ingram decided to adopt a new crew selection policy: they would only consider asking people they already knew and trusted. In the end, those people turned out to be their girlfriends. Ann-Marie Horne had been involved with the issue for almost a year and was keen to accompany McTaggart. Ingram's girlfriend, Mary Lornie, a young legal secretary with some sailing experience, was also up for the adventure. Apart from female companionship, McTaggart hoped that the presence of two young women would restrain the French and prevent them from getting rough. On that score, he would be sorely disappointed.[84]

In marked contrast to the previous year, the New Zealand authorities did not attempt to hinder McTaggart's preparations in any way, a fact that demonstrated how the climate for antinuclear activists had changed under the new Labour government.[85] The weather, however, refused to cooperate, and McTaggart had to postpone their launch as huge seas pounded the harbor entrance. Finally, in late July, they managed to set sail. Although there were no major problems of the sort that had occurred among the crew the previous year, the chemistry between the new crewmembers was not especially good. Horne, who was quite naïve and religious, did not get along very well with the more worldly and cynical Lornie, and a slight tetchiness developed between McTaggart and Ingram. In fact, while he was in Auckland, McTaggart had heard rumors that Ingram was a French agent and during the worst moments of tension, he began to wonder if there was not some truth to them. Fearing that they would arrive too late, McTaggart and Ingram pushed the little yacht hard, taking advantage of the strong southwesterly winds, and within a remarkable three weeks, they once again reached the French cordon at Mururoa.[86]

By this time, whatever patience the French military may have possessed had been well and truly exhausted. Millions of francs and countless valuable days had been wasted by the disruptive actions of the *Greenpeace III* the previous year and by the *Fri* a few weeks before, and the return of the *Vega* was a provocation they were not prepared to endure. Within two days of their arrival in the test area, a minesweeper approached them. By its side was a group of commandos in a zodiac, an inflatable motorized dinghy. The small craft zipped across the waves toward the *Vega*, and it was immediately clear to the crew that the commandos intended to board their yacht, just as they had done to the *Fri* a few weeks before. Unlike Moodie and his crew, however, McTaggart had neither the discipline nor the temperament for nonviolent resistance. As the first sailor attempted to board his ketch, McTaggart brought his arms up in a blocking motion and checked him hard as he tried to climb the rail. This was all the provocation the commandos needed. With truncheons at the ready, three of them jumped over *Vega*'s railing and attempted to subdue McTaggart, who kicked and punched them in an effort get free.[87] One of the commandos managed to pull McTaggart's T-shirt down over his head and another grabbed his arms and pinioned him. Then the batons began to flail:

> The first truncheon came down with a weight and force unlike anything I had ever felt on the back of my head and the second came down across my shoulders and the next blow landed on the back of my neck . . . and I was suddenly in the air being flipped over the railing and being yanked furiously into the inflatable, unable to catch a single breath or even find a way to make a sound. . . . With scarcely a pause, the truncheons were flailing again, each blow rattling my teeth so that it seemed they would be shattered and that my spine and ribs and skull would cave in any second. Back. Neck. Head. Kidneys. . . . [It was] as though they all had gone mad and were simply trying to smash me to death, stamp me out of existence like some loathsome bug . . . Then, with one man still pinning each arm, I was yanked upright and jammed in a sitting position

against the side of the inflatable. Something crashed into my right eye with such incredible force that it seemed to come right into the middle of my brain in an explosion so that I thought that half my head had been torn off. And then everything went black.[88]

Before the boarding, McTaggart had instructed Mary and Ann-Marie to remain calm, whatever happened, and to photograph as much of the action as possible, Mary with the movie camera and Ann-Marie with the Nikon. Somehow, amid their screams of protest, they managed to do just that. Ingram had gone below deck to send an SOS message and when he emerged, he was immediately set upon by more truncheon-wielding commandos. As Mary screamed in protest and briefly stopped filming, one of the sailors grabbed her movie camera and threw it overboard. Ann-Marie took as many shots as she could and then ran below deck to stow the camera in a prearranged hiding place. A commando followed her, determined to confiscate the camera. Instead, he found a partially hidden decoy, which he carried with him back to the inflatable. It was a simple trick, but it nonetheless sufficed to fool the French into thinking they had gotten rid of all the evidence of the beatings.[89]

For the next half hour, McTaggart slipped in and out of consciousness, at one point raising his hand to his blinded right eye and feeling nothing but a bloody pulp. While the *Vega*, with Ingram, Horne, and Lornie on board, was towed to

French commandos boarding the *Vega* and beating David McTaggart, August 1973. Reproduced by permission of Ann-Marie Horne/Greenpeace.

Mururoa, McTaggart was flown to Papeete to have his eye operated on. Initially, he refused to allow the doctors to touch him, demanding that he be allowed to call his brother. French naval officials refused this request for several hours. They claimed that the *Vega* had strayed into French territorial waters and that they had been within their rights to board it, an assertion that McTaggart and Ingram denied vehemently. Eventually, they allowed McTaggart to call his brother in Vancouver, to whom he dictated a telegram to Prime Minister Trudeau, before he finally allowed surgeons to perform the vital operation that would save his eyesight. Meanwhile, the other three crewmembers were to be flown to the French military base on Hao Atoll, some 300 miles away, while preparations were made for their deportation. Their major problem, however, was that the film was still in Ann-Marie's camera, which remained hidden behind a panel in the cabin and which the French were bound to find once they searched the yacht thoroughly. That film was the only evidence they had to counter whatever story the French might concoct about the incident. Without it, their voyage would once again have been, from a propaganda perspective, a bitter failure. With it, they stood a chance of countering the French version of events. As the commandos left them to pack their belongings, they discussed how they could smuggle the evidence out. Simply packing it in their bags would be too risky. Finally, Ingram grimaced and said, "There are precedents," before taking the film with him into the toilet and trying to insert it into his rectum. The pain, however, proved to be unbearable, and he was forced to give up. At that point, Ann-Marie realized that there was only one possible solution. Taking the film from Ingram, she went into the toilet, inserted the film canister into her vagina, and, with Ingram steadying her, managed to disguise her agony and walk from the yacht to the waiting jeep.[90]

In the expectation that the story of the event would be a matter of their word against the crew's, the French released a press statement claiming that McTaggart had been hurt while violently resisting the French sailors, who were peacefully trying to board a boat that had entered their territorial waters. His eye injury, they maintained, occurred when he slipped and fell over. Initially, it was this version of events that the Canadian government preferred to believe.[91] While McTaggart lay in hospital in Papeete, Ingram, wearing handcuffs and escorted by a French security officer, was put on a plane to London. He managed to give the officer the slip at Los Angeles International Airport and transferred to a Vancouver-bound plane, where he was met by McTaggart's brother, to whom he handed the film.[92] Drew McTaggart passed the film on to Greenpeace, and as soon as they had it developed, they knew they had a huge publicity coup on their hands. The photos clearly showed the French commandos boarding the yacht, throwing McTaggart into an inflatable, and beating Ingram with their truncheons. The photos and accompanying story were published in newspapers

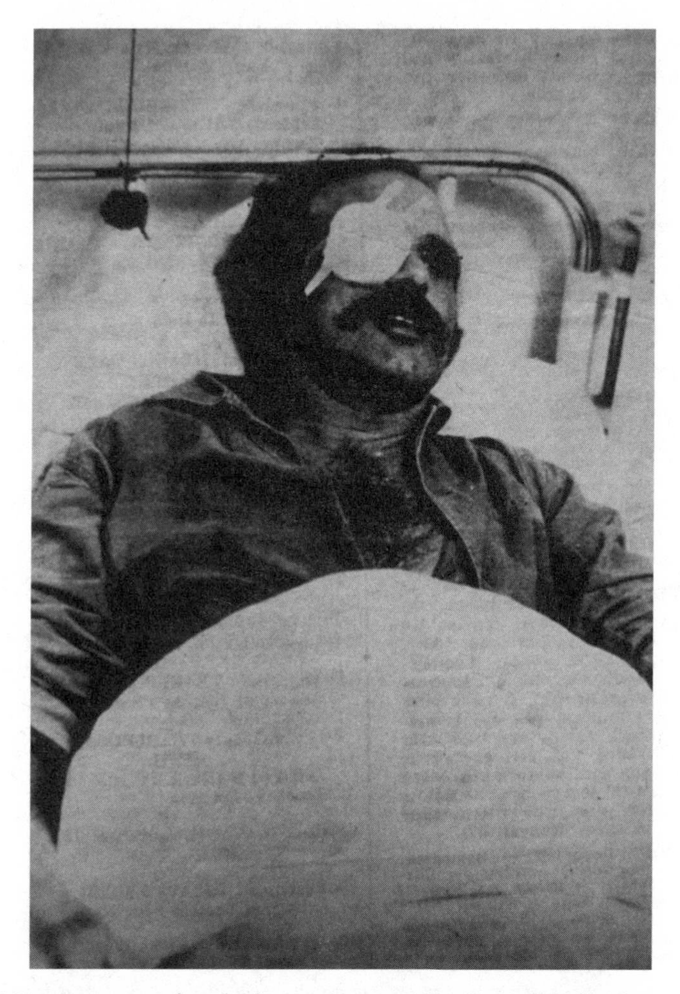

David McTaggart recovers from his beating, August 1973. Reproduced by permission of Ann-Marie Horne/Greenpeace.

throughout the world (except in France, where they were censored). Greenpeace finally had the publicity they craved, while McTaggart now had his evidence. If the French were capable of this, then surely they were also capable of lying about ramming the *Vega* and about the supposedly benign consequences of their nuclear tests. *Time* declared that the incident had "sunk Franco-Canadian relations to the lowest point since Charles de Gualle dropped his *Vive le Quebec Libre* clanger in 1967."[93] The French, according to the *Vancouver Sun*, were "startled to learn of the importance Canada attaches to resolving the Greenpeace III controversy." External Affairs Minister Mitchell Sharp spent almost half of a fifty-minute meeting with French Foreign Minister Michel Jobert discussing the case.[94]

Unlike the previous year, McTaggart made a triumphant return to Vancouver, with dozens of reporters and Greenpeace supporters greeting him at the airport. The operation in Tahiti had saved his eye, but there was a good chance he would develop glaucoma, which would require immediate medical attention to prevent the loss of his eyesight. Over the next few weeks, McTaggart and Ingram roved the various media circuits in Vancouver and Toronto, while McTaggart's book *Outrage!* was released by a small Vancouver publisher to modest sales.[95] Initially, McTaggart was pleased by his government's more enthusiastic attitude toward his case. This time, it seemed, there was a good chance that they would support his efforts to take the French Navy to court. Such support, however, waned as the headlines receded, and McTaggart once again grew frustrated as the government continued to come up with reasons why it could not espouse his claim. In December, the French allowed him to return to Tahiti to retrieve the *Vega*, although local authorities made things as difficult as possible for him once he arrived. In the end, the Canadian government, under pressure from the popular press, paid the $12,000 necessary to have the ketch freighted to Vancouver.[96]

Despite his frustration with the French and Canadian governments, there was one major consolation for McTaggart. Until 1973, the French had declared that, since their atmospheric tests were essentially harmless, there was no need to move them underground.[97] In November, however, they suddenly reversed their position and announced that in 1974, they would begin testing beneath the earth's surface. Greenpeace saw this as a significant victory and was not slow to claim its share of the credit. According to Hunter, they had achieved a "stunning victory" over the French military. "Who would have predicted," he wrote in his column, "that Europe's mightiest nation would so quickly do an about face in her policy of indifference to world opinion?" Greenpeace, he argued, created the public awareness that had caused the United States to reduce the number of tests they had scheduled on Amchitka, as well as forcing the French to move their tests underground. Groups such as Greenpeace and Peace Media, he continued, had "functioned as a flying wedge coming at France from beyond the usual diplomatic and governmental channels." Hunter believed that such organizations represented a political arena beyond the state and above the individual, empowering nonstate actors in an international system otherwise dominated by governments and corporations. "The lesson . . . is clearly that governments can only do so much. To abdicate our own individual responsibilities—to leave problems to be solved by governments alone—is to abandon the field. Greenpeace has shown that global village conditions do exist and that they can be bought to bear in politically forceful ways."[98]

Hunter's language foreshadowed the "civil society" concept that has become a popular analytical category in the study of social movements and nonprofit organizations. The political scientist, Paul Wapner, for example, has cogently argued that groups such as Greenpeace operate in the arena of "global civil society"—"that slice of associational life that exists above the individual and below the state, but also across national boundaries"—where they engage in what he calls "world civic politics."[99] In other words, the efforts to lobby and persuade governments represent only one dimension, albeit an important one, of global political activity carried out by international nongovernmental organizations. Greenpeace, Wapner argues, also targets "the global cultural realm," where it attempts to clue "into internationally shared modes of discourse, such as moral norms, symbols, and scientific argument, and . . . manipulates them to induce people to pursue . . . environmentally sound practices." By engaging in activities such as direct-action protests and media stunts, Greenpeace is "disseminating an ecological sensibility" that indirectly influences the behavior of both ordinary citizens and their governments. Thus, campaigns such as Amchitka and Mururoa represent efforts to "sting" people and governments with this ecological sensibility. They thereby constitute, in Wapner's words, "a type of governance" and "a mechanism of authority that is able to shape human behavior."[100] Although they may not have used the same language, the concept expressed by the phrase "world civic politics" would have been familiar to various Greenpeacers, particularly those such as Hunter and Metcalfe, whose Hegelian view of history predisposed them to such ideas. Even at this early stage of its existence, it seems clear that Greenpeace was moving, albeit somewhat haphazardly, into the arena of world civic politics.

In as much as Greenpeace was able to draw greater attention to nuclear testing than any organization previously, there was a ring of truth to their claims that their campaign had influenced the French decision to end atmospheric testing. None of the protests of the fifties and sixties, with the possible exception of the Aldermaston marches, had garnered as much media attention as the Greenpeace voyages. Although it is difficult to measure the impact this had on government policy, there seems little doubt that the French decision to move the tests underground after the two Greenpeace voyages was more than mere coincidence. Furthermore, by inspiring others to imitate their actions, Greenpeace raised the specter of an ever-expanding protest flotilla that threatened to invade the test zone year after year. The prospect of containing such protests, and dealing with their ever-expanding political fallout, must surely have weighed on the minds of French military officials and politicians.

While Greenpeace was happy to celebrate McTaggart's voyage and the press coverage it brought them, they remained reluctant to devote time and money to his legal battle. Some within the organization were already planning for another voyage in 1974, and they argued that what little money Greenpeace had raised should be spent on this upcoming campaign rather than on McTaggart's legal fees.[101] Feeling abandoned by his government and let down by Greenpeace, McTaggart began to consider the prospect of simply abandoning the case, turning his back on his confused and half-formed idealism and the movement that helped create it, and stepping back into the happy-go-lucky world of business. If only the French had been decent about the whole thing: if they had apologized, admitted they had lied, and paid for the Vega's repairs, McTaggart believes, he would have put the whole episode behind him and gone back to "normal" life.[102] Not surprisingly, the French would do no such thing. "The thought of the French getting away with what they had done," wrote McTaggart, "had filled me with a feeling that could only be described as outrage. . . . But was outrage enough to justify the enormous expenditure of energy and time and money that would be required to push my case through[?]" Given McTaggart's dogged, win-at-all-costs mentality, his outrage alone may, indeed, have fueled his motivation to proceed. But there were other factors involved, and during his moments of self questioning, his mind

> threw out dozens upon dozens of answers. There were principles involved. There were political realities involved. There were vast environmental factors at stake. The question at times seemed to come down to something so huge that I could barely grasp it. Although self-consciousness prevented me from discussing it openly with anyone, my sense of the issue was nothing less than the question of human survival. I knew that individuals aren't supposed to think in those terms, that history is supposed to be directed by swarms and masses and tides that take centuries to accumulate their momentum. But I could not shake off the thought that one single well-directed shot at the heart of the legal and political mechanism which was itself the force behind the machines which were being built to destroy the earth was worth more than everything my own life could add up to. I thought, too, of my own lost children and what hope they had of living out their time without being blasted or poisoned. And, inevitably, I wondered whether the whole tortuous business couldn't be reduced to something as simple as guilt for not having been able to fulfill my role as a father. Angrily then, I would go back to work.[103]

In May 1974, McTaggart, his luggage stuffed with files, affidavits, and photographs, left Vancouver and flew to Paris, where he would spend the majority of the next three years mounting a court case against the French Navy. The last thing on his mind was helping to start Greenpeace groups throughout Europe. Almost inadvertently, however, that would be one of the by-products of his time in Paris.[104]

Apart from the difficult to quantify role that Greenpeace played in the French decision to end atmospheric testing, the most notable result of the Mururoa campaign was the creation of Greenpeace New Zealand, which, on a per capita basis, has been, and remains to this day, the best-supported national branch of the international organization. Interestingly, however, neither McTaggart, Ingram, Metcalfe, nor any representative from the Vancouver group was directly involved in the founding of the New Zealand branch. Instead, it was Peace Media, the *Fri*, and their various supporters who were largely responsible for establishing the new outfit. In December 1973, Peace Media held its annual general meeting in Auckland. Among those who attended were David Moodie and the more committed members of the *Fri*'s crew: Francis Sanford, a leading Tahitian antinuclear activist and independence leader; and Peter Hayes, one of the organizers of the Greenpeace London to Paris march. Naomi Petersen, a twenty-three-year-old education student at Auckland University and a cook on the *Fri*, remembered the meeting: "Boat-owners and crew, Peace Media people and others were there. It was the first I had heard of the idea that Peace Media should change its name to Greenpeace, to signify the two groups' unity of purpose. Greenpeace was such a good name—it summed up what we were about."[105] Using the network of contacts built up by Peace Media, the new Greenpeace Foundation of New Zealand was officially registered in April 1974. Like the original outfit in Vancouver, the nascent Greenpeace New Zealand was primarily an antinuclear organization rather than a broadly conceived environmental group. Its first campaign, and the one that would occupy most of its time until 1977, was the appropriately named Peace Odyssey, in which Moodie and his crew would sail the *Fri* around the world, visiting the nuclear testing nations in order to publicize the threats posed by the tests, as well as trying to develop an international network of activists.[106]

The Mururoa campaign had, in many ways, been more successful than Amchitka. McTaggart had made it all the way to the atoll on both occasions and had directly confronted the French military, which in the end, had shown that they were willing to resort to violence when dealing with protestors, as well as being prepared to lie to the rest of the world to justify their nuclear testing program. Media interest had been strong in the areas more immediately affected by the testing, particularly in New Zealand, and with the release of photos of the beatings, Greenpeace garnered headlines around the world. In all likelihood, the campaigns seem to have played a role in the French decision to move their tests underground and in forcing the French media to focus on an issue on which there had, up until that point, been a broad consensus within French society. The establishment of Greenpeace New Zealand was another vital legacy of the campaign, albeit a somewhat indirect one. In the long term, however, the most important consequence may well have been that the campaign marked the beginning of David McTaggart's long association with Greenpeace.

For the founders in Vancouver, the period between 1972 and 1974 had been one of flux rather than consolidation. For the first nine months of Greenpeace's official existence, Stowe and Bohlen had scaled back their involvement in the organization, leaving Ben Metcalfe to run it as though it were a cross between a Madison Avenue public relations firm and an espionage agency. While Metcalfe's style may have been an effective method of creating pseudo-events and media images, it also alienated many in the group who favored a more democratic and participatory approach. By the end of the 1972 campaign, as Metcalfe began to lose interest in his role, Greenpeace was effectively leaderless, and its core members had drifted apart to pursue their own separate interests. Ironically, given his low opinion of Greenpeace, it was McTaggart who was largely responsible for rescuing the organization from its steady demise, albeit indirectly. The anger and the headlines generated by the French treatment of the *Greenpeace III* reawakened people's interest in the organization and helped put it back on a stable footing.

Despite this apparent consolidation, however, there remained a considerable degree of tension between various key players in the group, a divide that largely broke down along generational lines. The older, "straighter" members, such as the Stowes and the Bohlens, remained firmly committed to the antinuclear character of Greenpeace. For them, nuclear weapons still represented the ultimate threat to mankind's existence and to the health of the environment, and they were determined that Greenpeace should continue to focus its attentions on the nuclear arms race. Buoyed by the 1973 campaign (even though several of them had been lukewarm about supporting McTaggart), the antinuclear faction was preparing to send another vessel to Mururoa in 1974. They felt that they now had the French on the ropes and it was time to press home their advantage.[107] Bob Hunter, however, believed that the Mururoa campaign had "peaked" in 1973 and that the only way Greenpeace could draw more attention to the unfortunate atoll was if someone were killed during a protest. Moreover, Hunter felt that the Mururoa campaign had concentrated on traditional antinuclear concerns without placing enough emphasis on ecological issues. It was not that Hunter, or Greenpeace's countercultural faction in general, believed that other issues were more important than the antinuclear cause. Rather, there was a feeling among them that Greenpeace was focusing too heavily on the "peace" half of its name and not enough on the "green." It was time, Hunter felt, to move the organization in a different direction: one that emphasized not just the danger that humans were creating for themselves but also the threat that they posed to other species.[108]

The man who would lead Greenpeace in this new direction was a New Zealander working as an assistant professor of psychiatry at the University of

British Columbia. While the "straights" began to plan for the 1974 Mururoa campaign, and while McTaggart was in Paris preparing his court case against the French military, Paul Spong, armed with a recent PhD from the Brain Research Institute at UCLA, put Greenpeace on a path that would lead from antinuclear protest to a more broadly conceived form of environmental activism. It was a path that at times skirted close to the newly emerging philosophy of deep ecology. Spong's campaign would turn Greenpeace into darlings of the media, effectively making them the world's first "eco-superstars." By shrewdly exploiting Cold War tensions, it would also give Greenpeace its big break in the world's most lucrative NGO market—the United States.

7

Armless Buddhas vs. Carnivorous Nazis

By the end of 1973, from Bob Hunter's perspective, Greenpeace stood at a fork in the road. It could continue down the antinuclear path, as Bohlen, Stowe, and others among the older generation wanted. Or it could broaden its outlook and involve itself in issues that, at least in the public mind, were more overtly ecological. For Hunter, the fight to protect whales from exploitation by humans emerged as the perfect symbol of environmental survival. In his opinion, the antinuclear protests, particularly against the French, had run their course. The fickle world media, Hunter felt, was unlikely to be too excited about yet another antinuclear flotilla sailing across the Pacific. Furthermore, he had come to the conclusion that there was a limit to how much they could exploit people's fear of the bomb as a means of changing mass consciousness. The specter of nuclear annihilation was so mind-numbingly shocking that many preferred simply not to face it. Instead of using fear of death as a trigger for a consciousness revolution, Hunter felt, it would be more effective to appeal to people's reverence for life. Whales, he came to believe, constituted a vivid metaphor for humankind's wanton destruction. Sensitive, intelligent, and wondrous creatures, they continued to be slaughtered, as far as Hunter could see, because a small number of people could not conceive of a world where humans did not hunt whales.[1]

The split between Hunter and the older, Quaker-influenced activists mirrored other divisions within the unsettled organization. Some of these, such as the cultural differences between the younger countercultural crowd and the sober hair-shirt activists who had cut their teeth on the 1950s peace protests, had always existed but were beginning to intensify. Stowe, for example, argued that smoking was a form of pollution that should not be tolerated by environmentalists, and he moved to have it banned at all Greenpeace meetings. Drugs also constituted unacceptable pollutants of the mind and body, as well as a threat to the group's image.[2] Naturally, the younger generation, particularly the chain-smoking, acid-dropping hippies, such as Hunter and Marining, chafed at such stern paternal pronouncements.

In late 1973, from an organizational perspective, Greenpeace remained poorly defined, precariously balanced between a Vancouver-based antinuclear organization and an ad hoc international movement. It continued to officially exist as a registered nonprofit group under the British Columbia Societies Act, but it is unlikely that it would have passed inspection under the act's regulations. Metcalfe had resigned as leader, leaving Hunter and Marining in charge as "interim co-chairmen." No annual general meetings had been held, no audits conducted, and the books, such as they were, "amounted to an untidy pile of scribbled account numbers, bills [and] bank statements." According to Hunter, it was "a situation that would bring a glow of sheer joy to the cheeks of an anarchist."[3] Meanwhile, a Greenpeace New Zealand had sprang up independently of the Vancouver outfit, a group of anarchists in London were referring to themselves as Greenpeace, and all manner of groups and individuals around the world felt free to use the name at their own discretion. In Bonn, for example, a group of antinuclear demonstrators began to refer to themselves as Greenpeace Germany, while in Australia a band of daredevil parachutists had also adopted the name. In certain circles, an ad hoc gathering of antinuclear activists was sometimes simply referred to as a "Greenpeace activity."[4]

Beginning in 1974, and primarily under Hunter's leadership, Greenpeace experienced a series of dramatic changes in the makeup of its members, its campaigning style, and its organizational structure, so much so that by 1976 it bore little resemblance to the group that, five years earlier, had attempted to sail a worn-out fishing boat to Amchitka. From an antinuclear organization with an environmental emphasis, Hunter and those who supported him attempted to remake Greenpeace into an ecological strike force that would use daring and spectacular actions to draw attention to the rapid demise of the largest creatures that have ever inhabited the planet. Drawing on an eclectic and at times confusing mixture of biocentric thought, Eastern religion, New Age romanticism, and some controversial neuroscience, Greenpeace embarked upon a radically new way of protecting wildlife. In the process, they unleashed the full force of the counterculture onto the environmental movement.

The man who inspired Greenpeace's entry into the antiwhaling movement was Paul Spong, a physiological psychologist who had a history of conducting unorthodox cetacean research. Spong had worked at the Vancouver Aquarium until his boss, fed up with Spong's hippie demeanor and his calls to free the aquarium's whales, decided not to renew his contract. Spong grew up in a small town on the northeast coast of New Zealand and studied law at the University of Canterbury in Christchurch. After taking some psychology courses, however, he became interested in brain physiology and the emerging field of neuropsychology. A brilliant student, Spong crossed the Pacific in 1963 to attend graduate school at the Brain Research Institute at UCLA, one of the world's foremost neuroscience facilities. Spong's research involved high-tech computer analyses of brain-wave patterns in humans. His dissertation examined how the brain processed sensory stimulation, particularly perceptual and cognitive phenomena such as attention, concentration, and consciousness.[5]

In Los Angeles, Spong became deeply involved with members of the local counterculture. With little interest in politics, he gravitated toward the music and drug scene, frequently holding court among groups of devotees captivated by his drug-inspired monologues on human behavior and psychology.[6] An Asian colleague introduced Spong to the *I Ching*, which he came to regard as a profound and eminently useful philosophical guide to life. Like Bob Hunter and some of the other more countercultural members of Greenpeace, he turned to it unhesitatingly whenever he was faced with a dilemma.[7] The ancient Chinese text was originally a simple divination tool before Confucian scholars formalized it some 2,000 years ago. It then became the key text in Chinese cosmology, underpinning not only Confucianism, but also Taoism, Feng Shui, and traditional Chinese medicine. In the West, the *I Ching* was part of the broader interest in Eastern ideas that began to flower in the early twentieth century. It became widely accessible to German readers in 1923, when the Sinologist Richard Wilhelm published his influential translation. Carl Jung, who saw the *I Ching* as a tool that could help people access the collective unconscious, wrote the foreword. Wilhelm's version—complete with Jung's foreword—was translated into English in the early 1950s and quickly became part of the intellectual tool kit of the counterculture, sitting comfortably alongside influences such as gestalt, cybernetics, and systems theory.[8] Ken Kesey was one of its most famous aficionados. In 1971, as part of a supplement to the *Whole Earth Catalog*, Kesey described the *I Ching* as an "oracle" that worked on a "cybernetic gestalt principle" and that was "the most practical day-by-dawdling day tool" for accessing "the knowledge [that] is usually not permitted audience in the tight-assed regime of the courthouse of ego and attachment that we recognize, in a kind of diplomatic dither, as our consciousness."[9]

For those who might be skeptical of such putatively oracular devices, their attraction for devotees seems to reside in the "fit" between the broadly conceived advice and the particular set of circumstances being faced by the wisdom seeker. To the likes of Kesey, Spong, and Hunter, however, the *I Ching* was a conduit into some sort of deep and as yet unseen universal information structure. This was not all that unusual among scientists who embraced the sixties counterculture. Hewlett Packard, for example, offered an *I Ching* random-number generator on its new desktop calculator. When tech-savvy hippies, like *Whole Earth Catalog* founder Stewart Brand, threw the *I Ching*, they were self-consciously mimicking both the ancient Chinese and the Merry Pranksters, while also acting, in Fred Turner's words, "in concert with the probabilistic outlook of information theory." They were tapping into the "informational energies of the world to transform the 'system,'" indeed, the world itself. At the same time, they "could experience the ancient and the new, the Eastern and the Western, the literary and the technological, as mutually legitimating elements of [their] 'whole' experience."[10] Like Theosophy, Eastern religion, Native American spiritualism, and the broad interest in all things occult, the *I Ching* constitutes one of antinomies of modernity: it exists in a state of (sometimes fruitful) tension between seemingly irreconcilable forces and ideas such as scientific materialism and mysticism.[11]

Devotion to ancient Chinese oracles and acid dropping are probably not traits most people would associate with brilliant researchers at the forefront of neuroscience. The contradiction, however, is not as paradoxical as it may seem. On closer inspection, it turns out that LSD, neuroscience, cetacean research, and the military-industrial complex were strange but compatible bedfellows, and the Brain Research Institute was a very hospitable bedroom.

In 1967, after a one-year postdoctoral fellowship at UCLA, Spong was offered an unusual research job at the University of British Columbia. He would be allowed to split his time between working at the university's neurological lab—the kind of work that he was accustomed to—and conducting behavioral research on the Vancouver Public Aquarium's newly acquired killer whale. The whale was named Skana, a Haida word that meant something like "supernatural one." Despite knowing next to nothing about whales, Spong was intrigued by the job description. After a lengthy discussion with Kenneth Norris, a renowned cetologist at UCLA, Spong decided to accept the position. In 1967, he and his American wife, Linda, moved from sunny Los Angeles to the drizzly, provincial Canadian city a thousand miles to the north.[12]

Despite his countercultural proclivities, Spong remained a career-oriented scientist. He had no compunction, for example, about inserting electrodes into cats' skulls. How could one understand brain function without probing a few brains? He devoted himself to his new job, dividing his time between the UBC lab and the aquarium, where he developed a series of experiments designed to test Skana's visual acuity. Aware that there was considerable controversy regarding cetacean intelligence, Spong "deliberately adopted a conservative scientific strategy and approached [his] subject from the outset as if she were an unknown mammalian life form, perhaps as complex as the laboratory rat."[13] His experiments with Skana mostly involved placing two cards—one with a single line and one with two lines—in the whale's pool and rewarding her whenever she chose the double-lined card. Gradually, Spong reduced the distance between the two lines from four inches down to one-sixteenth of an inch, which appeared to be the limit of Skana's visual resolution. After repeating the experiment 200 times, Spong determined that Skana's rate of accuracy in distinguishing between two lines one-sixteenth of an inch apart and a single line had reached 90 percent. In July 1968, to verify his results, Spong decided to repeat the experiment again at one-sixteenth of an inch. Much to his surprise, Skana not only failed the test but also managed to choose the wrong card eighty-three times in a row. It was clear to Spong that such a result could not happen by mere chance: that would be the equivalent of getting eighty-three tails from eighty-three coin tosses. Skana, for whatever reason, was deliberately choosing the wrong card.[14]

At first, Spong became despondent with Skana's conduct. His visual acuity data had been ruined, and Skana's uncooperative behavior meant that she could not be trusted to perform further tests. Throughout the months he had been working with her, Spong had grown to enjoy Skana's company, but he was still fearful of her size and power and continued to maintain an objective distance between himself and his subject. Now, he decided, it was time to abandon the formal experiments and simply spend time with Skana, observing her, interacting with her, and getting to know her better. One day as he was sitting at the edge of the pool with his feet dangling in the water, Skana approached him slowly, as she often did, before suddenly slashing her open mouth across his bare feet. Her four-inch teeth, which could easily have severed his feet like twigs from a branch, merely grazed his skin with a gentle caress. He immediately pulled his feet out, gasping in astonishment. In short time, however, his curiosity overcame his fear, and he gingerly lowered his legs back into the water. Skana again raked her teeth across the tops and soles of his feet, and once more Spong instinctively jerked them out of the water. He repeated the procedure eleven times with the same result. Then, on the twelfth, he became determined to restrain his urge to flinch. This time, Skana delicately clasped his motionless feet in her mouth, let them go, and swam away making what sounded like contented vocalizations. Spong left his feet in the water, but Skana did not approach them again. Bewildered and excited, Spong felt like he had just undergone a role reversal: Skana was now the experimenter, and he was her subject.[15]

At that point, according to Spong, he and Skana entered a "joyous period of mutual exploration": "I dropped my posture of remoteness, opened my mind, and personally engaged myself in Skana's learning. . . . This whale was no big brained rat or mouse. She was more like a person: inquisitive, inventive, joyous, gentle, joking, patient, and, above all, unafraid and exquisitely self-controlled."[16] Spong postulated that whales' sense of hearing was far more important to their well-being than their sight. He abandoned the visual acuity tests and began bringing various musical instruments into the aquarium to see how Skana and Tung Jen, another captive whale, would react. The whales appeared to be stimulated by a wide variety of music, particularly if it was played live, prompting Spong to adopt increasingly unorthodox research methods, such as bringing live musicians into the aquarium. He also became bolder in his interactions with his subjects, and it was not long before he was swimming with the whales and riding on their backs. He began to see the whales as far more complex and intelligent creatures than was generally recognized, a view he cautiously proffered in his 1969 research report:

> *Orcinus Orca* (the killer whale) is a highly evolved, complex and capable creature, perhaps so different from us in the fundamental organization of its nervous system and the physical construction of its body, that despite our common mammalian ancestry, our capacity to understand even its most elementary characteristics is necessarily severely limited.[17]

Spong's reading and his interaction with the whales convinced him that they were primarily acoustic creatures for which sound was as important as vision was to humans. From this position, it was only natural to conclude the concrete pools were cetacean sensory-deprivation tanks. This was not only cruel to the whales but also severely limited people's understanding of what whales and dolphins were really like. Spong wanted to air these views publicly but was concerned about how his scientific colleagues and employers would react. He knew that many of his colleagues would be deeply skeptical of any claim that whales possessed a language and a highly evolved form of intelligence comparable to humans. So, for some time, he shared his views only with his family and

Paul Spong serenades a killer whale in the Vancouver Aquarium, 1974. Reproduced by permission of Rex Weyler/Greenpeace.

sympathetic friends. At the same time, he began to once again immerse himself in the local counterculture. He and Linda moved into an alternative community that had built a ramshackle settlement on the edge of some mudflats south of Vancouver. He grew his hair and beard and began turning up to work in what his boss, aquarium director Murray Newman, described as a "Che Guevara beret." Newman and the aquarium's directors began to view Spong's antics with increasing distrust, fearing that his behavior was becoming "weird and possibly hazardous to the animals." Their negative views were reinforced when an elderly potential benefactor arrived at the aquarium, "took one look at the hippie Spong, and that was that."[18]

By 1969, after working with Skana and Tung Jen for over a year, Spong had become utterly convinced that keeping cetaceans in captivity was both cruel and scientifically counterproductive. Whales, he felt, should be studied in their natural habitat, for our benefit as well as for theirs. At that point, Spong began to mull over the possibility of expressing these views in the talk he was due to present as part of his department's seminar series. But the threat this presented to his career made him hesitant. As always when faced with such a dilemma, he turned to the *I Ching*. He cast the coins, which added up to hexagram forty-seven. Spong had virtually memorized the *I Ching* and knew immediately that the number signified "Oppression." "When one has something to say," he read from the 2,000-year-old text, "it is not believed. Superior men are oppressed and held in restraint by inferior men." The commentary explained that the superior man "fears that he may have cause for regret if he makes a move. But as soon as he grasps the situation, changes this mental attitude, and makes a firm decision, he masters the oppression."[19] On this occasion, Spong's reading encouraged him to air his views at the staff seminar, though one wonders what action he would have taken had the *I Ching* urged him to remain silent.

Spong began his lecture in uncontroversial fashion, outlining some of the orca's general characteristics and describing his visual acuity experiments. Toward the end, however, he began to speculate on the sophistication of whale communication and their level of intelligence compared with humans:

> Orca vocal behavior displays sufficient variability, complexity, and capacity for modification and elaboration, that it seems likely to serve useful communication functions. Their capacity to send, detect, attend to, derive information from, and otherwise process and store auditory information is by any standards quite remarkable. . . . *Orcinus orca* has an enormous brain, with more convolutions per surface area than the human brain. The highly developed cerebral cortex leads me to conclude that the orca has evolved this brain through using it, just as primates did.[20]

Not content to merely imply that whales were comparable to humans, Spong took the next step and made the comparison explicit:

> *Orcinus orca* occupies a place in the oceans equivalent to the one which humans occupy on land: at the top of the food chain with no predators.... It is too early in my research to make any assumptions as to the exact nature of *Orcinus orca* intelligence, but I am quite excited about the possibilities.... My respect for this animal has sometimes verged on awe. [It] is an incredibly powerful and capable creature, exquisitely self-controlled and aware of the world around it, a being possessed of a zest for life and a healthy sense of humor and, moreover, a remarkable fondness for and interest in humans.[21]

At this point, in the eyes of some of his colleagues, Spong had crossed the line from scientific speculation to naïve anthropomorphism. To state that whales' large brains indicated a highly developed, if poorly understood, intelligence was one thing: to suggest that they possessed human-like *consciousness* was scientific blasphemy. Spong could see professorial eyes rolling and heard people shifting uncomfortably in their seats. His concluding remarks served only to exacerbate his audience's incredulity:

> My experiences with Skana and Tung Jen at the Vancouver Aquarium have led me to conclude that holding them there is not going to further our true knowledge of them. The captive environment, aside from being a sensory deprivation experience, is also socially depriving.... I believe now... that these whales should probably be freed, and that we should continue our studies with freed or semi-captive *Orcinus orca* in its natural habitat.[22]

Spong's words echoed those of the controversial neuroscientist, John Lilly, whose work with dolphins had led him to believe that they possessed a degree of intelligence and communicative complexity that was comparable to that of humans.[23] Such views reflected broader changes in people's attitudes toward whales and dolphins, a theme we shall return to shortly. In Murray Newman's eyes, however, Spong had allowed himself to be seduced by his subject. And it was Spong's immersion in the local counterculture that was largely responsible for this "corruption." For Newman, the whales served two purposes: to entertain the public and to act as subjects for scientific inquiry. By demanding that the whales be freed, Spong was questioning the aquarium's raison d'être. As a result, Newman felt he was left with no choice but to fire Spong by refusing to extend his contract.[24]

If Newman had any qualms about his actions, Spong soon helped him overcome them. After being refused further access to Skana, he set up camp under a tree behind the aquarium, with a big velvet pillow, a bowl of marijuana, a large mirror, a guitar, a flute, a zither, a tambourine, a mandolin, a sitar, a conch shell, a three-stringed gartwang, and $1,000 worth of sound equipment. "When I start playing this stuff," he told a *Vancouver Sun* reporter, "Skana is going to start dancing all over the place and there's not a bloody thing anybody can do about it. They'll be able to hear me all over the park. Everybody is going to be blasted right out of their minds." "I've always liked Dr. Newman," he said of his erstwhile employer, "but he's just a bit dull."[25] Spong then took his case to the *Georgia Straight*, where he adopted the counterculture argot in a drug-induced stream-of-consciousness interview with sympathetic *Straight* reporters. Sitting cross-legged among his instruments, his hair down to his shoulders and his beard halfway to his belly button, Spong proclaimed that the whale was "number one on this planet . . . the highest creature. Believe you me, baby, and I'll prove it to you . . . excuse me . . . I won't prove it, the whale will prove it. Let me tell you, as soon as I can get some liquid crystal in me hand, the whale will start talking to us in English. In words that we can see, like the *I Ching*." Spong freely admitted that he took drugs while working. "It helps me work. Helps me tune into the killer whale space. 'Cause when I take droogs [*sic*], the killer whale warms up her nose, and steam comes off her nose, and who knows. . . . They didn't understand, man, but that's ok, they'll come around, y'see, 'cause the I Ching says so."[26]

Although Spong was increasingly beginning to resemble a character from the *Electric Cool Aid Acid Test*, he remained, at heart, a scientist. He resolved to practice what he preached and started making plans to study whales in their natural habitat. In 1970, he and Linda, with their baby son, Yashi, drove up the BC coast to the Kwakiutl village of Alert Bay, the same place where, in the following year, the *Greenpeace* crew would be made honorary Kwakiutl tribesmen. After receiving advice from the local Indians, Spong selected a research site on Hanson Island, a few acres of lush temperate rainforest near the northeastern tip of Vancouver Island. Groups of orca were known to frequent the narrow channels in the area and could be easily observed from the beaches and cliff tops. With the help of some friends, he constructed a rudimentary shack, the first stage of a renowned whale-research project that Spong continues to run as of this writing.[27]

By this time, Spong was an expert on orcas and on whale behavior and physiology in general. Surprisingly, however, he was less well-informed about the plight of the world's great whales, many of which had been hunted to the brink of extinction. He only had the vaguest notion that Japanese and Soviet whaling ships were still hunting large numbers of sperm whales in the North Pacific, as well as more-endangered species, such as sei and fin whales, in the Antarctic. Only in 1972, after a meeting with the famous Canadian nature writer Farley

Mowat did Spong become aware of how decimated the world's great whale populations had become. Mowat was in Vancouver promoting his latest book, *A Whale for the Killing*, and was introduced to Spong at the book launch. Mowat was fascinated by Spong's research but was pessimistic about the future of the great whales. In a somber tone, he told Spong that he hoped that there would still be some whales left to study in a few years time. Spong was taken aback by Mowat's despair: "Do you think it's that bad?" According to Spong, Mowat leaned forward, lowered his voice and replied:

> Listen, Paul, there are only a few people who have seen what you've seen, who've had the experience. Most of the world still gets its ideas about whales from lingering images of Moby Dick and tales of vicious monsters of the deep. The factory ships are wiping out the whales as fast as they can. What do they care? It's all for a few dollars. When the whales are gone, the whalers can transfer their assets to another business. They make dog food and lipstick out of the whales. The Russians lubricate their ICBMs with whale oil. So do the Yankees.[28]

Mowat's words rang like a tocsin in Spong's ears. He suddenly realized that in focusing his attention solely on a few captive whales, he had almost completely missed the bigger picture: whales, as a species, were under grave threat from human exploitation. Mowat told Spong of Project Jonah, a San Francisco-based antiwhaling organization whose Canadian chapter he headed, and asked him if he would be willing to be the organization's representative in British Columbia. Spong, who had so far steered clear of activism, was wary at first, but Mowat soon persuaded him that he could be useful.[29]

Once he had made the decision to work for Project Jonah, Spong threw himself into his new activist role with all the zeal of a convert. His first task was to pressure the Canadian government to end the country's remaining whaling operation in the North Atlantic and to persuade them to stop casting their vote with the main whaling nations—Japan, the USSR, and Norway—at the International Whaling Commission (IWC) meetings. He and Linda energetically mimeographed fliers and petitions, collected signatures, and tried to get the local media interested in the plight of the world's whales. A few months later, Project Jonah's hard work appeared to pay off when the Canadian government announced that it would officially put an end to the nation's whaling industry. Spong insists that this was primarily due to Mowat, who knew Prime Minister Trudeau personally and was able to persuade him to support the antiwhaling movement.[30] It is doubtful, however, that Mowat's influence alone would have persuaded Trudeau and other politicians to shut down the nation's whaling industry. More likely, it was the activism of people like Spong that helped produce a groundswell of support to which politicians responded. Ultimately, however, the ban probably would never have been enacted were it not for the simple fact that by the 1960s, whaling played an insignificant role in the Canadian economy.[31]

In early 1973, Spong flew to San Francisco to visit Joan McIntyre, the founder and head of Project Jonah. McIntyre was a Friends of the Earth activist who had established Project Jonah with the help of Maxine McCloskey, the wife of Sierra Club director Michael McCloskey. She and Spong discussed their campaign for the coming year. McIntyre and other antiwhaling activists had been inspired by the resolution adopted the previous year at the United Nations Conference on the Human Environment in Stockholm, which recommended a ten-year moratorium on commercial whaling.[32] To maintain pressure on the whaling nations, McIntyre decided to mount a campaign to focus the world media's attention on the International Whaling Commission meeting to be held in London in June. She planned to organize "whale celebrations" throughout the United States and the United Kingdom and asked Spong to put together a similar show in British Columbia. By this time, Spong was becoming a well-known media figure in Vancouver. Combining scientific eloquence with an almost evangelical zeal for his cause, he was much sought after by radio and television talk show hosts. It was not long before his schedule began to fill up with interviews, lunch meetings, and dinner invitations. On the second of June 1973, Spong and Linda organized a "Whale Celebration" in Stanley Park, near the aquarium that still housed his beloved Skana. The celebration, which drew a sizeable crowd, attempted to combine educational activities with good clean fun, eschewing angry speeches and countercultural activities that were likely to put off ordinary middle-class citizens.[33] It coincided with similar celebrations throughout the world, including a large gathering in front of the London hotel where the IWC meeting was taking place. Despite the sympathetic media coverage of the show of support for the moratorium, the IWC failed to muster the three-fourths majority that was required to implement the resolution. Spong was particularly annoyed by the Canadian vote. While the United States and the United Kingdom voted in favor of the moratorium, Canada abstained, thereby implicitly supporting the prowhaling nations.[34]

Spong and his family spent the rest of the summer of 1973 on Hanson Island, where they continued to work on their whale research station. During these months, Spong began to plan a series of public "whale shows," consisting of films, slides, and anecdotes. Such shows, he hoped, would further the public's awareness of the unique nature of whale intelligence, as well as alert it to the fact that many species were facing imminent extinction. This time, however, Spong hoped to take his campaign well beyond the borders of British Columbia. He made plans to tour Scandinavia, Iceland, and the most obstreperous of all the whaling nations, Japan. In order to gain some sympathetic publicity for the show, as well as to learn how to run an effective environmental media campaign, Spong decided to contact Bob Hunter at the *Vancouver Sun*. Given their countercultural proclivities, their prominence in the Vancouver media world, and their converging interests, it was virtually inevitable that the two men would meet at some point. On a gray, autumnal day in 1973, Hunter took Spong out onto the choppy waters of English Bay in his converted fishing boat, where the two conducted what to this day must surely rank as one of the seminal conversations of Greenpeace's history.[35]

Unlike most of the journalists Spong had talked to, Hunter had read John Lilly's controversial work on cetacean intelligence and Farley Mowat's book on people's attitudes toward whales. After relating his work with Skana and his research project on Hanson Island, Spong, emboldened by Hunter's charm and sympathetic attitude, not to mention his beer, launched into his theory of whale intelligence:

> I honestly believe that the whales have a very highly evolved social structure and that they have evolved their large brains by using them for extremely complex social communication. They may not speak in words, but they may create sound pictures. In any event, their huge brains are perhaps the most sophisticated biological computers on the planet in certain ways. Humans don't even come close to whales in the ability to process auditory information. I was a classic behavioral scientist, content with manipulating behavior in animals, until I realized that the animal I was working with was as intelligent as and in some ways quite possibly *more* intelligent than I was. My work now is concerned with trying to reveal—or even just glimpse—a new *kind* of intelligence, something that humans aren't even aware of.[36]

Hunter assured Spong that some people were indeed aware of different forms of intelligence and would be willing to further explore Spong's ideas. He outlined his Hegelian notion of consciousness revolution and the vital role that ecological values would play in such a historic development.

> The industrial nations, blind to the laws of ecology, are coming up against their karmic debts, having ripped off the resources to the point that the machines are starting to grind to a halt. Industrial economies are in trouble, and a more ecological, coevolutionary paradigm is emerging. If what you say about whales is true—and I believe it is—then the whales are way ahead of us. They seem to have already learned how to live harmoniously within their surrounds, to control their

> populations, to live ecologically within their environment, and to manage their societies without aggression and violence. It sounds like the whales have a more *gestalten* language, not really a language at all as we know it, but a way of communicating about relationship. They intuitively understand systems theory. This puts them way ahead of human intellect.[37]

To most scientists then and now, Hunter's and Spong's ideas about cetacean intelligence would be controversial at best, if not flagrantly mystical. But for a hippie scientist who felt he was on the brink of a breakthrough in our understanding of intelligence and a countercultural journalist prone to cosmic levels of theorization, the whales appeared to represent both a model and a metaphor for ecological consciousness. By elevating whales to the top of the brain chain, Hunter and Spong were setting them up as ideal subjects for an animal rights campaign. Simultaneously, by portraying them as creatures that have evolved in a state of utter ecological harmony, they became exemplars of the deep ecological ethic that humans needed to adopt if they were to survive on the planet. The whale, in short, would be the messiah of Hunter's Whole Earth Church, while Greenpeace would be its apostle. The result, it was hoped, would be the ecological enchantment of the world.

As the conversation continued, Spong described his work for Project Jonah. He felt that antiwhaling activists were making some progress, but that it was too slow. By the time they succeeded in establishing a moratorium, most of the world's great whales would be hunted to the brink of extinction. Something more radical was required than merely collecting signatures and lobbying politicians. This prompted Hunter to explain his "mind bombs" theory to Spong: How the modern electronic media acted as an instant delivery system for images and ideas throughout the world. The theory sparked an idea in Spong: "What if we took a boat out and blockaded the whalers. Sailed right between them and the whales and didn't let them shoot their harpoons! Do you think that the media networks would cover that?" Hunter replied that they probably would, but that the logistics of such an action would be formidable. Where would Spong find a captain who was willing to place his boat in front of a harpoon? How would they find the whalers in the vastness of the Pacific? Spong then suggested that they could use kayaks to paddle between the whales and the harpoons. "We'll protect the whales with our bodies. If they fire, they have to kill us first." Hunter pointed out that Spong would never be able to keep up with a whaleboat in a kayak. Nevertheless, he felt that there was merit in the general idea.[38]

The thought of employing direct action against the whalers, however logistically difficult it may have been, excited Spong. Would Greenpeace be interested in such a campaign? he asked Hunter. Hunter thought the idea had potential and believed that Greenpeace was due for a change of direction. But he was skeptical about persuading others in the organization. They were, after all, an antinuclear group. Furthermore, in late 1973, Greenpeace was fragmented and in debt. The likes of the Stowes, Bohlens, and Will Jones were impressed with the 1973 Mururoa campaign and wanted to launch a similar voyage in 1974. Meanwhile, David McTaggart continued to demand that Greenpeace help finance what would undoubtedly be a lengthy and expensive legal case in the French courts. Under such circumstances, Hunter was not very optimistic about persuading others that they should direct funds, which they did not yet have, toward an entirely new campaign, especially one with such undeniably flaky overtones. Nevertheless, he resolved to give it a try and asked Spong to present his ideas at the next Greenpeace meeting.[39]

Spong's arrival at Greenpeace led to a further fracturing of the already fragmented organization. Although he respected Spong as a fellow scientist, Bohlen considered the whaling issue to be "soft" compared with the threat posed by nuclear weapons, and most of the older activists agreed with him.[40] Irving Stowe was perhaps the most implacable opponent of Spong's plan, a puzzling stance given his earlier pronouncements on whales. After listening to Roger Payne's recording of humpbacks, Stowe had written that the "thoughtless slaughter of the world's largest mammals is the perfect symbol for the most enormous mistake we could ever make; wiping out another species and destroying the chain of life."[41] Perhaps Stowe's change of heart was due to his proprietary feelings for Greenpeace. Whatever the case, his opposition would make it difficult for Spong and Hunter to succeed. On the other side were Rod Marining, Doc Thurston, and Hunter's friend, Hamish Bruce, the hippie lawyer who had first conceived the idea of a "Green Panthers" ecological strike force.[42] Bruce led something of a double life. A tough, radical left-wing lawyer during work hours, he was also, in Hunter's words, "a true hard-core West Coast freak . . . and a devotee of the *I Ching,* ancient Indian prophecies, Jungian concepts of synchronicity, and a believer in magic."[43]

Since Metcalfe had disappeared and neither Bohlen nor Stowe wanted to assume the responsibility of running Greenpeace, it appeared that Hunter was next in line to become the official chairman. Bohlen and Stowe, however, would not countenance such a move. Bruce was elected as a compromise candidate on the understanding that he would continue to give priority to Greenpeace's antinuclear campaign, while allowing Hunter and Spong to concentrate on the whaling issue. Hunter and Spong then formed the Stop Ahab Committee, which was set up, as Hunter sardonically noted, to explore "ways and means of getting an expedition together to save the whales despite Greenpeace."[44] As a central part of their working relationship, and as a sign of things to come, Hunter and Spong agreed to turn to the *I Ching* to resolve all differences of opinion.[45]

Spong's first order of business was to put together his public presentation on whales. He booked a medium-sized theater for the event and convinced Greenpeace to allow him to call it the "First Annual Greenpeace Christmas Whale Show." In exchange for the use of the Greenpeace "brand," however, Bohlen and Stowe insisted that Spong sign a notarized letter stating that he would personally be liable for all losses.[46] Spong hoped that the whale show would raise enough funds to enable him to travel to Japan, where he intended to stir up a public outcry against the Japanese government's support for whaling. If this failed, then he and Hunter would find a boat and launch a direct-action campaign against the whalers in 1975. The show, on December 28, 1973, was a sellout, and Spong's funds were boosted by a $5,000 donation from Canadian folksinger Gordon Lightfoot. He now had enough to finance his trip to Japan.[47]

Before Spong and his family flew to Japan in February 1974, he and Hunter began to plan the logistics of a nonviolent direct-action campaign against whalers. At that point, according to Hunter, "Apart from the idea of a boat going out to confront the whaling fleets somewhere in the ocean and a favorable *I Ching* reading, we lacked any specific plans."[48] After speaking with McTaggart and examining his photos, however, Hunter noticed that the French commandos had approached and boarded the *Vega* in inflatable motorized vessels they called "Zodiacs." A French invention, these craft were capable of speeds of up to thirty knots on a flat sea, as well as being extremely stable and maneuverable. Spong recalled that he had seen Canadian Fisheries Department scientists using the inflatables, as well as remembering that Jaques Cousteau had used them while filming footage of whales. Zodiacs, they resolved, would be the key pieces of equipment in their direct-action antiwhaling campaign.[49]

Shortly before Spong departed for Japan, he and Hunter decided to drum up some local publicity about Spong's traveling whale show. They managed to convince Murray Newman to allow Spong back into the aquarium in order to hold a press conference beside Skana's concrete pool. It proved to be another classic epiphany in Hunter's life. Spong and the whale greeted each other like long-lost lovers, with Spong down on his knees, patting her and pressing his face against her giant jaw, tears welling up in his eyes. He then motioned for Hunter to join him and introduced him to Skana. Initially, Hunter was fearful of the whale's enormous power:

> Yet the moment I got down on the platform and put my hands out to Skana, she moved lightly—like a woman—against me, allowing me to stroke the underside of her jaw. The fear fell away within seconds, to be replaced by a sense of marvel and excitement. Before I quite realized what had happened, I was rubbing my forehead against hers, stroking her, feeling nothing but sensuousness.[50]

Startled journalists and embarrassed aquarium officials looked on as Hunter and Spong continued their frolicking. Then, as Hunter was leaning out over the pool, Skana emerged from the water and took his head inside her mouth, holding it there for several seconds. "I could feel [her] teeth making the slightest indentation against the back of my neck," recalled Hunter.

> It was not possible for me to move a fraction of an inch. I was like a crystal goblet in a vice. I experienced a flash of utter aloneness, knowing that if she chose to chomp, there was nothing anyone on earth could do to save me. I was completely at her mercy. Fear exploded in my chest, yet the feeling of trustful happiness continued in my head. As though satisfied, she let go and sank away—ever so gently—with a handful of my hair snagged around two huge teeth.[51]

One way of understanding Skana's behavior might have been to compare it with a dog that will hold a child's hand in its mouth without feeling the need to chew off its fingers. But how many people could muster this kind of cool detachment after such an encounter? Probably very few, and certainly not people with personalities and worldviews like Hunter's. Instead, Hunter believed that Skana had intentionally revealed the path he should now follow. She had exposed his fear and illuminated the secret corners of his soul. Afterward, Hunter found himself

> standing among the trees alone, crying like a baby with relief because at last, *at last,* I could see exactly who I was, what my limitations were. . . . I had been through marathon t-group therapy sessions and emotionally exhausting workshops with the great Gestalt therapist Fritz Perls, but neither experience had been so far out of the framework of my understanding that it left me as shaken as I was now.[52]

If Hunter had had any doubts about embracing Spong's antiwhaling campaign, the experience with Skana wiped them away. She had revealed herself, in his mind, to be more than just a clever sea mammal, the oceanic equivalent of a higher primate. Instead, she came to represent a supreme form of power and intelligence rooted in oneness with nature, a state that humans, in their pathetic struggles to conquer the natural world, could never achieve.

> Could it be that there were serene superbeings in the sea who had mastered nature by becoming one with the tides and the temperatures long before man had even learned to scramble for the shelter of the caves, but who had not foreseen the coming of small vicious monsters from the land whose only response to the natural world was to hack at it,

> smash it, cut it down, blow its heart away? Had the whales enjoyed a Golden Age lasting millions of years, before their domain was finally invaded by a dangerous parasite whose advance could not be checked by any adaptive process short of growing limbs and fashioning weapons? What, indeed, could a nation of armless Buddhas do against the equivalent of carnivorous Nazis equipped with seagoing tanks and Krupp cannons?[53]

Hunter, it is clear, was ready for a crusade. He was also ready to end his increasingly unhappy marriage with Zoe and credited his encounter with Skana for giving him the courage to do so.

While Spong began his Japanese whale tour, the core Greenpeace members continued to hold meetings—mostly in Hamish Bruce's living room—where they argued over campaigns and, most importantly, funding. Greenpeace's fame was out of all proportion to its size and organizational competence. By now, the organization was becoming well known throughout Canada, Australasia, and the South Pacific, and among environmental and peace activists in Europe and the United States. Most of the group's meetings, however, descended into petty arguments over funding and expenses. What little money they possessed came mostly in the form of unsolicited donations and ad hoc fundraising. There was no advertising, mail-outs, or newsletter, and their address changed every time they chose a new leader. Their only significant asset was, in effect, their compact and evocative name which, as Hunter was well aware, fit snuggly into a one-column headline. The struggles within the organization, therefore, often came down to a battle for control of an increasingly valuable brand name. Bohlen and Stowe insisted that this brand name was inextricably associated with the antinuclear cause. To branch out beyond that would be to dilute its power, as well as misleading those who thought they were donating funds to an antinuclear organization. Hunter, Marining, and Spong, however, argued that the brand would be far more effective if it embraced a variety of ecological causes, thereby becoming a synonym for direct-action environmentalism.[54]

Under such circumstances, Hamish Bruce proved to be the ideal interim chairman. His feeling that Greenpeace was somehow serving a higher cause kept him above the petty and profane bickering. Furthermore, he viewed all donations as almost sacred, thereby retaining tight control over the few funds that were coming into the organization at the time. Not overly impressed by McTaggart's feat at Mururoa, Bruce was able to resist his calls to direct all of Greenpeace's funds toward his legal actions against the French Navy. In the overall cosmic scheme of things, as Bruce liked to view the world, McTaggart's actions were relatively minor, barely making up for the swashbuckling and profligate lifestyle he had led over the previous thirty years. Bruce kept the antinuclear faction happy by giving them funding priority for the 1974 Mururoa voyage, while also lending moral support to the whaling campaign. At times, such support reached a level of eccentricity that dumbfounded even hardcore acid droppers and *I Ching* freaks like Hunter and Spong. For example, Bruce wanted to organize a "call-in" for whales along English Bay. This would involve hundreds of people gathering on the beach at Kitsilano and using telepathy to attract whales from all over the world, providing them with sanctuary.[55]

In Japan, meanwhile, Spong was gathering considerable notoriety as he took his whale show throughout the country and courted the local media. His message centered on two themes which, to one degree or another, would define Greenpeace's antiwhaling campaigns for the rest of the decade. Firstly, he raised concerns about the ecological impact of a decimated whale population. Baleen whales, he pointed out, ate enormous quantities of plankton, while sperm whales fed mostly on squid. Would we have a dramatic increase in plankton levels if the whales were removed from the system? Would this have a drastic effect on ocean ecology? Would elevated plankton levels reflect more sunlight and alter water temperature? What impact would a squid population explosion have on fish populations? We did not yet know the answer to such questions, but at the rate that whales were being hunted, we would not be able to find out until it was too late to take remedial action.[56]

Secondly, Spong postulated that whales were the highest-evolved life-forms in the oceans, perhaps on the entire planet. This meant that we owed them a special duty of care as fellow creatures that perhaps possessed a level of intelligence and consciousness that, although different from ours, was comparably sophisticated. Instead of brutally slaughtering them and using them to manufacture products that could be produced in other ways, Spong insisted that we needed to learn from the whales. He compared the stability of the whale group structure with the volatile, fractured nature of modern human societies, noting that whale families stayed together for life and that a pod of whales "could very effectively mobilize group energy to satisfy individual needs." Such messages received a mixed reception among the Japanese press and public. Some saw Spong as an enlightened scientist whose warnings should be heeded, while others viewed him as a scientific imperialist and moralizer, the latest in a long line of Westerners intent on foisting their culture upon the Japanese. Spong's tour was not going to topple Japan's deeply vested whaling interests. It did, however, open a tiny crack in the Japanese consensus that whaling was an unquestioned good, as well as laying the foundations for a small but vocal antiwhaling movement in Japan.[57]

Just as it appeared that the Hunter-Spong faction was getting a strong foothold within Greenpeace, their influence suddenly melted away. While Spong was in Japan, both Hunter's and Bruce's marriages were falling apart, prompting the two of them to withdraw from the day-to-day affairs of the organization. With the main whale-campaign advocates temporarily out of the way, the antinuclear faction elected a Vancouver schoolteacher, Neil Hunter (no relation to Bob) as the new chairman. Hunter's straight, pedagogical style drove away many of the younger members who were more likely to have supported the antiwhaling cause, allowing the antinuclear group to concentrate Greenpeace's efforts exclusively on the 1974 Mururoa campaign.[58] This time, the campaign was based in Melbourne, where it was run primarily out of the local Friends of the Earth office. The nascent Greenpeace New Zealand group, Barry Mitcalfe's Peace Media, and the French branch of FOE (Les Amis de la Terre), were also involved.[59] Since McTaggart decided to remain in France to fight his court case, Greenpeace chose a German-born sailor, Rolf Heimann, to replicate the *Vega*'s voyages. His small ketch, renamed the *Greenpeace IV*, sailed from Melbourne in the middle of the southern winter. He and his crew eventually made it to Mururoa, but a week after the French had ended their tests for the season. Heimann was a skilled and experienced sailor whose antinuclear credentials were far more impressive then McTaggart's. But he lacked McTaggart's maniacal drive and stubbornness. The voyage, as Bob Hunter had predicted, failed to capture the attention of the world's media. Since the French government had already committed itself to ending aboveground testing, few journalists or politicians could see any practical point to the protest.[60]

The somewhat disappointing 1974 campaign precipitated a series of events that, in a relatively short period of time, would see Greenpeace depart from its antinuclear roots. Neil Hunter's faction, perhaps chastened by the Mururoa campaign, began to lose interest in Greenpeace, and their influence waned accordingly. Jim and Marie Bohlen bought a parcel of land on Denman Island, near the town of Courtney, British Columbia, where they set up the Greenpeace Experimental Farm, a model of alternative, sustainable living. Using their own labor and local materials, they started building a house composed of two geodesic domes, which Bohlen considered to be the optimum design for a space- and energy-efficient home. They established a large organic garden, developed various forms of alternative energy, and attempted to lead a lifestyle that verged on subsistence. Despite the fact that they clung to the Greenpeace name, however, the Bohlens had effectively withdrawn from the organization and would play no further role until the mid-1980s.[61]

The clearest and most tragic sign that Greenpeace's predominantly antinuclear phase was coming to an end was the news that Irving Stowe, a vegetarian, antismoking activist, and health food devotee, had been diagnosed with terminal stomach cancer. Greenpeace's pater familias endured a drawn-out, painful death, understandably embittered by the caustic irony of the situation.[62] Fate, it must have seemed, was mocking him. Stowe had been Greenpeace's deepest and most direct link to Quakerism and to the American peace movement of the 1950s. His influence over the younger activists had been considerable, even if they had not always appreciated his paternalistic tone and his lengthy, self-righteous speeches. His death, in October 1974, finally cleared the way for the sixties generation to assume control of Greenpeace. In Bob Hunter's words:

> There was no one left to resist any further Greenpeace's transformation from nuclear vigilantism to whale saving. And there was no one left to prevent us from dropping the hard brick-by-brick logic of the normal political world completely, seizing our *I Ching*s and allowing signs and visions to determine our course.[63]

The signs and visions would lead them out into the vastness of the Pacific Ocean once again. Within a year, they would make the antiwhaling movement one of the most high-profile and emotive environmental causes in the world. In the process, the name "Greenpeace" became synonymous with that cause. In media terms, it was a powerful synergy, one that greatly benefited both whales and Greenpeace. Nobody should doubt the courage, inventiveness, and sheer persistence of Hunter and his cadre of swashbuckling eco-warriors. At the same time, we should not exaggerate their influence. Their achievement was the end product of a series of broad cultural shifts that occurred over the previous decades. Unorthodox scientists and countercultural intellectuals were preoccupied with cetacean brains several years before Greenpeace began lobbing mind bombs at the world's media and blasting whales to the forefront of Western ecological consciousness. Before launching into Greenpeace's first antiwhaling campaign, therefore, it will be helpful to take a detour into the fascinating and sometimes bizarre history of Hunter's "armless Buddhas."

8

The Reenchanted Whale

In October 1954, *Time* magazine, with no apparent disapproval, described how a group of American soldiers enthusiastically slaughtered a pod of one hundred killer whales off the coast of Iceland. The Icelandic government considered the whales—which *Time* described as "savage sea cannibals up to 30 ft. long and with teeth like bayonets"—a menace to the local fishing industry and appealed to the U.S. soldiers stationed at a lonely NATO airbase on the sub-Arctic island for help. Seventy-nine bored GIs responded with enthusiasm, firing thousands of machine gun rounds at the whales until "the sea was red with blood." "It was all very tough on the whales," the report concluded somewhat jovially, "but very good for American-Icelandic relations."[1] Though the incident was particularly bloody, it was hardly unique. Throughout the 1950s, the U.S. Navy routinely used whales for target practice, pretending that they were Soviet submarines.[2] Meanwhile, the Norwegian whaling industry, hoping to tap into a potentially huge market, attempted to inculcate in Americans a taste for whale meat. Bemused housewives, however, were rarely tempted by "Capt. Seth's Frozen Tenderloin Norwegian Whale Steak" and similar cetacean delights. Nevertheless, the product itself did not outrage anyone. As late as 1962, a major U.S. publishing company released a popular book, written by a former whaler, celebrating and romanticizing the exploits of whalers.[3]

Within little more than a decade, the Western public's attitude toward whales would change completely. By the mid-1970s, as the first whale-watching tours began off the coast of Massachusetts, the idea of machine gunning whales or hunting them for sport, a not uncommon activity among wealthy big-game hunters in the early twentieth century, would have been seen by most Americans as nothing short of barbaric.[4] What explains this dramatic change in the way many people in Western democracies came to view cetaceans? Answering this question requires understanding how people thought about whales in the decades prior to the save-the-whales movement of the 1970s, before examining the cultural changes that helped facilitate a new attitude toward cetaceans in the mid-twentieth century.

Whales have furnished humans with an astonishing variety of useful and arguably frivolous products: Meat and fat for hungry populations; oil for burning lamps and lubricating machinery; bones and teeth for grinding into fertilizer or carving into works of art (scrimshaw); and baleen for the painful, constricting corsets that were popular among bourgeois European women in the nineteenth century. Until the nineteenth century, whaling did not, on the whole, endanger the world's whale population (with the exception of Right whales hunted to commercial extinction in the North Atlantic). In 1865, however, a Norwegian whaler, Svend Foyn, invented a device that suddenly, and radically, gave whale hunters an enormous advantage over their pelagic quarry. Foyn developed a barbed harpoon that could be fired from a shipboard canon that, on impact, would detonate an explosive charge inside the whale. The primitive and dangerous practice of hunting whales with hand-thrown spears was quickly replaced by the deadly, high-tech harpoon, completely shifting the odds in the whaler's favor.[5]

The Norwegians continued to be at the forefront of whaling technology well into the twentieth century. In the 1920s, they brought the assembly line to the high seas with the development of enormous factory ships. Giant stern slipways enabled crewmen to drag whales aboard where they could be flensed, boiled, rendered, and packed into barrels. While it was not unusual for nineteenth-century whalers to spend months, or even years, at sea, their efficiency and productivity was limited by the difficulty of processing large whales on the open ocean. The factory ships, with their fleet of nimble hunting vessels, allowed whalers to pursue the previously underexploited (from the whalers' perspective) Antarctic whale populations on a massive scale. By the 1930s, whaling was an almost completely unregulated multinational industry, fueled by U.S., British, and Scandinavian capital.

After the Second World War, Japan and the Soviet Union became the two major whaling nations in the world. For the Japanese, whales constituted a useful supply of protein and fat, particularly during the hungry years immediately after the war. In the Soviet Union, scientists came to see whales as the ultimate marine resource, and whale products were used in the production of medicine, leather, perfume, oil, and fertilizer, as well as animal feed, particularly for the growing Soviet fur industry. Aided by massive state subsidies, Soviet and Japanese fleets killed hundreds of thousands of whales in the postwar era. The combined take of Japanese, Soviet, and Norwegian fleets, along with whales killed by renegade operations, such as those of the Greek shipping-tycoon Aristotle Onassis, ensured that the world's whale population continued to plummet alarmingly.[6]

In an effort to prevent the extinction of several whale species, the United States helped establish the International Whaling Commission (IWC) in 1945. The IWC turned to science as an objective arbiter, hoping that all whaling nations would agree to abide by the recommendations of cetologists and population biologists. Whale research in the mid-twentieth century was dominated by what historian Graham Burnett calls the "hipbooted" scientists who would accompany commercial whaleboats to sea and stand thigh-deep in whale corpses to learn more about cetacean anatomy. Not surprisingly, therefore, there were very close ties between cetacean science and the whaling industry.[7] To compound the problem, many scientists, particularly those from Japan and the Soviet Union, saw their job as justifying whatever number of whales their nations were able to catch. Accordingly, they developed the spurious notion that hunting whales actually benefited the whale population, since fewer whales meant more food for those remaining. As a result, they argued, these whales would grow faster and larger, breed earlier and more frequently, and would ultimately develop into a fitter and more productive population. The whales could draw little comfort from the fact that the IWC had replaced laissez-faire hunting with the econometric logic of scientific conservation, a form of natural resource management that relied on concepts such as maximum sustainable yield and carrying capacity. For example, the measuring device that whale scientists developed to allocate and keep track of whaling statistics was known as the Blue Whale Unit (BWU). Each BWU was equal to one blue whale, two fins, two-and-a-half humpbacks, and so on. Since it was generally easier and more cost-effective to kill one blue whale rather than six smaller ones, the BWU ensured that it was more efficient for whalers to concentrate their efforts on the largest, and most endangered, species, and then work their way down the list as their numbers dwindled.[8]

Beyond the confines of the whaling industry and the IWC meeting rooms, however, whales were coming to represent far more than a mere marine resource. Until the mid-twentieth century, Burnett notes in his exhaustive study, few would have suggested that whales possessed "intelligence" or "beauty." One would need to "comb closely through marginal materials to catch even a hint of such preoccupations." Similarly, bottlenose dolphins were more likely to be viewed as annoying "herring hogs" that stole fish from the nets of hardworking honest fishermen rather than as the oceanic equivalent of higher primates.[9] However, by the early 1970s, whales and dolphins were rapidly becoming cultural icons, particularly within the North American and Western European counterculture. As well as representing a unique form of intelligence, cetaceans came to symbolize an idealized form of ecological harmony, particularly among those whose environmentalism was infused with countercultural mysticism. Such an outlook represented a dramatic shift from the Moby Dick-inspired image of whales as ferocious leviathans of the deep.

The transformation in the way Americans and, given the broad influence of American popular culture, most Westerners viewed whales can be traced to the late 1930s, when a group of scientists and entrepreneurs opened the Marine

Studios (later Marineland) aquarium in St. Augustine, Florida. The aquarium's curator, Arthur McBride, perfectly represented its twin missions of conducting cetacean research and turning a profit. McBride was undoubtedly a talented scientist, being the first to deduce that dolphins used their acoustic senses for navigation.[10] He was also a keen promoter who was not above complementing scientific rigor with sentimental anthropomorphism if he felt it would attract more people to the aquarium. In a 1940 article in *Natural History*, McBride introduced dolphins as our "most 'human' deep-sea relatives" whose "astonishing habits, observed at Florida's Marine Studios, reveal an appealing and playful water mammal who remembers his friends and shows a strong propensity for jealousy and grief."[11]

In 1954, another Marineland was opened in Palos Verdes, California, and its success spurred the development of several mammoth Sea World theme parks throughout the country. The stars of these aquariums were the clever and playful bottlenose dolphins (*Tursiops truncates*), whose tricks, vocalizations, and apparent delight in interacting with humans won the hearts of millions. In 1964, the Seattle Aquarium exhibited the world's first captive killer whale, an event that spawned a succession of articles in major magazines and newspapers, as well as a series of captive killer whale exhibits at the various Sea Worlds. Sea World was primarily run as a business and devoted little time and few resources to scientific studies. Instead, it concentrated on training dolphins and whales to entertain the crowds that poured through the theme parks' doors. Its scripted dolphin and killer whale shows emphasized the putative emotional similarities between humans and cetaceans, while eliding the differences, in the process creating a neutered, desexualized, domesticated pastiche. Such performances did little to educate people about how cetaceans lived in the wild, but they undoubtedly contributed to an increasingly Disneyesque view of dolphins and whales, one that antiwhaling groups such as Project Jonah and Greenpeace would later benefit from.[12]

Happily for the numerous aquatic theme parks across North America, cetacean Disneyfication was a shared cultural project. In fact, the great impresario of schmaltz was involved from the start.[13] In 1946, Walt Disney released *The Whale Who Wanted to Sing at the Met*, a short animated film starring one of America's most famous crooners, Nelson Eddy. Produced by the same people responsible for *Bambi*, the film did for marine mammals what *Bambi* did for their terrestrial counterparts: it created a sympathetic creature whose only goal was to please humans and live in peace and harmony with the rest of nature. The hugely popular 1963 movie, *Flipper*, and the subsequent television series of the same name, featured a tame dolphin as a clever and courageous pet—an aquatic version of Lassie—who frequently saved the day whenever his human friends got into deep water. In his fantasy novel, *The Day of the Dolphin* (which in 1973 was turned into a Hollywood film directed by Mike Nichols), French author Robert Merle created a scenario where dolphins were trained to speak with humans and to save the world from nuclear devastation.[14]

Congenial Flipper and Merle's delphinoid green berets were more than mere Disneyesque fantasies. They were also the cultural offspring of John Lilly, a brilliant neuroscientist whose controversial cetacean research ultimately destroyed a promising scientific career. For Lilly, as for Paul Spong a few years later, it was a worthwhile sacrifice. Born in St. Paul, Minnesota, in 1915, Lilly was a precocious youth with an aptitude for science. He majored in physics and biology at Caltech, moved on to the Dartmouth Medical School in 1938, and completed his medical training at the University of Pennsylvania in 1942. During the war, he conducted research on the physiological impact of high-altitude flying on pilots. It was the first in a series of affiliations that would deeply enmesh Lilly in some of the more fantastic and creepy military research projects developed at the height of Cold War paranoia. After the war, Lilly worked for the Public Health Service Commissioned Corp, the military branch of the Public Health Service, before becoming a section chief at the National Institute of Mental Health (NIMH) in 1953. As part of his research, Lilly wanted to study the human brain in the absence of all external stimuli. To that end, he developed an isolation tank, a sort of enclosed, soundproof bath filled with warm salty water. The subject—frequently Lilly himself—would lie in the tank for hours like a fetus in a sac of amniotic fluid. Such experiments were somewhat useful, but Lilly wanted to delve further into the human brain. However, there were limits as to how far he could probe: the prohibition on human vivisection, for a start. Perhaps there were other large-brained mammals that might tolerate being suspended for hours in saline isolation, but who were in no position to complain about having their skulls cracked and their brains prodded.[15]

Lilly began conducting invasive cortical research on dolphins in the mid-1950s (he had been doing it with monkeys for many years). This was not exactly neutral science. At the height of the Cold War, the U.S. military was keen to explore the possibilities of manipulating the human brain. Various branches of the military funded research into brainwashing (or "reprogramming" as it was then called), sleep deprivation, and "operant control" as part of the ongoing battle against the enemies of capitalism and, to a lesser extent, democracy. Animals whose brains were most similar to humans were naturally seen as useful subjects, and Lilly's dolphin experiments offered a promising avenue of research. Ironically, though perhaps not surprisingly, much of Lilly's early dolphin work was conducted at the Florida Marine Studios. So while one group of dolphins splashed about in the Studio's pools and balanced beach balls on their noses, their less fortunate conspecifics were in the lab next door getting their skulls drilled and their brains probed by Lilly's electrodes, before succumbing to a mercifully swift death (dolphins cannot breathe when anaesthetized). That is until Lilly had his eureka moment. As another of his experimental subjects was expiring in the service of science, it emitted a series of wheezing noises that, to Lilly, sounded like human speech.[16]

Cetacean-inspired revelations are not to be taken lightly. Lilly left his position at the NIMH and set up his own dolphin research station, the Communication Research Institute (CRI), first in Miami and then in St. Thomas in the Virgin Islands. He left his wife of twenty years and married a fashion model from St. Croix. He received lucrative offers to publish a book on his work with dolphins, resulting in the best-selling *Man and Dolphin*, published in 1961. The opening sentence set the course for the rest of the book and the rest of Lilly's life: "Within the next decade or two, the human species will establish communication with another species: nonhuman, alien, possibly extra-terrestrial, more probably marine, but definitely highly intelligent, perhaps even intellectual."[17]

Initially, Lilly was able to take this new direction without having to cut his ties with his generous benefactors; the Office of Naval Research, the Office of Space Sciences' Bioscience Program (part of NASA), and the National Science Foundation, among others, continued pouring money into his research. Lilly was not the only cetacean researcher to benefit from the largess of Cold War bioscience; indeed, he was part of a fairly extensive network of scientists who found themselves at a lucrative nexus: the sudden and unexpected convergence of cetacean research, neuroscience, and Cold War bioscience. The curator of Florida Marineland, for example, was appointed the head of the newly formed Marine Biosciences Facility of the Naval Missile Center. This was the forerunner of the navy's Marine Mammal Program, which throughout the 1960s trained bottlenose dolphins to perform numerous military tasks, many of which had been suggested years before by Lilly. Ronald Turner, a recent PhD graduate from UCLA's Brain Research Institute, was hired by the Naval Ordnance Test Station (NOTS) to work on the operant conditioning of dolphin vocalization. Kenneth Norris, the renowned whale researcher whom Spong consulted before moving to Vancouver, was on Turner's doctoral committee and was also the recipient of Naval Research funding, as were several other scientists at the Brain Research Institute.[18]

Despite its considerable promise, Lilly's scientific career began to unravel. To some extent, this was due to the psychotropic transcendence of LSD. Throughout the 1950s, countercultural intellectuals like Aldous Huxley and Alan Watts had been experimenting with Albert Hoffmann's chemical synthesis, frequently in the drawing rooms of Los Angeles' cultural elite. Lilly had not been part of this milieu, but he soon would be. His first LSD experience took place in Los Angeles in 1963. He shared it with Constance Tors, whose husband, Ivan, had directed the movie *Flipper*. Through his links with his former employer, the NIMH, Lilly was able to procure as much LSD as he needed. Over the next several years, he took the drug regularly, frequently while naked and floating in his isolation tank. He also began to inject it into dolphins. Given the tenor of the times, this was not as bizarre as it seems today. The U.S. military was keen to see if LSD could be a useful tool in controlling and manipulating behavior, and it encouraged certain scientists to experiment with it. Many took the drug themselves before administering it to volunteer subjects. And Lilly's dolphins were certainly not the first nonhumans to have LSD flowing through their bloodstream: numerous animals, from rats to chimpanzees, had also received doses.[19]

Nevertheless, as word of Lilly's increasingly peculiar exploits spread through the world of cetacean research, he incurred the wrath of some of his more intellectually and morally orthodox colleagues. By the early 1960s, in Burnett's words, "The CRI facility in the Caribbean began to look less like an outtake from a wonkish version of *Flipper*, and more and more like a bachelor crib for randy scientists seeking sun and surf."[20] Numerous nonconformist luminaries, including Aldous Huxley, Carl Sagan, and the anthropologist Gregory Bateson (formerly married to Margaret Mead) spent time at CRI, speculating on the possibilities of interspecies communication and other matters of cosmological significance while frolicking in the pool.[21]

Among Lilly's many unorthodox "experiments," one stood out for its flagrant flouting of both the rules of science and the widely acceptable boundaries of human-animal interaction. In 1964, Lilly convinced Margaret Howe, a young woman whom Bateson had gotten to know at a local resort, to spend several weeks cohabiting with an eager adolescent male dolphin in a thigh-deep pool. Howe wore a skin-tight leotard and bright red lipstick (to help the dolphin read her lips) and was excited at the prospect of participating in what Lilly hoped would be a breakthrough moment in his communication research. Peter the dolphin, however, seemed more excited by Howe's leotard. This was not the clever and clownish asexual dolphin of Disney fantasy: this was an adolescent male with strong urges and no inhibitions. Peter's increasingly aggressive overtures began to alarm Howe, but Lilly urged her to remain calm and to meet the dolphin half way. Here is Howe's description of what followed: "When Peter was upstairs in the Fiberglas tank he would occasionally become aroused, and I found that by taking his penis in my hand and letting him jam himself against me he would reach some sort of orgasm, mouth open, eyes closed, body shaking, then his penis would relax and withdraw. He would repeat this maybe two or three times, and then his erection would stop and he seemed satisfied."[22]

Many unusual experiments have occurred in the service of science, but it was difficult to know what to make of Lilly's cohabitation project. Far from being circumspect about the experiment, he happily published the explicit details in his next book, *The Mind of the Dolphin*, a follow-up to his 1961 bestseller, *Man and Dolphin*. While his bohemian lifestyle and questionable research methods were clearly a liability, it was his insistent claims about the transcendent possibilities of dolphin intelligence and communication—claims that most scientists found outlandish—that ultimately derailed his career. *Man and Dolphin* was a hit with the general public and would become a countercultural favorite, but it received numerous critical—even dismissive—reviews in scientific journals, where Lilly was criticized for his sloppy thinking. For example, he simplistically assumed a direct correlation between brain size and intelligence. Since sperm whale brains are about six times the weight of a human brain, Lilly proposed that they must have far more surplus gray matter than humans. Among other things, Lilly argued, this would enable sperm whales to recall past experiences with perfect clarity and learn from them. Burnett aptly describes this approach as a form of "flat-footed cortical reductionism." Nevertheless, this obsession with brain size would continue to appear in the antiwhaling literature for many years.[23]

By the late 1960s, just as Paul Spong was beginning his visual acuity experiments with Skana, Lilly found himself cut off from the world of Cold War bioscience that had sustained him for two decades. Indeed, these two events were probably not unrelated. UCLA's Kenneth Norris was among those who ensured that Lilly would receive no further funding from the likes of NOTS and NASA.[24] He also, in all likelihood, advised Spong to avoid the crowded and increasingly tendentious world of cetacean communication research and instead to concentrate on studying the visual acuity of killer whales.[25] His scientific career in tatters, Lilly was free to indulge his psychedelic urges and become a full-blown countercultural guru. He became affiliated with the Esalen Institute and began advocating the therapeutic potential of his isolation tank.[26] He struck up a friendship with Timothy Leary, who shared a similar career arc and penchant for LSD. Lilly also continued to develop and propagate his ideas about cetacean intelligence and interspecies communication. It mattered little that his former scientist colleagues no longer paid attention: a whole new generation was ready to tune in. Lilly almost single-handedly laid the groundwork for the creation of a new mammalian construct. Had he thought to give it a name, he might have called it *Cetaceus intelligentus*.

Virtually all the pop-cultural and pop-science cetacean literature that emerged from the early 1960s onward was inspired by Lilly. In fact, the first such book preceded the publication of *Man and Dolphin*. In 1960, Leo Szilard, the brilliant Hungarian physicist who had been one of the architects of nuclear fission, published *The Voice of the Dolphins*, a satirical novel in which highly intelligent dolphins help guide scientists through the complex process of nuclear disarmament. Szilard developed the idea after several conversations with Lilly at the National Institute of Health in the late 1950s. The book sold over 35,000 copies in the United States alone and was translated into six other languages.

Lilly's work was also the direct inspiration for both Robert Merle's *Day of the Dolphin* and Farley Mowat's *A Whale for the Killing* (1972). The latter sold several million copies and was a major catalyst in the antiwhaling movement of the 1970s.[27]

Among the writers who sought to promote the idea that whales were a uniquely intelligent species, and one that was under threat from human exploitation, few were as articulate or prolific as Scott McVay. An administrator at Princeton University, McVay had become fascinated by whales as a result of studying *Moby Dick* in college. In 1961, he attended a lecture that Lilly delivered to the Princeton psychology department. The two became good friends, and McVay went to work for Lilly at CRI from 1963 to 1965. He then returned to Princeton, where he worked as the assistant to the university president. A firm believer in Lilly's *Cetaceus intelligentus,* he devoted virtually all his spare time to alerting the world to the plight of whales. In a 1966 article in *Scientific American,* McVay outlined the history of the IWC and wrote a stinging critique of its practices, essentially accusing it of being a cozy club for whalers rather than a regulatory organization that was genuinely interested in conservation. "If essentially unrestricted whaling continues," he wrote, "the only surviving stock . . . is doomed to become a monument to international folly."[28] McVay attended several IWC meetings as an observer, and his presence only served to confirm his pessimistic view of the commission: "The most desirable goal of all, a ten-year moratorium . . . seems beyond the capacities for cooperation and restraint of the nations present at the (IWC) meeting."[29] By 1971, McVay was couching his antiwhaling arguments in a language that reflected the holistic outlook and moral ecology that would suffuse many elements of 1970s environmentalism:

> Our survival is curiously intertwined with that of the whale. Just as all human life is interconnected . . . so have we finally begun to perceive the connections between all living things. The form of our survival, indeed our survival itself, is affected as the variety and abundance of life is diminished. To leave the oceans, which girdle seven-tenths of the world, barren of whales is as unthinkable as taking all music away . . . leaving man to stumble on with only the dryness of his own mutterings to mark his way.[30]

Such pronouncements, linking human survival with that of whales, became increasingly common throughout the late 1960s and early 1970s and were instrumental in casting whales as the great symbol of 1970s environmentalism. Folk legend Pete Seeger summed this view up perfectly in his ballad, "The Song of the World's Last Whale":

If we can save
Our singers in the sea
Perhaps there's a chance
To save you and me.[31]

Roger Payne, a scientist at Rockefeller University and a friend of McVay's, supplied further evidence to bolster the theory that whales and dolphins had sophisticated communication systems similar to humans. Using a hydrophone, Payne recorded the vocalizations of humpback whales near Bermuda. His analysis of the recordings led him to conclude that the sounds were, in the truest sense, songs: discrete phrases repeated over and over and sometimes lasting for up to thirty minutes. Payne produced a record, *Songs of the Humpback Whale*, which introduced millions of people to the animals' haunting sounds, which, in the context of the rapid demise of the species, could easily be interpreted as cries for help.[32] In effect, Payne and McVay had realized one of Lilly's central goals: to provide people with a meaningful acoustic encounter with intelligent aquatic beings.

In 1972, an eminent Swedish scientist, Karl-Erik Fichtelius, decided that his colleagues had dismissed Lilly's work too hastily. With the help of a journalist, Sverre Sjölander, he published a book, provocative titled *Smarter than Man?* Fichtelius and Sjölander systematically compared the human brain with those of whales and dolphins in an effort to draw some broad conclusions about their comparative intelligence. The dolphin's cerebral cortex, they found, is larger than ours, has twice the number of convolutions, and 10 to 40 percent more nerve cells. The section of the cortex devoted to motor skills was considerably larger in humans than in dolphins, but this merely meant that "the dolphin has more cortex left over for the higher mental processes than we do. . . . The surprising conclusion of [our] comparison is that the dolphin brain *could* be superior to ours."[33] This did not mean that whales were more intelligent than humans, at least in the sense that we understand the concept of intelligence. However, the authors speculated, it was possible that our definition of intelligence was simply too narrow.

> A little thought will tell us that what we take such pride in—our capacity to adapt to new situations and to use previous experience to solve new problems—is an ability that harsh selection called forth in a naked ape turned predator. And this ability is suited to the life of such an ape, even if it has also lent itself to the invention of whaling ships, hydrogen bombs, and concentration camps. With a little imagination one can actually conceive of a different type of intellectual activity than ours. And it need not necessarily be the case that our form of intelligence is the most appropriate form for whales as well.[34]

The ultimate ode to *Cetaceus intelligentus* was *Mind in the Waters*, a collection of articles, essays, and poems assembled by Project Jonah's Joan McIntyre and published by the Sierra Club. The volume included a variety of works by people in various fields. Sterling Bunnell, a medical doctor who also taught evolutionary ecology at the California College of Arts and Crafts, echoed the work of Fichtelius and Sjölander, arguing that since humans had evolved primarily to recognize and avoid danger, it was "difficult for us to understand intelligent and non-manipulative beings which are so well adapted to their habitat that the survival considerations of finding food and avoiding danger have been much less of a problem for them than they have been for us." Bunnell also offered some thought-provoking ideas about the difference between human and cetacean communication. The cetacean auditory system, he argued, is predominantly spatial, like human eyesight, and is designed to process much simultaneous information. Thus whales and dolphins could communicate a whole paragraph of information in one elaborate instantaneous hieroglyph. Therefore, Bunnell reasoned, pouring a little cold water over Lilly's interspecies communication experiments, "for them to follow our pattern of speech might be almost as difficult as for us to study the individual picture frames in a movie being run at ordinary speed." Nevertheless, he continued, it was possible that the cetacean communication system was even more complex than our own, only in a very different way. According to Bunnell, another important indicator of intelligence was the extreme playfulness and humor exhibited by both captive and wild cetaceans. "Despite its low status in puritanical value systems," Bunnell explained, "play is a hallmark of intelligence and is indispensable for creativity and flexibility. Its marked development in Cetaceans makes it likely that they will frolic with their minds as much as with their bodies."[35]

John Sutphen, a Connecticut physician with more than a passing interest in whales, argued that whales and dolphins probably had a far more sophisticated emotional intelligence than humans. Since echolocation is three dimensional, "one dolphin scanning another dolphin does not just receive an echo from the other's skin but from the interior body as well." Therefore, apart from immediately recognizing if another animal was ill, cetaceans would also "be constantly aware of a considerable portion of each other's emotional state." "What sort of candor," Sutphen mused, "might exist between individuals where feelings are instantly and constantly bared? It would be irrelevant to hide, to lie, or to deny one's feelings." Cetaceans, it seemed to Sutphen, were not only as intelligent as humans, but quite possibly morally superior. Sutphen's analysis, like Spong's, abandoned the conservative caution of much scientific discourse, even going so far to suggest that whales and dolphins possessed a "culture."[36]

Paul Spong's contribution to *Mind in the Waters* summarized his experiences with Skana and his work on Hanson Island. Scott McVay attempted to disabuse his readers of the "old square-rigged notions about whaling" that continued to "linger like a gauzy pink haze" and which "abound[ed] in contemporary writing." The romantic image of the whale hunter, he declared, "has begun to pall, for the whale has no more chance than a bull in the ring as it is scouted by helicopter, scanned by sonar, and run down by mechanized ships designed to travel three knots faster than a finbacks's top speed." The renowned marine biologist, Victor Scheffer, who was chairman of the presidentially appointed Marine Mammal Commission of the United States, urged the IWC to look beyond mere conservation measures and to consider whaling as an ethical issue. "The esthetic and educational values of whales alive," he contended, "are greater than the values of (the products) which might be derived from their carcasses." Scheffer espoused views that were not dissimilar from those emerging from the fledgling animal rights movement. "Morality," he argued, "extends beyond ordinary humaneness, or the prevention of pain and terror in the animal, to a consideration of the simple right of the animal to live and to carry on its ancestral bloodline." Even an unreconstructed traditional conservationist such as Lee Talbot, the senior scientist on the president's Council of Environmental Quality and the scientific adviser to the U.S. delegation to the IWC, was willing to flirt with the rhetoric of holistic ecology in the service of the whales. "The time is past," he declared, "when we can equate conservation with maximum sustained yield, or when we can base management of a living resource simply on our economic 'need' for its products. We are slowly coming to the realization that maintenance of the health of the habitat is a prerequisite to the survival of a species."[37]

The most eloquent contributions to *Mind in the Waters,* though it could also be argued the most speculative and sentimental, came from Joan McIntyre. With equal parts holistic ecology and New Age romanticism, McIntyre decried the Cartesian worldview that denied feelings, imagination, awareness, and consciousness to other creatures. "It seems that in our craze to justify our exploitation of all non-human life forms," she declared, echoing Rachel Carson, "we have stripped from them any attributes which could stay our hand." Try, she urged her readers, "to imagine the imagination of a whale, or the awareness of a dolphin. That we cannot make these leaps of vision is because we are bound to a cultural view which denies their possibility."[38] According to McIntyre, the plight of the whales needed to be understood as part of a broader trend of human beings' relationship with the natural world; indeed, it was this very bifurcation between nature and culture, and between the mind and the body, that lay at the root of the problem. In the water, the cradle of cetacean consciousness, the distinction between the mind and the body had been dissolved: "Without the alienating presence of objects and equipment, with only the naked body encasing the floating mind, the two, split by technological culture, are one again. The mind enters a different modality, where time, weight, and one's self are experienced holistically." In the sea, she continued, "the world can be *thought* and *experienced* simultaneously—not broken down into categories that stand for experience rather than experience itself."[39]

After reading McIntyre and the other contributors to *Mind in the Waters*, one is left with the image of whales and dolphins as exemplars of ecological virtue and holistic consciousness. These are creatures who are totally in tune with their environment and with each other; who possess advanced systems of communication and construct "thoughts" from acoustically derived images; whose brains are larger than ours and have a greater degree of gray matter leftover for higher mental processes, rather than for simply manipulating objects. With this in mind, it is all the more shocking for the reader to learn that a whale was killed every twelve minutes, "the living tissues blown into agony by explosive harpoons." Compounding this brutality was the fact that almost all the products made from whales—things such as chicken feed, cattle fodder, fertilizer, car wax, shoe polish, lipstick, cosmetics, margarine, cat and dog food, and feed to raise minks and foxes for fur coats—could be synthesized or substituted from other sources. The future of the whales, McIntyre insisted, was inextricably bound together with our own: "in saving them we can create a model of international action that can demonstrate a way to save ourselves and the rest of the earth we cherish."[40]

This, clearly, was not the language of scientific conservation. In the face of *Cetaceus intelligentus*, mere conservation become irrelevant, indeed, abhorrent. The only way to save the whales was to abolish whaling, not merely control and regulate it. For the likes of McIntyre, Mowat, Spong, and Bob Hunter, the logic of scientific conservation could be as absurd and harmful as laissez-faire whaling. Take, for example, the ideas of Gifford Bryce Pinchot, a marine biologist and the son of the man commonly viewed as the father of twentieth-century conservation. Pinchot felt that both world hunger and the extinction of the great whales could be prevented by simply turning whales into the cattle of the sea. His ambitious plan involved pumping deep-sea water into tropical lagoons in the Pacific Ocean. This would spur the growth of phytoplankton—great masses of aquatic algae—which would in turn be eaten by zooplankton such as krill. The most efficient way to convert this mass of stored energy into protein and fat for human consumption was to "farm" blue whales in the lagoons. In this way, the great whales, like the American bison before them, could be simultaneously saved and savored.[41]

There is little doubt that Pinchot was genuinely concerned with the possibility of the blue whale's extinction. Nevertheless, to those subscribing to the *Mind in the Waters* worldview, the idea of turning whales into semi-domesticated stock was possibly even worse than hunting them on the open sea. If the Blue Whale Unit and raising whales in lagoons for human consumption were the best that scientific conservation could do, then clearly, scientific conservation was grossly inadequate. How could one, after all, equate an intelligent and sensitive creature such as the whale with the doltish cow?[42] Yet, it was exactly here that the tension between *Cetaceus intelligentus* and ecology became apparent. If whales deserved to be spared the harpoon because of their intelligence, their awareness, their consciousness, then abolition was the only solution. If, however, one also accepted that whales were part of a broader ecosystem, then no matter how harmoniously they lived within that system, it was hard to argue that humans should not be allowed to hunt them. After all, whales themselves were hunters, and one of the key tenets of holistic ecology is that no individual creature, neither human nor whale, should be viewed as more important than the ecosystem to which it belongs. From this perspective, arguing that humans should never kill whales, particularly if they were not endangered, was as absurd as arguing that it was morally wrong for orcas to kill seals or rip apart baby humpbacks, which they did regularly. It was a philosophical dilemma that groups such as Greenpeace would grapple with for years to come.

Nevertheless, despite this rather thorny paradox, the process of cetacean reenchantment was well under way by the time Greenpeace arrived on the scene. In a matter of a few decades, whales had been transformed from blubber and baleen to Buddhas of the deep. Once a mere natural resource, by the early 1970s they had been symbolically reconstituted into icons of ecological holism and holistic consciousness. In the process, particularly within the counterculture, they served as an all-purpose emollient that would heal the intellectual and ecological wounds that Western culture had been inflicting upon the world since Descartes separated the mind from the body and insisted that all creatures other than humans were mere machines. The periodic revolt against dualism, as philosopher Arthur Lovejoy called it, was once again gathering momentum.[43] But if Hunter's armless Buddhas were going to complete the consciousness revolution, someone would have to save them from the carnivorous Nazis that were driving them toward extinction.

9

Stop Ahab

By late 1974, Hunter and Spong's Stop Ahab Committee had, in effect, become Greenpeace. Although technically a mere subunit of the broader organization, the departure of the Quaker and antinuclear members left the Stop Ahab Committee as the only functional and active element within the group. Despite that they continued using the "Stop Ahab" letterhead on their official correspondence, all those working on the campaign referred to themselves as Greenpeace, and the media rarely used the more cumbersome Moby Dick-inspired name. In addition to Hunter, Spong, Marining, and Bruce, Patrick Moore had returned to the core group after an eighteen-month hiatus, during which he had remarried and completed his doctorate in ecology. Along with Hunter and Spong, Moore played an important role in authoring several key documents that constituted a kind of Greenpeace manifesto. The efforts to sketch the underpinnings of a new ecology-based religion—Hunter's Whole Earth Church—paralleled the efforts of various philosophers who were working on similar projects, most notably deep ecology. Greenpeace's entry into the antiwhaling movement, therefore, was an exercise in both theory and praxis.[1]

The first of these documents arose from the original Greenpeace voyage to Amchitka in 1971 when Moore had delivered his spontaneous "a flower is your brother" lecture. Hunter had then whimsically proclaimed everyone present a member of the Whole Earth Church. Despite the element of parody and gentle self-mockery in Hunter's pronouncement, he and Moore did indeed set up a Whole Earth Church. The members, drawn mostly from Vancouver's countercultural ghetto, were all considered ministers, and each received an official certificate bearing a Greenpeace moniker and a reminder of their sibling responsibility to flowers. The entire venture was a not entirely unserious effort to establish a New Age religion with ecology as its deity. According to their manifesto, members of the church believed that "all forms of life are inter-related" and that any form of life that went "against the natural laws of inter-relatedness and interdependency" had "fallen from the State of Grace known as ecological harmony." Slipping into traditional Christian argot, the manifesto insisted that it was "mankind's duty to transform the earth into a heavenly garden" and that it was "a sin to cause any degradation to the land, air, water, plants . . . or animals because this causes harm to humans."[2]

The Whole Earth Church manifesto was followed by the "Three Laws of Ecology," which were then incorporated into a "Declaration of Interdependence." Based on a similar document written by Gary Snyder and others in the very early 1970s, it reflected a variety of influences, from Barry Commoner and the Club of Rome, to Leopoldian expressions of biocentric thought.[3] Abandoning the hint of anthropocentrism that had characterized the Whole Earth Church, Hunter, Moore, and Spong declared: "As suddenly as Copernicus taught us that the earth was not the center of the universe, ecology teaches us that mankind is not the center of life on this planet." Accordingly, the First Law of Ecology stated: "Life is a great web, continually regenerating, evolving, interchanging, and converting energy, programming its own reproduction and survival, and becoming conscious of itself." Inspired by theories of biospheric holism, the First Law also insisted that the biosphere itself was a living organism. The "Second Law" asserted that the stability of an ecosystem was directly proportional to its diversity, while the "Third Law" proclaimed that all resources were finite and that there were "limits to the growth of all living systems." The Declaration combined Moore's ecological training with Hunter's penchant for grand theorizing. By "embark[ing] on a quest for the great systems of order that underlie the complex flow of life on our planet," ecology bore the hallmarks of religion as much as of science. "Like religion," Hunter and Moore continued, "ecology seeks to answer the infinite mysteries of life itself. Harnessing the tools of logic, deduction, analysis, and empiricism, ecology may prove to be the first true science-religion."[4]

The idea of ecology as a religion for the New Age fit well with Hunter's Hegelian theory of consciousness revolution. Politics, in Hunter's reckoning, took a backseat to philosophy and psychology. The goals of direct action should not merely be the narrow ones of ameliorating present conditions or replacing one regime with another. Rather, activism had to become a tool in the struggle to change people's fundamental perceptions of the world around them. This would require an almost cosmic paradigm shift. Familiar systems of knowledge and belief such as science, Christianity, capitalism, and Marxism would have to yield to the new deity of holistic ecology. In 1974, Hunter elaborated on these thoughts in an unpublished essay titled "The Politics of Evolution." Writing in the midst of Greenpeace's preparations for the whale campaign, Hunter attempted to demonstrate the organic connections between the self—the interior being that centuries of Western thought had severed from the outside world—and nature:

> Once, in a heightened psychedelic state . . . a voice from deep within told me: "I am the Earth." And I realized later that it was quite true, at

> every level, from the biological to the psychic, that the earth and its ecosphere is not only metaphorically but *literally* "my body," in as much as I am the equivalent of a cell or microbe within it.[5]

In such a "Cosmic Ecology," as Hunter referred to it, the awareness of our true nature and our proper place in the world already existed, albeit in a dormant state. It was like an *I Ching* reading; all that was required was "the consciousness to perceive the meaning." The message, according to Hunter, was simple: "Change consciousness, not the things around you. For, by changing consciousness, the things around you automatically change too. . . . True revolution is revelation." Hunter spurned the traditional politics of the Left and the Right, arguing instead that human society needed "to move to the centre, to the very core of politics: Man's relationship to life itself." Such musings led Hunter to the language of ethical extension that was a trademark of biocentric thought from John Muir and Aldo Leopold to Gary Snyder and Christopher Stone. "We must seriously begin to inquire," Hunter insisted, "into the rights of rabbits and turnips, the rights of soil and swamps, the rights of the atmosphere itself, and, ultimately, the rights of the planet."[6]

Despite Hunter's appeals for an ecological egalitarianism, however, there was a contradictory thread within his argument, one that implied the kind of hierarchical taxonomy that biocentrism supposedly repudiated. Not surprisingly, this ecological elitism surfaced in Hunter's discussion of whales and dolphins. Science fiction, Hunter argued, had prepared humans for the arrival of intelligent beings from outer space but not for the discovery of such creatures within our very midst. Whales and dolphins possessed "an alien form of intelligence approximately equal to our own." The mind of a dolphin, he continued, "might be as strange in relation to our own as that of a Martian, yet it bears all the hallmarks of higher consciousness." Hunter's ecological elitism was somewhat mitigated by his insistence that flowers, too, possessed a form of consciousness. Nevertheless, Spong's influence on his thought was growing. Despite the biocentric rhetoric throughout Hunter's essay, there was also a strong implication that whales and dolphins deserved special consideration based on the fact that their intelligence and consciousness was of a similar sophistication to our own.[7]

Hunter's resolve to mount a Greenpeace antiwhaling expedition was cemented in July of 1974 after a close-up encounter with a pod of wild whales. Disappointed by Greenpeace's decision to mount yet another Mururoa campaign and grieving over the recent breakup of his marriage, Hunter decided to camp out on a lonely beach on the west coast of Vancouver Island to contemplate his future. Every summer for as long as locals could remember, a pod of gray whales would appear offshore and venture into the shallow waters near the beach to feed. Hunter had bought a small plastic child's dinghy along, and, after a favorable *I Ching* reading, decided to paddle some two miles offshore to visit the whales. At one point, one of the forty-ton giants approached to within a couple of yards of the tiny and wholly inadequate dinghy. A wave of fear rippled through him. The whale hovered just below the surface, and Hunter could see its giant eyeball staring up at him. "My trembling ceased," Hunter recalled, "and a feeling came over me that I normally associated with spectacular sunsets. It was not that I was eyeball to eyeball with *something*, but rather, with *somebody*." Unlike his encounter with Skana, however, this particular episode did not leave Hunter feeling as though he had just encountered some aquatic guru with insight into the human condition. Rather, he climbed very slowly up the beach "feeling very normal and real and ordinary, aware that I had been close to a normal, real, ordinary whale, and that we both faced the normal, real, ordinary world—one in which his kind were being hunted to death. It was time I got on with the business of doing something about it."[8]

By September of 1974, Irving Stowe was near death and Jim and Marie Bohlen had moved to their new home on Denman Island. David McTaggart had decamped to France to continue his fight against the French military, and Neil Hunter and the rest of the antinuclear faction had largely disappeared from the scene. According to Hunter, "The Greenpeace Foundation had effectively ceased to function."[9] All that remained was the Stop Ahab Committee, which consisted of Hunter, Spong, Marining, and Hamish Bruce. Booking a small hall in a community center in Kitsilano, the heart of Vancouver's countercultural scene, Hunter and his cohorts set up a meeting to plan a direct-action campaign against whaling. Some seventy people attended, most of them under thirty and representing the spirit of their generation. From the very outset, the meeting, and those that followed in the forthcoming months, was characterized by a tone of joyous celebration and good humor, in marked contrast to the "heavy atmosphere of moralistic purity," which, according to Hunter, had pervaded earlier Greenpeace meetings. As Patrick Moore recalled, the "sober suffering" of the Stowes and Bohlens was replaced by a joie de vivre, a fact that could partially be explained by the positive nature of the campaign. As long as Greenpeace's raison d'être had been to oppose nuclear weapons, there was little to celebrate. But now, in Hunter's words, "instead of fighting death, we were embracing life. It was not just that we wanted to save whales, we wanted to meet them, we wanted to engage them, encounter them, touch them, discover them. For the first time there was a transcendent element lying at the center of the undertaking."[10]

Another distinctive feature of the new Greenpeace and its supporters was their general adherence to the biocentric views outlined by Hunter, Moore, and Spong. In the early 1970s, the Norwegian philosopher Arne Naess published a seminal article in which he drew a distinction between "deep" and "shallow" ecology. Hunter had never heard of Naess, but he was clearly swimming in the same intellectual waters. His version of the shallow/deep dichotomy distinguished different forms of environmentalism based on various levels of "planetary consciousness." While conceding that members of Greenpeace's antinuclear faction were concerned about the impact radiation would have on nonhumans as well as humans, Hunter felt that "their moral and spiritual revulsion was still essentially centered on their concern about the impact on people, not whales or sea lions or lizards." So while Irving Stowe was a "planetary patriot," he was still, in Hunter's opinion, "an agent of the human race, looking out for the human race's long-term interests, especially his own great-great grandchildren." People such as himself and Spong, on the other hand, accepted that nonhuman species had an intrinsic value that made their salvation a worthy goal in itself, regardless of the impact on humans. While such characterizations may not do justice to the likes of Jim Bohlen, whose environmentalism was informed by both Leopoldian ideals and Buddhist thought, they indicate a concerted effort on the part of Hunter to distinguish the new Greenpeace, which was, after all, *his* Greenpeace, from the old. For Hunter, the ecology of the new Greenpeace was, to use Naess's language, "deeper." Although Greenpeace certainly was not molding itself to fit Naess's philosophy, it seems fair to assume that the views expressed by people such as Hunter, while not always rigorously consistent, closely approximated what Naess had in mind when he referred to deep ecologists.[11]

In addition to drawing new people to Greenpeace, the antiwhaling campaign also gave some of the organization's more peripheral members the opportunity to play a greater role. One of the most important and controversial of these was Paul Watson. Watson had been on board the *Greenpeace Too* back in 1971 and had participated to a minor degree in the Mururoa campaign. Like McTaggart, he has become something of a larger-than-life figure in the environmental movement, and his biographical details have been shaped to better serve the legend of "Captain" Watson, the world's most fearless and uncompromising eco-warrior. Hunter recalls that Watson, "because of his impetuousness, unpredictability, and a marked tendency to brag too much," was frequently out of favor with others in the organization. He also suffered from a persecution complex and found it impossible to attribute any of his misfortunes to his own actions. For Hunter, however, these flaws were outweighed by Watson's courage and total commitment to the cause, as well as his physical strength and sharp reflexes.[12] Of French-Canadian extraction, Watson was born in Toronto in 1950. The oldest of six children, his mother died when he was a teenager, and he frequently clashed with his father. According to Watson, the greatest influence on his early life was his maternal grandfather, Otto Larsen, a Danish-Canadian artist who liked to paint pictures of animals. As a child, Watson spent countless hours at Toronto's Riverdale Zoo watching his grandfather paint. It was here that he first developed his affinity for animals: "I got to know all of them," he claimed somewhat mawkishly, "and the love seeded then has flourished ever since."[13]

In the mid-1950s, Watson's family moved to a small fishing village in southern New Brunswick. The village housed a government-run marine biology station, and Watson spent much of his free time observing the station's wildlife and fishing in the bay. By his own account, his first act of animal liberation occurred when he was a mere ten-year-old: "I had a little friend, a beaver, who was killed in a trap line," Watson recalls. The beaver's death prompted him to begin destroying trap lines and releasing snared animals.[14] In 1967, he left home to work in Montreal, before "hobo-ing" his way across to British Columbia in railway boxcars. In Vancouver he found work as a merchant seaman aboard a Norwegian carrier, a job that was his ticket to the world. On returning from his first lengthy voyage, he studied linguistics and communication at Simon Fraser University, with a special emphasis on "interspecies communication."[15] He also claims to have been "an active member of the Society for Pollution and Environmental Control," the antipollution organization founded in Vancouver, and his desire to protect wildlife from pollution and radiation led to his involvement with the Don't Make a Wave Committee in 1971. Watson insists that he had little interest in the Vietnam War and the other major protest movements of the era, though others remember him as a radical antiestablishment figure who frequently wore a North Vietnamese flag on his jacket.[16]

Like many radicals and environmentalists of the time, Watson felt a deep affinity with the indigenous peoples of North America. Particularly among the counterculture, Indians symbolized an ecological harmony and courageous resistance in the face of the overwhelming might of European conquest. After reading Dee Brown's *Bury My Heart at Wounded Knee*—a catalog of injustices visited upon North American Indians—Watson decided to drive to South Dakota to offer his services to the members of the American Indian Movement who, in February of 1973, had seized the trading post and church at Wounded Knee and were involved in an armed standoff against the FBI. Accompanying Watson on the journey was his good friend David Garrick, a lanky, long-haired underground journalist with a drooping moustache. Garrick, a twenty-eight-year-old Ontario native, had studied anthropology at Trent University and was committed to the entire pantheon of countercultural causes, particularly aboriginal rights and holistic ecology. Like Watson, he was a member of the Vancouver Liberation Front. Throughout the 1970s, as a gesture of solidarity with Native peoples and the environment, he adopted the Tolkienesque name "Walrus Oakenbough." Walrus, as he was known within Greenpeace, also had a deep knowledge of Native American spiritualism. Through a combination of persistence, daring, and luck, he and Watson managed to pass through the FBI lines and join the members of the American Indian Movement, mostly Oglala Sioux, who were organizing the standoff. For several weeks they worked as medics, which mostly involved lancing children's boils and bandaging the odd bulletwound.

Watson described the standoff as a "combat situation," with FBI agents, federal marshals, and members of the U.S. Army's 82nd Airborne Division firing thousands of rounds of ammunition in the general direction of the Wounded Knee trading post. As a reward for their service, the two were inducted into the Oglala Sioux tribe as full-fledged members. In a traditional sweat lodge ceremony, Watson was given the tribal name "Grey Wolf Clear Water," while Walrus added "Two Deer Lone Eagle" to his already fanciful nom de guerre.[17]

On returning to Vancouver, Watson and Walrus enthusiastically joined Hunter and company in the preparations for the antiwhaling expedition. Another important new member of Greenpeace's inner circle was Rex Weyler, a twenty-eight-year-old American who had moved to Vancouver after fleeing the draft. Weyler grew up in Colorado and Texas, before moving to California to attend Occidental College. He planned to major in mathematics and physics, but in the late 1960s, he began subordinating his education to his activism; at one point he was suspended for a semester for participating in a blockade of the college's administration building. After dropping out of college, he spent time at various countercultural communities throughout California, including a stint at Joan Baez's Institute for the Study of Nonviolence in the Palo Alto hills, where he got to know people such as Baez and leading countercultural theoretician Theodore Roszak. Influenced by writers such as Roszak and Gary Snyder, Weyler began to gravitate toward Buddhist thought and ecology, which he fused with his ideas about nonviolent activism. He then hit the hippie trail, traveling throughout North America, then moving on to Amsterdam, North Africa, the Greek Islands, India, and Katmandu, before returning to the United States with his new Dutch-Indonesian bride. In June 1971, FBI agents visited his sister's house, informing her that Weyler was being indicted for violations of the Selective Services Act. If caught, he would be taken back to Texas to face trial at the federal court in El Paso, where he faced a possible twenty-five-year maximum prison term. Like many before him, Weyler fled to Canada, settling in Vancouver in the summer of 1972.[18]

In Vancouver, Weyler was hired as a journalist by the *North Shore Shopper*, an advertising circular that the publisher was preparing to turn into a more substantive newspaper. Weyler had considerable freedom to pursue whatever stories he felt were newsworthy. Since he had a strong interest in ecology, much of his reporting concentrated on environmental issues in Vancouver. In 1973, he wrote a story on a group of people protesting the city's plan to cut down some large cottonwood trees in North Vancouver to make way for a new development. A woman by the name of Bree Drummond had "occupied" one of the trees, setting up a perch high in its branches, while her boyfriend, Rod Marining, offered moral support from below. Weyler became friends with Drummond and Marining, and through them, met Hunter, Spong, and the rest of the Greenpeace crowd. Like Hunter and Metcalfe before him, Weyler's initial involvement was ostensibly as a journalist reporting on the organization's affairs rather than as an active member. However, he soon found himself drawn more tightly into the circle. Once the Hunter faction, to whom he was most sympathetic, had taken control of the organization, he decided to commit himself to the antiwhaling campaign. His primary responsibility was to be the photographer for the voyage, and many of the most spectacular early Greenpeace photographs were shot by his camera.[19]

By late 1974, Greenpeace's plans for their antiwhaling direct-action campaign had, at least in their broad outline, been confirmed: A crew of activists would sail into the Pacific and track down the Soviet or Japanese whalers. They would then launch their Zodiacs, follow the killer boats to the whale pods, and attempt to position themselves between the harpoons and the whales. The assumption was that in such a situation, the harpooner simply would not dare fire for fear of killing the protestors. The whole episode would be filmed and photographed, providing the media with irresistible images of a David versus Goliath conflict between the unarmed activists in their tiny Zodiacs and the whalers with their enormous factory ships and explosive steel harpoons.

The next few months were a whirlwind of activity. Greenpeace needed to find a boat, choose a crew, formulate a media strategy, and, above all, raise funds for the campaign. For Hunter, in particular, the period was marked by a bizarre series of events and coincidences that served to bolster his perception that Greenpeace was blessed with excellent karma. First, a man whom Hunter described as a "renegade Brahman" from India moved in next door to Rod Marining. The would-be guru walked barefoot in the snow and carried a hand mirror "into which he stared with admiring rapture like a God gazing upon himself." Clad only in a filthy blanket, he would play a flute at any given moment while rolling his eyes beatifically. He was barely competent enough to tie his shoe laces, "yet he wagged his dark bony finger in our faces and raged about how he had come to North America to guide us." The "renegade Brahman" showed up at the oddest times and places, "glaring malevolently one moment and giggling at some great cosmic secret the next, chanting to himself." One of Hunter's interests was phenomenological psychology, which posited, among other controversial views, that schizophrenics were perhaps the only sane people among us. The leading exponent of this view, R. D. Laing, visited Vancouver at this time, and Hunter vividly recalled a passage from his lecture: "The light that illuminates the madman is an unearthly light. He may be irradiated by light from other worlds. It may burn him out." The fact that Hunter, Marining, Weyler, and Bruce, among others, were open to such views meant that people who others may have dismissed as mere lunatics were allowed—perhaps even encouraged—to hang around on the fringes of Greenpeace.[20]

Another similar character was Henry Payne, a self-proclaimed shaman who lived in a nearby forest. Payne donated five acres of land that Greenpeace could raffle off to help fund the antiwhaling campaign. In return, he insisted on being "allowed to stand up at meetings and roar out his poems with thunderous passion." Like the Brahman, he demanded disciples and claimed to be an authentic voice of God, warning that he had within him the power to control the winds and the tides. He would frequently utter enigmatic statements such as: "I am the I of eternal I-am." The presence of such eccentrics, according to Hunter, played on people's minds and kept them open to all sorts of views, beliefs, rituals, and magic: "You couldn't climb into bed at night with any assurance that you wouldn't be awakened by the sound of chants or mantras, magical rituals or tinkling bells. What was worse was the suffocating feeling that, like it or not, these two basket cases might be genuine gurus, and they just might be casting spells, spinning webs, fiddling with events in sly, transcendent ways, despite the reluctance of any of us to worship them."[21]

Typical of the bizarre series of coincidences at this time was the appearance of Melville Gregory, a hippie musician and composer from North Vancouver. Spong had suggested that Greenpeace take some musicians with them on the voyage to try to attract the whales to their ship. However, they would need a special kind of musician: one with an affinity for interspecies communication. Hunter and Hamish Bruce were discussing this idea at their favorite haunt, the beer parlor at the Cecil Hotel in downtown Vancouver. Just as they were wondering where on earth they might find a musician with animal affinity, "an impish, bushy-bearded man in his early forties" threaded his way through the tables toward them. He introduced himself as Melville Gregory. The name itself was redolent with whaling culture: Herman Melville, of course was the author of *Moby Dick,* while Gregory Peck had played the role of Captain Ahab in the Hollywood version. Mel Gregory sat down next to Bruce and Hunter, saying that he had "had a flash you guys wanted to talk to me." Gregory, it turned out, was a professional musician who claimed to be able to communicate with animals. Hunter and Bruce were astonished by Gregory's arrival at the very moment they had been thinking of just such a person. Gregory himself, however, was unperturbed: "That's just magic, man. Happens all the time."[22]

A similar incident occurred in connection with Hunter's plan to capture the expedition on film. He realized that the voyage held the potential to yield an exciting, feature-length documentary, but he balked at the cost of hiring a film crew. Just as he was mulling the problem over, a cinematography graduate from Vancouver, Michael Chechik, called him and asked if Hunter would mind if he and his crew accompanied them on the journey to make a feature-length documentary. Such coincidences continued to occur on an almost daily basis. Spong wanted to bring along a high-tech Moog synthesizer that could produce whale-like sounds. Once again, unsolicited, a Moog musician from San Francisco wrote to Greenpeace asking if he could join the crew. Similar events occurred with the sound technician, Zodiac expert, and the appearance of Japanese and Russian speakers who could act as interpreters. Hunter's interpretation of these events demonstrates his inner struggle with New Age mysticism and scientific rationalism:

> A computer could no doubt have made sense out of all the good karma and coincidences, showing how they were simply mathematical probabilities whose time had come. The objective truth was undoubtedly that so many human billiard balls had been set in motion that they were beginning to intersect at angles that would not have been possible, or at least would have been highly improbable, with fewer balls on the table. It was only our relative position at the center of the network that made the feedback look so magical and effortless. Be that as it may, the subjective impression was that some force greater than mere human will was at work, and this impression gave the campaign the flavor of a crusade, or *jihad*, a sacred undertaking.[23]

The paradox at the center of this freakish, hippie-inspired vortex of events was that, in spite of all the apparent chaos, Greenpeace was becoming more organized than it had been at any other stage in its five-year history. Despite the surreal influence of the "renegade Brahman" and Henry Payne, the prominence of unreconstructed hippies such as Marining and Bruce, and the leadership of the visionary but perennially disorganized Bob Hunter, Greenpeace began to take on the trappings of a traditionally run nonprofit organization. The first, and perhaps most important step, was setting up an office in a small building on Fourth Avenue in the heart of Kitsilano. Such an apparently trivial event was a vital stage in Greenpeace's evolution. At last, there was an actual address where people could reach the organization, rather than having to contact individual members at their homes. Furthermore, there was a comforting sense of bourgeois legitimacy in the act of leaving home and going to an office.

Setting up an office also led to the adoption of the paraphernalia that one normally associates with offices: bookkeeping procedures, mailing lists, organized filing systems, and letterhead stationery. The buzz created by groups of people working together in a shared space contributed to a general sense of comradeship and to a more inspired and efficient work ethic. Malingerers could be politely escorted from the premises. The move from the meeting hall to the office also impacted the interpersonal relations within the organization. As Hunter noted: "open up physical space where everyone can move around, interact, mingle, press up against one another; fill it with people whose average age is in the mid-twenties; add an extraordinarily high percentage of beautiful, intelligent, liberated, radiant, positive females, most of them single; introduce them to young men and older men, men who are alive with the excitement of embarking upon a dangerous mission—and the juices cannot help but begin to flow." Despite the high level of sexual tension, Hunter insisted, promiscuity was not the order of the day: "There were too many spiritual people," as well as "an overriding prohibition on any overt displays of male chauvinism." Furthermore, "the rhetoric and to a large degree the actual spirit of women's liberation had penetrated to our core." Not surprisingly, perhaps, women occupied many of the chief organizing and administrative roles in the new Greenpeace office with Hunter's new partner, Bobbi MacDonald, particularly influential.[24]

The task of finding a boat for the campaign was left to Hunter, who spent several weeks scouring the docks and spreading the message that Greenpeace was looking for a sturdy vessel that could keep up with a whaling fleet. It was not long before word of this reached John Cormack, whose rickety old halibut seiner had been the original *Greenpeace* in 1971. Cormack had experienced a good fishing season the previous year and had invested heavily in repairing and upgrading the *Phyllis Cormack*. But the outlook for the 1975 season was not too encouraging, prompting his colleagues to suggest that he should "try to get one of those Greenpeace deals again." Though not exactly thrilled with the thought of spending another two months in the tight confines of the *Phyllis Cormack*, Hunter decided that it was better to proceed with a known quantity than to take a chance on finding something better. The original *Greenpeace*, therefore, became the *Greenpeace V*. Then, much to Hunter's surprise, a retired law professor by the name of Jacques Longini turned up to offer his yacht and himself at Greenpeace's disposal. The yacht was called the *Vega*, and he had purchased it from David McTaggart. Thus the *Greenpeace III* was reincarnated as the *Greenpeace VI*.[25]

By early 1975, Greenpeace had two boats, an office, and a rapidly growing base of support in Vancouver from which they could choose a crew for the campaign. The raffle for Payne's five acres of land was producing a reasonable level of funds, while money and support also started to come from a variety of sources. Timber giant MacMillan Bloedel, for example, donated two tons of paper for brochures, which the Vancouver School Board allowed Greenpeace to distribute throughout the city's schools. Donations also came from groups as disparate as the Animal Liberation League, the Lower Mainland Automotive Dealers' Association, the Federation of Labor, the British Columbia Association of Indian Chiefs, and the British Columbia Teachers' Association. In addition, they received the blessing of the province's socialist premier and were able to arrange the cooperation of important official bodies, such as the Port of Vancouver and the Vancouver Police Department. Hunter also took the liberty of "borrowing" the symbolic killer whale crest of the Kwakiutl Indians. The crest formed the base upon which Greenpeace superimposed their ecology and peace emblems, and this new logo was painted on the *Phyllis Cormack*, as well as printed on T-shirts, buttons, and the Greenpeace stationery letterhead. Hunter reasoned that this was permissible, feeling that Greenpeace had a special relationship with the native peoples of North America. After all, he and his fellow Greenpeacers had been made honorary Kwakiutl tribesman in 1971. Furthermore, Spong was on good terms with the Kwakiutl group near Hanson Island, the sight of his whale observation facility, and Watson and Walrus were full-fledged members of the Oglala Sioux.[26]

These developments were encouraging, but there was one vital factor still missing. It was entirely implausible to expect that two small boats bobbing around on the sea would simply bump into a whaling fleet. Where, in the vastness of the Pacific Ocean, might Greenpeace actually encounter Soviet or Japanese whalers? Paul Spong was given the mammoth task of obtaining some sort of useful data about the location of the fleets. But where would he find such information? Certainly the Japanese and Soviet whalers were not going to disclose their intended whereabouts nor would their governments. Spong concluded that his best bet would be to find a way to gain access to the heavily guarded files of the Bureau of International Whaling Statistics in Norway. He decided to combine this task with the job of spreading the Greenpeace anti-whaling message throughout Canada and Europe. So, in early 1975, he, Linda,

Paul Spong (*left*) enlists the support of BC Premier Dave Barrett, 1974. Reproduced by permission of Rex Weyler/Greenpeace.

The Kwakiutl-inspired Greenpeace symbol adorns the *Phyllis Cormack*'s sail, Sydney Harbour, British Columbia, 1976. Crew (*left to right*): Michael Manolson, Bob Hunter, Rex Weyler, and Kazumi Tanaka. Reproduced by permission of Rex Weyler/Greenpeace.

and Yashi revived the whale show of the previous year and took it on the road, visiting over a dozen Canadian cities from Edmonton to St. John's. Apart from educating Canadians about the plight of the world's great whales, the Spongs were also acting as apostles of Greenpeace. At each of the towns they visited, they left behind a small cadre of whale lovers, some of whom formed the nucleus of what would later become Greenpeace branch offices.[27]

In Iceland, Spong met the first of a series of contacts that eventually led him to Einar Vangstein, the director of the Bureau of International Whaling Statistics in the southern Norwegian town of Sandefjord. The pragmatic Norwegians, having concluded that whaling was unlikely to remain a lucrative economic activity for much longer, had largely exchanged their whaling fleets for crude oil tankers. Nevertheless, there were many people in the country who had devoted their lives to whaling, which continued to play an important role in Norwegian culture. Vangstein was among those who hoped that Norway would one day resurrect its proud whaling tradition. Spong managed to convince him that he had a purely scientific interest in observing sperm whales in their natural habitat. The director was hesitant at first, but a combination of Spong's polite deference, Linda's charm, and Yashi's playfulness soon led the avuncular Vangstein to change his mind. While Vangstein chatted with Linda and played with Yashi, Spong was allowed to enter the office containing the most detailed whaling data in the world. He scribbled down thirteen pages of figures, including the geographical coordinates, dates, and kills of the major Soviet and Japanese fleets during the 1973 and 1974 seasons. By this time, Vangstein was so comfortable with the Spongs that he had his secretary type up the data. The information did not guarantee that Greenpeace would be able to locate the fleets, but it improved their odds considerably. After translating the rows of figures, Spong figured out that the Soviet fleet had come within fifty miles of the California coast in its pursuit of sperm whales. It was this fact, rather than any broad geopolitical considerations on Greenpeace's part, that ensured that the Soviets would be the most likely target of their campaign.[28]

Spong's research-cum-espionage indicated that the best time to begin the antiwhaling voyage would be sometime in early June, in anticipation of intercepting the Soviet fleet off the California coast in late June or July. While Spong was away, however, Hunter and the others had virtually at random chosen April 27 as the launch date. They had already organized a huge send-off for the occasion, replete with music and speeches, and expected thousands of people to see them off at the dock. Not for the last time, public relations got the better of logistics, and Hunter decided to stick to the April 27 launch date. In the meantime, Hunter also decided to quit his job as a columnist at the *Vancouver Sun* in order, he told readers in his farewell column, "to commit myself fully to what is loosely called the environmental movement." A new age was dawning in which "a vivid blending of spirituality and technology" was taking place. While Hunter enjoyed studying, documenting, and exploring this phenomenon, he felt it was time for him "to get into it and work with it and be part of it, rather than trying to remain on the outside, looking in."[29]

Hunter, however, was not being entirely honest with his readers. As he later explained, the task he set himself was to be a "flack" for the whales. To achieve this, he was prepared to "use every trick I had ever learned from Ben Metcalfe and be ready to invent quite a few on top of that." Subjective pieces written by columnists, Hunter believed, were rarely picked up by the international wire services, while the "objective" stories from reporters were treated as legitimate news. In order to attain this level of "objectivity," while still acting essentially as "news manager" for Greenpeace, Hunter was quite prepared to "invent quotes, place them in the mouths of various agreeable crew members, then 'report' to the outside world what they had said." His overall aim was to "censor any unflattering realities, control the shaping of our public image, and when things got slack . . . arrange for events to be staged that could then be reported as news." Instead of reporting the news, Hunter proclaimed proudly, he was "in fact in the position of inventing the news—*then* reporting it."[30]

By April 27, the planned launch date for the voyage, the frantic preparations had taken their toll on Hunter, who was suffering from fatigue, a poor diet, and overconsumption of alcohol and other stimulants. To compound his misery, Watson had spilled a jar of formaldehyde in the *Phyllis Cormack*'s hold on the morning of the voyage, and Hunter spent over an hour helping him clean up the noxious liquid before becoming violently ill. He spent much of the day doubled over in pain and dry retching, while a motley crew of well-wishers gathered at a Kitsilano beach to see the boats off. It was certainly not the most auspicious beginning, but Hunter's problems turned out to be relatively minor compared to the psychological condition of Hamish Bruce. Bruce had endured a stressful period during the campaign's preparation, as well as coping with problems in his personal life. This, combined with the ingestion of some highly potent fungal matter, left the hippie lawyer in a trance-like state. For five straight hours after the *Phyllis Cormack*'s departure, he stood at the bow, his salt-caked hair blowing in the cold sea wind while icicles formed on his moustache and beard. Neither the ship's doctor, Myron MacDonald—who was, incidentally, the former husband of Hunter's new partner, Bobbi—nor Hunter were able to coax him away from the bow, and he continued to simply stare at the sea like the statue of the ancient mariner.[31]

Apart from Hunter, Bruce, Cormack, and MacDonald, the core of the crew included Watson, Weyler, Walrus, Moore, and Mel Gregory. Marining, along with Bobbi MacDonald, would coordinate the campaign from Greenpeace's Vancouver office, while Spong flew to London to try to attend the IWC meeting as a Greenpeace representative. For the first month of the campaign, the makeup of the crew was constantly changing as different people embarked and disembarked at various ports, sometimes according to plan, sometimes not. They included the well-known Indian actor and singer Don Francks; a San Francisco oceanographer named Gary Zimmerman; several musicians; Nicholas Desplats of Les Amis de la Terre in Paris; a Kansas City doctor named Jim Cotter; a cameraman named Fred Easton; and Ramon Falkowski, who had sailed to Mururoa aboard the *Fri*. The more permanent and influential new crew members included George Korotva, a giant blond Czech who had been shipped to a Siberian labor camp for taking part in a student uprising in Prague. After escaping from the camp, he had managed to find his way to Western Europe, before immigrating to Canada. He had considerable sailing experience and was also supposed to be Greenpeace's Russian translator, though some were later skeptical of his putative Russian language skills. The ship's engineer and all-round electronic wizard was Al Hewitt, a tall, thin thirty-seven-year-old who could operate, repair, or, if necessary, construct just about any electronic or mechanical gadget that Greenpeace needed. Also aboard for the duration of the voyage was Carlie Trueman, a twenty-four-year-old scuba diver and Zodiac operator from Victoria, British Columbia, whose skills persuaded Cormack to abandon his "no women on board" policy.[32]

Because they had at least a month to kill before they could start searching for the Soviet whaling fleet off the California coastline, the two boats sailed along the west coast of Vancouver Island looking for whales. With their musicians, hydrophones, and Moog synthesizer, Hunter and the crew hoped to conduct some interspecies communication experiments. Rex Weyler reflected the positive and, at times, self-righteous attitude of the crew in his journal: "We sail against greed—and greed will lose, because in greed is self-destruction . . . we are going into a war we cannot lose, because life is always the winner, and we are fighting for life."[33]

After two days at sea, Hamish Bruce's condition had not improved. It was becoming increasingly clear that Greenpeace's erstwhile leader was in dire need of psychiatric help. Although Hunter had managed to coax him away from the

Canadian actor and musician Don Francks plays the flute to a killer whale near Bella Bella, British Columbia, with *the Phyllis Cormack* in the distance, April 1975. Reproduced by permission of Rex Weyler/Greenpeace.

bow, thus saving him from almost certain hypothermia, Bruce refused to eat or talk. He appeared on deck the next day with war paint on his cheeks and began to chant in a language nobody else on board had ever heard. With his piercing eyes he would stare at other crew members with a mixture of pity and contempt. Dr. MacDonald, something of a New Age physician, felt that Bruce was either having a schizophrenic episode or was some kind of "awakening being" or Buddha. For Hunter, with his interest in phenomenological psychology and Eastern mysticism, both possibilities were intriguing.

By the fifth night, however, Bruce's behavior had become too psychotic for even the most tolerant mystic to endure. While the rest of the crew slept during a stopover in Patrick Moore's hometown of Winter Harbor, Bruce ran amok in the woods, howling and raving like a man possessed, terrifying the local villagers. Journalists, including one from the *New York Times*, were turning up to write stories about the campaign, and Bruce was hardly likely to inspire confidence in a group that already had more than its fair share of flaky overtones. Lyle Thurston, the doctor on the original *Greenpeace* and a friend of Bruce's, was summoned from Vancouver, but his arrival only seemed to add to Bruce's explosive and increasingly violent rage. MacDonald and Thurston crushed some tranquilizing tablets and mixed them up with mayonnaise, which they then spread on a cheese and pickle sandwich. Bruce, in the meantime, had barricaded himself in the ship's hold, where Hunter and Korotva were trying, unsuccessfully, to subdue him. The doctors passed the sandwich down with instructions to feed it to Bruce, who had not eaten all day. Bruce waved it away dismissively, but Hunter and Korotva, who also had not eaten, devoured it ravenously, before collapsing to the floor in a drug-addled stupor. Fortunately, before the situation reached an even more cartoonish level of absurdity, MacDonald managed to calm Bruce down enough so that he could jab a syringe into his posterior. Then, using a sleeping bag with a giant Stars and Stripes design as a straightjacket, Bruce was tied up, put on a plane, and flown to a psychiatric hospital in North Vancouver. As if that were not enough, the next day, Paul Watson came down with acute appendicitis and had to be flown by helicopter to the town of Campbell River for an appendectomy. Suddenly, the fact that the expedition had set out a month earlier than Spong had deemed necessary began to seem like a blessing in disguise.[34]

As the voyage progressed and the boats dropped into small towns along the BC coast, it became increasingly obvious that the current crop of activists, like the crew that had sailed to Amchitka in 1971, could be divided into "mechanics" and "mystics." On the one hand, there were the hardcore *I Ching* devotees, who were influenced by Indian mythology and Eastern religion, took drugs freely, and tended to see magic at work in every event that was the slightest bit unusual or coincidental. The main mystics included Hunter, Weyler, Walrus, and Gregory. The mechanics included, to one extent or another, Cormack, Korotva,

Moore, Hewitt and Carlie Trueman. To be sure, the categories were not mutually exclusive. Hunter, for example, felt that the world was entering an age where Eastern religion and Western science were merging together. Like the New Age physicist, Fritjof Capra, he believed that "without losing what we have learn[ed] from quantum physics, we may comfortably adapt ourselves to the idea of the Tao."[35] Nevertheless, the phrase "the mystics versus the mechanics" served well as a general description of the clashes that occurred regularly among the crew, particularly when they needed to make important campaign decisions. At such times, the mystics would always bring out the *I Ching* and insist that its advice should at least be considered, if not followed.[36]

For the mystics on board, there were countless inexplicable events and mysterious coincidences to validate their belief system. Rainbows, one of the chief symbols of the self-described Rainbow Warriors, appeared everywhere. Given that they were sailing along a coastline that was commonly referred to as the "rain coast," this should not have been too surprising. Nevertheless, for the mystics, the rainbows were signs of magic at work. They noticed them at night during a full moon, in the spray from waves crashing into the boats, and, most symbolically, in the mist sprayed out of the whales' blowholes. On remote Moresby Island off the coast of Vancouver Island, they were greeted by, of all things, two Buddhist monks who gave them a flag from a Tibetan monastery. This had particular resonance for Hunter, who wore around his neck a strip of red cloth he had received from His Holiness Gyalwa Karmapa XVI, the Grand Lama of the Oral Tradition of Tibetan Buddhism. Soon thereafter, Hunter's amazement was compounded by the sudden appearance of his beloved and long lost Uncle Ernie, who had disappeared from Hunter's native Winnipeg when Hunter was a child and whom he had long given up for dead. While Hunter was weeping and embracing his uncle, the two monks excitedly directed his attention to the sudden appearance of a "stunningly vivid, complete, double-ringed rainbow." It was enough to shake the skepticism of even the most diehard "mechanic."[37]

In mid-May, Hunter moved from the *Phyllis Cormack* to the *Vega*, which he sailed to the edge of the continental shelf in order to monitor the radio for any sign of the whalers. Patrick Moore was left in charge on the *Cormack*, which sailed down to an area known for its gray whale population. Moore, however, was not as well-liked as Hunter and did not have his leadership qualities. As Weyler wrote in his journal: "Bob is the leader by instinct, simply capable of making and carrying out decisions. Pat just wasn't able to do that as easily; it was a struggle for him." Hunter, in Weyler's opinion, sensed "the grander scheme, the totality of the event," while simultaneously seeing "absurdities and incongruencies [*sic*] that others miss."[38] Moore's rather didactic and patronizing style alienated many on board and foreshadowed the problems he would have in future years when he would succeed Hunter as Greenpeace's leader. His relationship with Watson was particularly difficult. Watson, having quickly recovered from his appendectomy, had turned up to rendezvous with the *Phyllis Cormack* in the small port of Tofino. Moore, however, refused to let him rejoin the crew. But Watson was not to be denied, swimming out after the departing ship and forcing Captain Cormack to rescue him and bring him back on board.[39]

The tension on board the *Phyllis Cormack* was eased once they reached the gray whale pod, which, for many on board, was their first encounter with live whales. The crew took turns in the Zodiacs, zigzagging among the whales while attempting to "communicate" with them through various musical instruments. The mystics in particular felt that the whales were responding positively to their overtures. Weyler wrote that they "loved the music—you can call it curiosity, intelligence, appreciation, communication, anything you want. They came! They came and they stayed, and they *spoke* to us." At one point, a whale came up underneath Weyler and his fluke "gently rapped the underside of the Zodiac." "It was no mistake," wrote Weyler: the whales were responding to Greenpeace's good vibes and attempting, in their own way, to communicate with them.[40] Will Jackson, one of the musicians on board, spent several hours meditating in a Zodiac, trying to reach a Zen-like state of mental purification in an effort to communicate with the whales. For the mystics, this approach was validated by the appearance of two whales who lingered near Jackson's boat. Meditating, it appeared, worked even better than music. Despite the different ways that various crewmembers chose to interpret the whales' behavior, the experience had the effect, in Hunter's words, "of 'converting' everyone into whale freaks."[41] For Weyler, the lesson was clear: "We realize[d] that if we were to get closer to understanding other creatures, and developing an appreciation for the relationship among all creatures, we must learn to trust and respect them. We must allow them to take an active role in the encounter. They must *want* to be with us. That is a certainty."[42]

For virtually everyone involved with the campaign, the encounter with the gray whales cemented the feeling that they were on a mission of cosmic importance. Nevertheless, the fact that they had set out so early meant that Hunter was faced with the difficult job of maintaining media interest in the voyage. By late May, as the *Phyllis Cormack* sailed toward the California coastline, even the *Vancouver Sun* was relegating Hunter's dispatches to the back pages. In a desperate bid to grab the headlines, Hunter concocted a story about Greenpeace's plans to observe the battle between a sperm whale and a giant squid. To give the story an air of scientific authenticity, Hunter placed his words in the mouth of "Dr. Patrick Moore, an ecologist." Moore explained that at the time of the full moon, millions of tiny organisms would float up from the sea floor to absorb the additional lunar light. These were followed to the surface by giant squid which normally lurked in the depths of the ocean. The squid was the major food source of the sperm whale, and Moore stated that Greenpeace hoped to film an encounter between two of the ocean's giants: "Sperm whales grow up to 80 feet long. Giant squid can reach 50 feet. The struggle between these two immense creatures is probably the most awesome encounter to take place in the natural world."[43] Hunter admitted that a perceptive editor should have seen in the story "the thrashings of a desperate publicist." Nevertheless, the fanciful tale proved to be the most widely reported story of the campaign to that point, thereby confirming Hunter's faith in his ability to manipulate the media.[44]

Up until the *Phyllis Cormack* had reached a point some fifty miles from the northern California coastline, Hunter had not told anyone else of the existence of the coordinates that Spong had found at Sandefjord. Instead, he kept them carefully hidden at the bottom of his duffel bag to prevent anybody from accidentally leaking the information to the media. A contact at the Soviet embassy in Ottawa had warned Korotva that an agency of the Canadian government had been feeding the Soviets regular reports of Greenpeace's location. While Hunter could not verify this, there seemed little reason to doubt the story's veracity. After all, Greenpeace had already proven to be a major headache for the Trudeau government. McTaggart, who was in Paris at that very moment, was continuing to demand that his government aid him in his court case against the French Navy. It was reasonable to assume that Trudeau would be eager to avoid a similarly awkward incident with the Soviet Union. Hunter, therefore, decided to give the impression that Greenpeace had no idea where the whalers might be: that they were merely sailing blindly through the Pacific in the hope of stumbling across them. To solidify this impression, Hunter stopped at the Canadian military base in the Queen Charlotte Islands, where he had arranged for a private interview with the major in charge of the post. He pretended to con the officer into giving them the coordinates of the Soviet fleet, faking desperation and acting dejected when the major refused to reveal any details. He also spoke of Greenpeace's plans to search for the whaling fleet off the coast of British Columbia, an area far to the north of the Soviet hunting grounds. Convinced that his counterespionage effort had been successful, Hunter felt he had done his utmost to throw his own government off Greenpeace's trail. Soon thereafter, he shared Spong's coordinates with the rest of the crew, and Cormack set course for an area of the Mendocino Ridge some one hundred miles off the coast of Eureka, California.[45]

For several days in early to mid-June, the *Phyllis Cormack* plied the waters west of California, searching for the elusive Soviet whaling fleet. Supplies were running low and a fuel spill had contaminated much of the ship's food, giving everything a distinctly diesel flavor. The ship's radio continued to pick up snatches of Russian conversations, but they were always too brief for the radio directional finder (RDF) to establish their exact coordinates. Fed up with the fruitless search and with the constant argument about whether the RDF or the *I Ching* was a more useful tool for tracking down the whalers, Cormack, the lead mechanic, decided to confront Hunter, the chief mystic. Do you really believe, he asked Hunter, fixing him with his steely gaze, "in this here book from abracadabra two-humps land?" Hunter insisted that he did. With the rest of the crew watching intently, Cormack picked up the *I Ching* coins and threw them on the table in the prescribed fashion. The result was hexagram 19, "Approach," a positive, if somewhat obscure prophecy that translated as "becoming great" and that alluded to "the approach of something highly placed in relation to what is lower." Cormack remained unconvinced, while even the mystics were loath to read too much into the vague prediction. Within a few hours, however, Russian voices again appeared on the radio, and this time the signal was strong enough for the RDF to gauge the fleet's general direction. Clearly, it was a victory for the mystics.[46]

While the *Phyllis Cormack* continued its search for the fleet, Spong was in London trying to gain entry into the IWC meeting. British Columbia's socialist premier, Dave Barrett, had promised to help Greenpeace, and he put the BC representative in London at Spong's disposal. Spong's announcement that Greenpeace would be attempting to stop the whalers through "nonviolent intervention" had caused considerable excitement among the many protestors who had gathered at the London meeting. It did little, however, to endear him to the IWC delegates, who refused to give him a pass to attend the sessions. Spong's frustration was compounded by the Canadian delegate's decision to vote with the Soviet Union and Japan for an increase in the sperm whale quota. Several days into the meeting, however, Spong's luck changed. After eating lunch with a group of delegates at a nearby pub, Spong found a red IWC entry badge under a table. Upon entering the meeting room, the surprised and rather concerned looking Canadian delegate asked Spong who had let him in. "God," he replied with an enigmatic smile. Over the next few days, Spong attended all the IWC meetings and lobbied frantically on Greenpeace's behalf. In the evenings he gave lectures and speeches on the plight of the whales, showing films and slides of the orcas he was observing at Hanson Island. However, as the meeting entered its last days, it began to look increasingly likely that the intended high point of Greenpeace's campaign—the confrontation with the whalers—would not materialize. By June 26, the second to last day of the meeting, the *Phyllis Cormack* had still not found the Soviet fleet. It seemed as though Spong would have to leave his supporters disappointed and his opponents smirking with delight.[47]

By midnight of June 26, the *Phyllis Cormack* was some forty miles southwest of Cape Mendocino. A heavy sea and a particularly diesely dinner had left several of the crew moaning in their bunks. The five chain smokers on board, Hunter among them, were almost out of cigarettes, while only seven cans of beer remained in the hold. Meanwhile, Korotva continued to monitor the radio for a sign of the whaling fleet. At one point, he heard the word *Vostok*, which was the name of one of the major Soviet factory ships, and Hewitt set about tracking the voices on the RDF. There were also Soviet fishing fleets in the area, making it difficult for Korotva and Hewitt to distinguish the fishermen from the whalers. At ten o'clock the next morning, a stark naked Mel Gregory was at the wheel with firm orders from Cormack to maintain a steady course forward. A few weeks before, Gregory had incurred Cormack's wrath when he had steered toward the full moon, rather than according to the compass, thereby taking the ship some thirty degrees off course. The experience, however, had done little to instill Gregory with a sense of nautical discipline. Soon after Cormack left the wheel to take a nap, Gregory spotted a rainbow and turned the ship toward it. Some thirty minutes later, the towering figure of the *Dalniy Vostok* appeared on the horizon. Certainly, without the radio and the RDF, Greenpeace would have had next to no chance of finding the whalers. The mystics, however, took heart from the fact that, ultimately, it was a rainbow that led them to their goal.[48]

The sight of the whaling fleet instilled the crew with a sense of orderly discipline as they went about their tasks. Several donned wetsuits as they prepared to enter the Zodiacs, while others checked their cameras one last time. As they approached the fleet, Cormack noticed a red triangular flag protruding from a gray-blue lump on the ocean surface. When they reached it, the crew was shocked to find that it marked the body of a dying baby sperm whale. After their interactions with the gray whales several weeks earlier, stumbling across the harpooned whale was akin to finding a pet dog that had been shot with a crossbow at a suburban park. Hunter went even further than this, implying that the act was comparable to murder:

> The wave of emotion that hit us then—revulsion, rage—was so concrete that it staggered us all in our tracks for several seconds, so that there was a sharply focused instant when nothing seemed to move, neither the boat, the people on the decks, nor the dead whale-child in the water. From Walrus Oakenbough came a short howl or wail such as an Indian might have made coming back to his camp to find his children massacred and his world in ruin.[49]

Paul Watson took his Zodiac out to the whale and climbed onto its body, caressing it lovingly and staring into its dying eye. While Weyler and others took photos in order to measure the size of the whale against Watson's body, Watson leaned over and gently pulled the creature's eyelid closed. To some extent, the incident helps explain the rage that Watson has directed toward whalers ever since. To many of those on board the *Phyllis Cormack*, the whales had ceased to be mere animals in need of protection; they had become something closer to humans who were the victims of genocide.

Paul Watson sitting on a whale calf recently harpooned by Soviet whalers, June 1975. Reproduced by permission of Rex Weyler/Greenpeace.

The crew's first close-up view of a factory ship also made a vivid impression. Its massive rear slipway resembled a giant mouth into which the whales, "looking no bigger than sardines," were fed. Hunter's description aptly captures the impact the sight had on the crew:

> From an opening the size of a sewer outlet halfway along the length of the massive hull, blood flowed as casually as oil from the bilges of ordinary boats, fountains of thick red blood that poured out and kept pouring out, enough blood every minute to fill a bathtub. Who had ever seen so much blood flowing before? The smell, as we came downwind, left half the *Phyllis Cormack*'s crew retching over the sides. The peculiar obscenity of the *Vostok* came into focus the moment we realized that here was a beast that fed itself through its anus, and it was into this inglorious hole that the last of the world's whales were vanishing—before our eyes.[50]

Hunter's enraged, melodramatic reaction is instructive. Here was a man who had worked in a slaughterhouse and had written in visceral, gory detail about the experience. Yet the sight of, indeed, his participation in, the slaughter of cattle and pigs incited nothing like the moral opprobrium that he felt toward the factory ship.[51] Hunter's differing reactions could be interpreted in a biocentric light: cattle were not in danger of extinction, therefore it was permissible to participate in their slaughter and to eat their meat (Hunter was no vegetarian). But Greenpeace's rhetoric, both at the time and subsequently, illustrates that whales were on a different moral plane from cattle. According to Patrick Moore, there were very few among them who did not share Spong and Hunter's view that whales, because of their putative intelligence, deserved special treatment:

> The whale was the largest animal that has ever lived on earth, with a brain bigger than ours. It's been here sixty million years living in peace. The line, "if mankind was born a step below the angels, the whales are somewhere in between," more or less summarized our thinking on that. I still think that way. I don't see any reason not to think that way. For one thing, there is no logical reason to think that we are the highest form of life in anything other than a very specialized way. We are the highest specialized life in that we know how to build automobiles, but not the highest form of life than can swim in the sea and be peaceful for sixty million years and have a brain larger than ours. I think they are pretty amazing. I always saw them as a perfectly legitimate sacred cow. Sustainable development doesn't have to apply to whales.[52]

As the *Phyllis Cormack* came close to the *Dalniy Vostok*, Weyler and cameraman Fred Easton set out in their Zodiacs to film the harpoon boats that were transferring dead whales to the factory ship. Hundreds of perplexed workers lined the ship's railing, wondering what to make of this motley group of people zipping around in their tiny boats. On the *Cormack*, Mel Gregory and Will Jackson took out their guitars and megaphones and began singing a whale song they had written during the course of the trip. This led to the surreal sight of Russian whale flensers clapping and swaying to the chorus:

> We are the whales, living in the sea
> Come on now, why can't we live in harmony?
> We'll make love, above the ocean floor
> Waves of love come crashing on the shore.[53]

The next stage of Greenpeace's aural assault on the whalers was to play Roger Payne's recordings of the humpback whale songs at full blast. The haunting sounds echoed back and forth between the *Cormack* and the giant steel hull of the *Vostok*. Once those on board the factory ship realized what they were hearing, friendly waves were soon replaced by shaking fists. Will Jackson began to play his synthesizer in an effort to simulate whale songs, although, according to Hunter, the painfully loud feedback sounded more "like a crashing jet full of terror-crazed cats." If the whalers still had any doubts about Greenpeace's attitude

A Greenpeace Zodiac approaches the *Dalniy Vostok*, a Soviet whaling factory ship, and two harpoon boats, June 1975. Reproduced by permission of Rex Weyler/Greenpeace.

toward their livelihood, they were soon set straight by Korotva, who, in his broken Russian, explained to them that Greenpeace was there "to stop you from killing the whales."[54]

With the Soviets now well apprised of Greenpeace's mission, it was time to attempt the core element of their plan: to place live human beings between the whales and the harpoons. The *Phyllis Cormack* followed one of the harpoon boats, the *Vlastny*, as it headed off in search of its quarry. Some time later, Walrus, who was on the mast, spotted the *Vlastny* in pursuit of a pod of sperm whales. The crew was astonished to realize that the whales were "coming directly, unerringly, straight as an arrow, toward the *Phyllis Cormack*." For the mystics in particular, this was too much to be a mere coincidence. "Out of the 360 degrees on the compass that the whales had to choose from in their flight," Hunter observed, "they had somehow picked the one particular bearing that would bring them to our side, dragging the whalers in their wake, making our whole effort to protect them possible. That moment on deck, as the significance of what was happening hit us, was worth enduring a whole lifetime of meaningless struggle just to experience once."[55]

The crew then set about launching their three Zodiacs. One was piloted by Korotva and carried Weyler and his photography equipment. Moore drove the other one and was supposed to maneuver it into a position that would allow Fred Easton to film the action. Easton, however, had some bad news: he was down to his last ten or so minutes of film, which was of poor quality. Furthermore, the light was beginning to fade and the battery on his camera was playing up. Hunter and Watson, meanwhile, were in the "kamikaze" Zodiac, the one that would attempt to place itself between the whales and the harpoon. The men were flushed with a sense of nervous excitement as they headed toward the killer boat. At one point, Hunter, wearing a multicolored Peruvian cap and the red cloth given him by the Karmapa, turned to Watson, who had a kamikaze-style white cloth tied around his head, grasped his hand in a revolutionary handshake, and yelled above the roaring engine: "We're doing it, Paul! We're doing it!"[56]

Watson approached the bow of the *Vlastny*, where he and Hunter had their first close-up view of the harpoon. On noticing them, the harpoon operator let them know that he was not amused by their presence. He shook his fist and spat at them in anger as Watson powered the Zodiac through the choppy sea to a position some forty feet in front of the harpoon boat's bow. Then, some sixty feet in front of them, they saw the fleeing pod of sperm whales breaking the surface for air. For the next several minutes, they maintained their position between the harpoon and the whales. Hunter recalls that "rather than being terrified at this point, we were exultant. We did not believe they could fire. We did not believe they would take the political risk of killing two human beings in international waters." This feeling was reinforced by the actions of the harpoon gunner, who kept taking aim at the whales and then walking away from the harpoon in frustration, unable to get a clear shot. Hunter could only recall one clear thought going through his mind at that point: "Gotcha, you bastards! Gotcha!" Then, all of a sudden, the Zodiac's outboard engine shut down, bringing Hunter and Watson to a rapid halt with the *Vlastny* bearing down on them. As Watson pulled frantically on the ripcord, Hunter saw the captain of the *Vlastny* hanging over the bow, "laughing wildly and slashing his forefinger back and forth across his throat." The boat kept coming toward them, as though determined to crush them under the weight of its rusting hull. Hunter feebly flashed a peace sign at the captain and prepared to dive off the Zodiac. Watson, however, refused to give in and continued to flail away at the motor. The *Vlastny*'s bow was now some ten feet away with Hunter frozen to his spot in the Zodiac, unable to jump unless Watson jumped. At literally the last second, the bow wave from the harpoon boat swept them aside and the *Vlastny* continued past them at full speed, close enough for Hunter to reach out and touch it. Suddenly, Hunter was no longer so confident in his belief that the Soviets would not risk killing them.[57]

Korotva had observed the entire incident from his Zodiac and within a matter of seconds had pulled up beside Hunter, yelling at him to swap places with Weyler. With the fifty-horsepower engine at full-throttle, Korotva and Hunter soon caught up to the *Vlastny* and resumed a position between the harpoon and the fleeing whales. While Korotva concentrated on driving, Hunter watched as the *Vlastny*'s captain spoke with one of his sailors, who then strode purposefully along the catwalk and relayed the message to the gunner. Korotva later told Hunter that he could tell the moment the gunner was about to fire by watching Hunter's face, "which suddenly went ashen and sick-looking." Weyler, who was in another Zodiac with Moore, described what happened next: "We heard a deafening blast as the gunner fired his explosive harpoon over the heads of George and Bob. I was afraid that the rope would cut them in half as it came lashing down."[58] By the time Hunter and Korotva had instinctively ducked, the harpoon had already pierced the body of an exhausted sperm whale, exploding like a hand grenade deep inside the whale's body. The harpoon cable whipped the water less than five feet from the port side of the Zodiac. The gunner, showing great skill and taking a considerable risk, had waited for the Zodiac to float down into a trough between two waves, firing over it as the whale came up for air.[59]

Watson, in the meantime, had managed to restart the stalled Zodiac and was piloting it alongside the *Vlastny*'s bow as Hunter and Korotva were blocking the harpooner's shot. By this time, Easton, who was now in the Zodiac with Watson, realized that he only had thirty seconds of film left and that his battery was probably dead. Feeling he had nothing to lose, he lifted the camera to his shoulder and began to film. Much to his surprise, the battery came back to life. He panned between the whales and the *Vlastny*, but was too far away to see the harpooner preparing to fire. Through sheer luck, or, as the mystics preferred to think of it, some sort of divine intervention, Easton, with his last few feet of film and the dying breath of his battery, managed to capture the exact moment when the harpoon was fired. When played in slow motion, the harpoon can actually be seen flying just a few feet above Hunter's and Korotva's heads before smashing into the whale's body. At the time, Easton was not entirely sure if he had gotten the shot. But if he had, he knew that Greenpeace would have the "mind bomb" that Hunter had so desperately wanted.[60]

As soon as the whale was hit, Korotva turned the Zodiac around and headed back toward the *Phyllis Cormack*. A cetacean expert in Vancouver had told them that whalers would usually shoot a female first, thereby enraging the lead bull, who would often turn back to try to rescue her. This would leave the rest of the pod confused and leaderless, allowing the harpooner to pick them off at will. The expert had also told them that the bull might try to attack whatever he felt was threatening the pod. The Greenpeace eco-commandos in their small rubber boats would present easy targets for a vigilante sperm whale. The bull did indeed turn, but not on the Zodiacs. Instead, "as though knowing perfectly who his real enemy was," he turned his anger on the *Vlastny*, throwing his huge body out of the water and snapping his enormous jaw at the harpoon gunner on the bow. The gunner watched the charging whale calmly, as though expecting exactly this response. On the whale's second leap, he aimed the harpoon steadily and fired it as the whale's body hit the water. Everyone on board the *Phyllis Cormack* was astonished by the bull's behavior. Once more, the whales had apparently demonstrated that they could distinguish between those humans who wanted to slaughter them and those who were trying to save them. Although there was no scientific proof that this was the case, the behavior fit into a broader anecdotal history of whales being able to make distinctions between "friendly" and "unfriendly" humans.[61]

Though saddened by the death of the two sperm whales, Hunter and the crew were elated with their efforts. Weyler had snapped dozens of shots from his Zodiac, while another photographer, Ron Precious, had taken a similar number of photos from on board the *Phyllis Cormack*. If Easton's film also turned out to be viewable, and he was confident that it would, then Hunter had the story of his dreams and the images to accompany it. "No network would be able to resist such footage," thought Hunter, "just as no wire service would be able to ignore the story." Greenpeace, Hunter felt, had achieved its immediate goal:

> Soon, images would be going out into hundreds of millions of minds around the world, a completely new set of basic images about whaling. Instead of small boats and giant whales, giant boats and small whales; instead of courage killing whales, courage saving whales; David had become Goliath, Goliath was now David; if the mythology of Moby Dick and Captain Ahab had dominated human consciousness about Leviathan for over a century, a whole new age was in the making. Nothing less than a historic turning point seemed to have occurred. From the purely strategic point of view of changing human consciousness, there was little more that we could hope to achieve.[62]

Hunter somewhat overstates the achievement. People's consciousness regarding whales, especially in the North America, had been gradually changing for a number of years. For many Americans, the Sea World-inspired image of the friendly and intelligent killer whale had already displaced Moby Dick from their consciousness. Those who had read books such as the highly popular *Mind in the Waters* were even further along the path to "enlightenment." Nevertheless, the heroic and exciting nature of Greenpeace's actions, combined with the vivid images of whale slaughter, would provide a huge boost to the antiwhaling movement. The major significance of the action was not so much that it had massively changed people's consciousness regarding whales, but rather that it represented a courageous and highly publicized effort on the part of a group of human beings who were deliberately putting their lives at risk to protect nonhumans. This was the truly revolutionary aspect of Greenpeace's campaign.

By now, the *Phyllis Cormack* was running low on fuel and supplies, and Easton had run out of film. By their own count, Greenpeace had been responsible for saving eight whales that had escaped from the harpoon boats during the confusion caused by the Zodiacs. They tried to chase the fleet for another day but were unable to keep up. So Cormack turned the boat around and headed toward San Francisco. By this stage, all the crewmembers were fully convinced of both the righteousness of their cause and of the fact that whales were unique and extraordinary beings that needed to be saved at all costs. On this point, comparisons with nineteenth-century abolitionists would not be invidious.[63] According to Hunter, the mystics were "utterly convinced that the gods had intervened so often, not only before our eyes but before our cameras, that there could be no further debate: we *were* blessed."[64] The proof of this was all around them: in the rainbows, in the way Easton's camera had suddenly come to life at the vital moment, and in that the whales appeared to recognize Greenpeace as their saviors. And it did not stop there. The morning after the encounter with the *Vlastny*, Weyler awoke to the site of a large cloud, shaped unerringly like a sperm whale, hovering above the horizon.[65] The entire experience sent Hunter into a rapturous meditation on the meaning of it all:

> We had entered a zone of profound mystery more fascinating in itself than even the wonder and travail of the whales. We had brought the ancient influences of both Tibet and China out of the water off the coast of America to confront a fleet from the Asian mainland, led by a ship, the *Dalniy Vostok*, whose name meant "Far East." Here, surely, was a yin-yang situation of global proportions, a true meeting of East and West, except that the East was coming from the West, and the West from the East, which suggested that some tremendous reversal, a shifting of the axis of the spirit of the world, had taken place. The West now was the East and the East itself had been turned into its opposite. What had just occurred in the waters off the Mendocino Ridge was a microcosm reflecting all of this, signaling, in our minds at any rate, a change whose outlines dwarfed any of our previous lesser fantasies. It was as though the world's collective-unconscious mind was making one of its periodic efforts to render itself conscious, and we were the immediate instruments of the transformation.[66]

As the exhausted crew settled in for the overnight trip to San Francisco, it suddenly occurred to Hunter and several of the others that they had no plan what to do next. Their efforts had been entirely focused on tracking down and confronting the whaling fleet. Against considerable odds, that mission had been accomplished. What now? Ironically, as Hunter and the crew were soon to find out, the most exhausting, strenuous, and antagonistic part of the campaign still lay ahead. In London, meanwhile, the media had picked up Hunter's report about the confrontation with the whalers and reported the story widely just as the IWC meeting was winding up. As the Greenpeace representative in London, Spong was suddenly besieged by reporters wanting to know more about the campaign. All the major London newspapers carried Hunter's story, some on the front page, suddenly providing Spong with a newfound credibility among both the media and the antiwhaling protestors. The Soviet and Japanese delegates, irritated by the story, referred to Greenpeace as "pirates." The conservation forces had once again failed to muster the numbers for a moratorium, but stories of activists willing to risk their lives to stop whaling must have sent alarm bells through the various prowhaling delegations. The fact that several of those involved in the action had been Americans and that the events had taken place off the California coast was of particular concern. The United States, after all, was the strongest proponent of the moratorium. The Soviets could not help but have interpreted Greenpeace's actions within the broader context of the Cold War. The Japanese, for their part, feared that increased scrutiny of their whaling industry could trigger the United States to invoke the 1971 Pelly Amendment to the Fisherman's Protective Act of 1967, a law which threatened to ban the importation of seafood from nations whose practices threatened endangered species. It was obvious to all concerned that the antiwhaling forces had succeeded in raising the struggle to a new level.[67]

As the *Phyllis Cormack* pulled into San Francisco, a throng of reporters lined up along the Embarcadero to greet the crew. Immigration officials had to restrain the clamoring journalists, who leaned across the boat's gunwale with their cameras and microphones, impatient to talk to the heroic, if somewhat fanatical, environmentalists who had risked their lives to save the whales from the Soviet hunters. Various media envoys picked up Weyler and Easton in taxicabs and drove them to their film processing labs. Hunter talked to virtually every TV and radio station in the Bay Area, and the story, complete with Weyler's photos and Easton's film, was printed and broadcast throughout the United States and the world. According to one study, the whale campaign garnered more media coverage in the United States than all of Greenpeace's previous four years of antinuclear actions combined.[68] The *San Francisco Chronicle* reflected Greenpeace's hitherto status as a virtually unknown environmental group, introducing them as "an antiwhaling organization from British Columbia."[69] Walter Cronkite, the doyen of American news anchors, introduced them to a massive TV audience, featuring them on his "That's the Way It Is" evening broadcast. The *New York Times* published a lengthy and overwhelmingly positive feature on the organization. As well as describing the clash with the whalers, the *Times* cited Spong's experiments with killer whales as proof of whales' unique intelligence, thereby adding scientific credibility to Greenpeace's list of virtues. As a media event, the campaign was successful beyond Hunter's wildest dreams.[70]

In retrospect, it is clear that the media was primed for a whale campaign of the sort run by Greenpeace. For years the antics of captured whales and dolphins had been shown on television documentaries, while sentimental stories of beached whales regularly made the news. In January 1973, CBS News ran a story about the migration of the gray whales along the west coast of America. Images of tens of thousands of delighted whale watchers clustered on cliff tops and beaches were juxtaposed with footage of the slaughter of whales on the high seas. News programs also ran negative stories about the Japanese consumption of whale meat. Whales were depicted as intelligent and friendly creatures that were under threat from overhunting by the likes of the Japanese and the Soviets. The fact that whaling was of absolutely no significance to the United States' economy meant that there was little incentive for journalists to "balance" their stories by showing the whalers' point of view. From a media perspective, all that was missing were some compelling images that could further highlight the case against whaling. Greenpeace provided those images and the necessary element of spectacle and conflict. The mix was made even more irresistible by the fact that the "enemies" were the Soviet Union and Japan, nations that had inspired fear and suspicion in two generations of Americans.[71]

Greenpeace did little to distance itself from the Cold War overtones of much of the reporting and frequently resorted to military metaphors when describing its actions. Weyler, for example, began one of his articles with the observation: "There's no doubt that it was warfare. . . . We were armed with cameras; it was a guerilla encounter in an information war."[72] In the *New York Times*, *Vega* skipper, Jaques Longini, also resorted to the rhetoric of warfare in describing Greenpeace: "Protest means standing with a picket sign, being against something. That's not what we're doing. We are more like monkey wrenches, throwing ourselves into a huge machine. We're trying to stop whaling, not just complain about it. This is a nonviolent battle, not a protest."[73] Furthermore, Hunter frequently relied on military metaphors to describe his own theories on the workings of the media, referring to television, for example, as a "delivery system" for "mind bombs."[74] Although such language was problematic for an ostensibly pacifist organization with Quaker roots, neither Greenpeace nor the media were inclined to dwell on the paradox. The Cold War overtones, whether deliberate or not, suited both groups' purposes.

The crew spent a total of nine days in San Francisco, during which they were wined, dined, and generally celebrated by the local media and, to a lesser extent, local environmental organizations. After spending the better part of two months in the claustrophobic confines of the *Phyllis Cormack*, the glamour and polish of the San Francisco media world and the opulent houses of many of the city's environmentalists proved to be something of a culture shock. A somewhat jaded Hunter recalled one of Ben Metcalfe's favorite aphorisms: "Fear success." It was not long before the meaning of Metcalfe's words became clear. Less than twenty-four hours after their arrival, Hunter was contacted by the New York-based movie production company, Artists Entertainment Complex (AEC), the maker of such blockbuster films as *Earthquake* and *The Godfather, Part II*. The next day, an AEC agent, Amy Ephron, and a scriptwriter flew into San Francisco to meet with the crew to discuss a multimillion dollar movie about Greenpeace's exploits. Whatever tensions had existed on the *Phyllis Cormack* paled in comparison with the schism created by Ephron's visit. Her brusque New York style put off most of the Greenpeacers right from the start. She was prepared, she said, to offer them $25,000 for the movie rights to their story, with 10 percent down and a promise for the rest once the film was made. Although Hunter was no entrepreneur, he nonetheless knew that $25,000 was peanuts compared with the amount that Ephron's company stood to make from a successful film. Still, as far as Hunter was concerned, the objective was to raise "whale consciousness" around the world. The film, he felt, would contribute to this goal, as well as provide Greenpeace with a great deal of free publicity. Others, however, were deeply suspicious. Watson and Walrus were particularly upset and accused Hunter of being a "sell out." The contract required every crewmember to sign a release, giving the movie company the right to portray them as it saw fit. Watson and Walrus refused to sign, which infuriated Hunter and Moore, who accused them of grandstanding. The division over the movie contract, according to Hunter, "was never to fully heal itself and was to lead to divisions that would plague us for years." The acrimony was reflected in an exchange between Moore and Walrus on the final day in San Francisco. Moore tried one last time to persuade the rest of the crew to sign the movie release contract. In a fit of rage, Walrus climbed up the mast to escape Moore's overtures, screaming at him to go away and refusing to come down until the ship had left the dock.[75]

The sour taste created by the movie dispute was not improved by the welcome that Greenpeace received from local environmental groups. San Francisco had a reputation for being America's environmentalist capital, and there were over a dozen antiwhaling groups of one sort or another based in the Bay Area. Instead of cooperating, it seemed to Hunter, many of these groups operated alone and viewed other organizations as competitors for funds and publicity. Greenpeace's media coup seemed to make them a target for resentment as much as praise, particularly since they had been battling whalers virtually on San Francisco's doorstep. As Hunter discovered, "Rather than being welcomed as brothers and sisters by our fellow conservationists, we were greeted with token smiles and congratulatory remarks that barely masked the underlying mood of resentment and suspicion.... It was very much as though we had usurped someone's turf." Hunter and the others were also irked by the contrast between the hand-to-mouth existence that their dedication to Greenpeace entailed, and the opulent lifestyle that many of the San Francisco environmentalists seemed to lead. "It was impossible to avoid noticing," Hunter observed, "that the practice of environmentalism was the preserve of an elite." The reaction of Joan McIntyre, the head of Project Jonah and America's most well-known antiwhaling crusader reinforced this impression. "I want to see the whales saved," she told Hunter reproachfully, "but not *that* way." Greenpeace's confrontational and, from her perspective, macho style, did not sit well with McIntyre's genteel sensibilities. This new form of environmental direct action, it seemed, did not suit the tastes of many of the old-line conservationists, nor, for that matter, of sensitive New Age environmentalists such as McIntyre. It was clear to Hunter that if Greenpeace was going to make inroads into the American environmental scene, they would not be able to rely on San Francisco's environmental establishment for help.[76]

The rigors and excitement of the campaign, the San Francisco media circus, the schism created by the movie contract, and the overconsumption of alcohol and drugs finally took their toll on Hunter. On the journey back from San Francisco to Vancouver, he saw a "fog rainbow," a perfect rainbow that arched over the *Phyllis Cormack* from bow to stern like a multicolored half-nimbus. Convinced that this was another in a long line of miracles, Hunter stripped off his clothes and dived off the deck into the icy waters of the North Pacific, where he swum "in a state of bliss" toward the slowly drifting rainbow. By the time the others got him back on the boat, he was shuddering so hard that he could not speak, and it was clear that he had only narrowly avoided hypothermia. A couple of nights later he fell asleep with a lit cigarette in his mouth and awoke to find his bunk on fire. On a wooden ship at sea such an accident was unforgivable. The rest of the crew punished Hunter by not speaking to him for several days. By this time, Hunter admitted, he was so "blown out" that he could not distinguish between water and white wine. Despite the problems in San Francisco and the impact of the campaign on Hunter's sanity, the voyage ended on a note of triumph when some 10,000 people turned out at Vancouver's Jericho Beach to welcome the crew home.[77]

The voyage, overall, had been a startling success. The fact that Greenpeace had been able to find the Soviet whaling fleet was itself remarkable. Though their nonviolent direct-action tactics had, somewhat to their surprise, not prevented the whalers from firing their harpoons, they nonetheless provided Greenpeace with a series of potent images that could influence world opinion on whaling. As a staged media event, the campaign had succeeded beyond everyone's expectations. Their pictures and footage had been shown all over the world, and there can be little doubt that they played a major role in focusing public and media scrutiny on the IWC and on the plight of the world's remaining great whales. The media interest in the campaign, especially in the United States, ensured that Greenpeace was now in a position to enter the world's largest "environmental

Left to right: Patrick Moore, Rex Weyler, and Bob Hunter arrive in Vancouver after the inaugural Greenpeace save-the-whales campaign, 1975. Reproduced by permission of Greenpeace.

marketplace." Suddenly, the fund-raising possibilities and the potential for organizational expansion appeared limitless. Furthermore, Greenpeace's protest style, which combined the Quaker notion of bearing witness with a McLuhanesque media strategy and a good dose of guerrilla theater, could be adapted to a variety of causes, thereby providing frustrated environmentalists the world over with a new paradigm of ecological activism.[78]

Despite all the successes, however, there remained at the heart of Greenpeace's campaign a degree of philosophical tension and logical confusion. Those who were primarily responsible for framing Greenpeace's antiwhaling stance—Spong, Hunter, and Moore—were trying to reconcile two strands of environmental thought which, although closely related in some respects, held differing attitudes toward the value that humans should place on nonhuman life. Certainly, from both an ecological and an animal rights perspective, there were strong arguments for demanding the abolition of whaling. But there were also important differences in the philosophical foundations on which their respective arguments were based. For holistic ecologists, whales, like any other species, ought to be preserved because they play a vital role in the planetary eco-system and because, like nature in general, they have an intrinsic value that is independent of whatever values humans place on them. From an ecological perspective, however, there are no a priori reasons why humans should not hunt a species that has a healthy and stable population. Ecology, therefore, left the door open for the artisanal hunting of nonendangered cetaceans and a potential resumption of a hunt for larger whales if their numbers rebounded to their pre-nineteenth-century levels. From an animal rights perspective, however, whales, as intelligent beings capable of feeling pleasure and pain, should not be hunted under any circumstances. Neither the size of a given species' population nor the role that whaling played in any particular culture was reason enough to allow humans to hunt whales. The cessation of whaling, like the abolition of slavery, would last forever.

There were certainly compelling ecological reasons to oppose whaling, and Greenpeace as often as not resorted to the rhetoric of ecology when presenting their case to the public. Frequently, however, ecology played second fiddle to *Cetaceus intelligentus*. In press conferences and talks, for example, Hunter repeatedly emphasized that Greenpeace had been directly responsible for saving the lives of at least eight whales. From Greenpeace's perspective, whales, like humans, could be spoken of as individuals rather than merely bundled together as a collective or species.[79] Greenpeace's resident "inter-species communications expert," Mel Gregory, also reflected this rights-based approach in an article written shortly after the *Phyllis Cormack*'s return to Vancouver. The first paragraph invoked the standard portrayal of whales as exemplars of ecological harmony: "Somewhere in time, the whale decided to return to the water, and live in complete accord with nature." The great whales, he continued, "have never overpopulated the oceans" and have "kept in balance the delicately balanced ecosystem that provides the major supply of oxygen for all of the earth's creatures." The rest of the article, however, took the "mind in the waters" approach, insisting that whales possessed both higher intelligence and consciousness, while implying that it was these traits that made the whale worthy of being saved. When listening to Payne's recording of humpback whales, Gregory argued, one "can't help but read purpose and intellect into the thirty minute songs . . . that would burn out the circuits of any sound synthesizer. . . . The humpback whale, who is now at the point of extinction, has devoted his existence to perfecting the highest form of communication—MUSIC." The whale's "massive brain," Gregory reasoned, "has probably been developed through millions of years of introspection and meditative reflection." Where "man has gone out of himself with his hands to create, and sometimes destroy . . . the whale realizes completion within himself."[80]

Gregory's sentiments, like those of Spong and Hunter, demonstrate the eclectic and frequently conflicting array of Greenpeace's intellectual influences. The need to fashion a compelling popular image of the whale, combined with Hunter's and Spong's various philosophical and scientific predilections, led to the development of a fractured environmental philosophy based on various elements of biocentrism, neuroscience, animal rights, and New Age romanticism. It blended cutting-edge brain science with Eastern mysticism, ecological holism with sentimental anthropomorphism, and hitched them to a theory of consciousness revolution that was equal parts Hegel and McLuhan. It was a potent mix, one that proved attractive to both the media and environmental activists around the world. Its flaws and inconsistencies, however, would become more evident in Greenpeace's next major campaign: the effort to end the harp seal hunt on the ice floes of eastern Canada.

10

On Thin Ice

Since the beginning of the 1970s, Bob Hunter's overriding ambition had been to lead an international revolutionary ecology movement. By mid-1975, it appeared his dream had been realized. As well as exposing the actions of the Soviet fleet off the coast of California, the antiwhaling campaign had also exposed Greenpeace, and its spectacular nonviolent direct-action style of protest, to much of the world. The American media in particular seemed fascinated by the organization's tactics. The results, as far as Greenpeace's media strategy was concerned, were momentous. The *New York Times* and Walter Cronkite were now delivering Hunter's mind bombs into homes across the United States. Throughout middle America, newspapers such as the *St. Louis Globe Democrat* were decrying "the unconscionable extermination of many of God's most magnificent creatures" and urging President Ford to employ the Pelly Amendment against whaling nations.[1] Greenpeace, it seemed, even had the United Nations' imprimatur, with Maurice Strong, the head of the UN Environment Program, thanking the organization for highlighting the plight of the world's whales and urging them to continue their efforts.[2] For Hunter, all this was an opportunity too good to miss. A new age was dawning, he declared in one of his final columns for the *Sun*, in which "a vivid blending of spirituality and technology" was taking place. Vancouver, in his opinion, had replaced San Francisco as "the cradle of a whole revolution in consciousness."[3]

While the notion of being the Greenpeace Foundation's president may have seemed romantic and exciting, Hunter was now forced to confront the arduous and dreary chores that come with running an NGO. The intellectual frisson that Hunter derived from speculating about consciousness change and mind bombing was soon tempered by the mundane reality of bookkeeping, legal technicalities, and the fact that Greenpeace was in debt to the tune of $40,000. The sense of magic that had been present throughout the whale voyage quickly began to fade. As the Vancouver operation grew and new Greenpeace groups popped up throughout North America, Europe, and Australasia, Hunter was faced with a degree of factionalism that made him long for the simple "mystics versus mechanics" dichotomy that had characterized life aboard the *Phyllis Cormack*. Furthermore, Greenpeace soon decided to launch a protest against the annual harp seal slaughter on the ice floes of eastern Canada, a campaign that would attract an array of supporters, from hardcore animal rights activists to naked political opportunists, each with their own view of how campaigns should be run, why seals should be saved, and who, if anyone, should be allowed to hunt them. Perhaps not unexpectedly, the most serious factional disputes occurred not between Vancouver and the new Greenpeace chapters, but rather, within the Vancouver group itself. One such dispute—which took place between Paul Watson on the one side and Patrick Moore and Bob Hunter on the other—was particularly important in defining the acceptable limits of Greenpeace's nonviolent action.

The antiwhale campaign could still be plausibly framed as an attempt to preserve an endangered species rather than as an animal rights issue; after all, there was no doubt that certain species of great whales were threatened with extinction. However, the fate of the harp seal, from an ecological standpoint, was less clear-cut. Although Canadian and Norwegian hunters continued to slaughter hundreds of thousands of seal pups every year, the harp seal population, though considerably smaller than it had been during previous centuries, was nonetheless relatively stable. Furthermore, unlike whales, seals did not seem to be possessed of an extraordinary intelligence: thus there was no possibility of constructing a campaign around a "mind on the ice floes." What was undeniable, however, was that baby harp seals were utterly adorable, a fact that made their slaughter—usually achieved by a skull-crushing blow from a homemade club—seem excessively brutal. Herein lay a dilemma: in a world in which millions of cute animals are butchered or subjected to cruel experiments on a daily basis, would it be possible to adopt the harp seal as a symbol of humankind's wanton ecological destruction without running a gratuitously emotional campaign that demonized the mostly impoverished Newfoundlanders who slaughtered them?

Greenpeace's seal campaign, more than their antinuclear or antiwhaling protests, exemplifies a conflict that has been repeatedly played out during the past several decades between environmentalists intent on preserving a part of the natural world they value and the working people whose livelihoods depend on exploiting it. From the forests of the Amazon to the North Atlantic fisheries, environmentalists have often found themselves in the unenviable position of being pitted against rural working-class or indigenous people struggling to make a living, a conflict that is usually complicated by the actions of corporations and governments. Recently, various environmental groups have gone to considerable trouble to try to work with, rather than against, the people whose livelihoods depend upon the destruction of highly valued natural environments.[4] In Canada in the 1970s, however, there were few precedents for campaigns such as those run by Greenpeace. It is perhaps surprising, therefore, that the organization did indeed try to form an alliance with sealers against the large corporations who benefited most from the harp seals' slaughter. The fact that they ultimately failed in this endeavor was due as much to the intransigence of the sealers and the Canadian government as it was to Greenpeace's shortcomings.

The most pressing item on Greenpeace's agenda after the whale campaign was their $40,000 debt. Although their medium- to long-term prospects of paying this off and raising more money appeared positive, the short-term situation looked grim. At this point, Rex Weyler persuaded two of his colleagues at the *North Shore News* to offer their managerial and financial expertise. Peter Speck was the newspaper's owner and publisher and Bill Gannon was his savvy young accountant.[5] Gannon had moved to Vancouver from Winnipeg, and he immediately turned what had seemed to Hunter an insurmountable difficulty into a manageable problem. Soon T-shirt sales, donation boxes, and ad hoc gifts—such as those of the forest mystic Henry Payne—were supplemented by budgets, cash-flow projections, guarantors, and lines of credit. When Gannon learned that Greenpeace had held a lottery earlier that year, he asked if anyone had kept the receipts. After some searching, they were discovered in a garbage bag in the corner of the office. Gannon used the names and addresses on the receipts to start a direct-mail data bank. The first mail-out generated an impressive response rate of 8 percent, well above the average direct-mail response rate of 2 percent. Gannon also drew up a detailed budget and persuaded fifteen of Greenpeace's most devoted supporters to sign up as guarantors for $5,000 a piece. He took this plan to the Royal Bank of Canada and convinced them to loan $75,000 to an organization whose primary objective was to use direct action to disrupt what many a banker would have seen as legitimate economic activity.[6]

As well as venturing into the world of mainstream finance, Greenpeace also adopted a more traditional organizational structure. Apart from Hunter's presidency, Moore, Korotva, and Marining were appointed as vice presidents in charge of policy, operations, and communications respectively. The board included these four plus Cormack, Hunter's partner Bobbi MacDonald, Moore's partner Eileen Chivers, film producer Michael Chechik, Paul Spong, Paul Watson, and Gary Gallon, the executive director of Vancouver's other significant environmental organization, SPEC. Peter Speck demonstrated various managerial techniques that helped Greenpeace cope with their rapid turnover of personnel. He also helped them draw up realistic projections of future campaigns. Not everyone, however, was entirely pleased with such developments. Some of the hardcore mystics, such as Mel Gregory and Walrus, and radicals such as Watson, began to worry that Greenpeace might become just another mainstream environmental organization rather than the fluid, unstructured social movement they envisioned. Just as they had refused to sign the film contract in San Francisco, several of these people grumbled about such unheroic notions as cash flows and bookkeeping. From their perspective, it was hard to see what the "eco-revolution" had to do with contracting T-shirts to a distributor. Despite these murmurings of discontent, the situation remained, in Hunter's words, "within the bounds of acceptable comedy, because one of the worst of the flipped-out mystics—namely myself—was now the chief advocate of organization, fiscal responsibility, and the budget system itself."[7]

In addition to allowing professionals to transform the Vancouver office into something resembling a well-run nonprofit group, Hunter was also Greenpeace's leading proselytizer. In the fall of 1975, he and Bobbi traveled throughout Canada, visiting virtually every major university campus in the country. Hunter would present a slide show and lecture about the Greenpeace whale campaign, while Bobbi would sit at the back selling T-shirts and buttons and signing people up for membership. After each show, they would be approached by at least a dozen volunteers interested in setting up a Greenpeace group, many of whom had also been to one of Spong's whale lectures the previous spring. By Christmas of that year, there were approximately a dozen Greenpeace branches throughout Canada. Some were made up of merely a handful of people selling T-shirts, while others, such as those in Toronto and Montreal, were more substantial organizations that were soon contributing to Vancouver's campaigns, as well as mounting their own.[8] The most important office outside Vancouver, however, was undoubtedly the one established in San Francisco in the fall of 1975. This was to be Greenpeace's American beachhead. While the various Canadian branches were largely left to themselves, Hunter and his cohorts set up the San Francisco office in a more deliberate fashion. It would be the focal point for Greenpeace activity in the United States, providing them with access to the American media and an ideal base from which to plan further whale campaigns, as well as placing them at the center of California's lucrative fund-raising market. While some locals would help to run the branch, it was clear to Hunter, Moore, and the others in Vancouver that San Francisco was a subordinate office rather than an independent operation.[9]

Greenpeace's first action after returning from the antiwhaling voyage was to blockade the loading of a Soviet ship that they believed was heading off to supply the whaling fleet with fresh fruit and vegetables. Since such picketing was illegal, it soon drew the attention of the Vancouver Harbours Board police. The officer in charge took Hunter aside and suggested they cooperate: "Look, you're after publicity for your cause. All I'm after is removing you people without any trouble. How many do you want arrested?" Hunter asked the officer to give him a moment to check with his treasurer, Bobbi, and reported that Greenpeace could afford to pay bail for six people. He then returned to the group of protestors and asked all but five to move away. In return, the police allowed them to blockade the ship for a short time before leading them away one by one to be arrested. Only Watson broke the rules by stiffening and resisting, forcing the police to carry him to their squad car. The incident revealed Hunter's pragmatism and willingness to negotiate, while confirming Watson's status as the most radical and uncompromising member of the group.[10]

Another important development in the fall of 1975 was the publication of the first edition of the *Greenpeace Chronicles*. Rex Weyler was given the task of producing this wide-ranging journal, and the first edition appeared as a supplement in the *Georgia Straight*. Rather than merely a mouthpiece for Greenpeace, the *Chronicles* was "intended as the beginning of an international communications network, the main aim of which is to make environmental groups around the world more aware of each other's actions and contributions, ideas and plans." Weyler welcomed correspondence "from any and all environmental, wildlife, conservation and ecology groups" provided their activities emphasized "actions and campaigns rather than rhetoric or speculation."[11] To those less familiar with the organization, Weyler described them as "a prototype United Nations peace force that will eventually devote itself to the preservation of a habitable world." Politically, Greenpeace was "beyond left or right in the conventional . . . sense," operating as a nongovernmental force "in areas where governments themselves will not or cannot at the moment intervene." Those who were interested in helping the organization could join in one of two categories: "Supporters" were those who wished to restrict their activity to donating money, while "members" were defined as those people who wished to "take an ACTIVE role in defence of the environment." The latter group was invited to become involved in the general activities of Greenpeace and to "submit an outline of a plan of action." From Vancouver they would receive a campaign kit that contained examples of successful actions and "step-by-step instructions on how to apply innovative communications theory to draw attention to a particular issue."[12]

To illustrate its intentions of becoming an environmental clearinghouse, the *Greenpeace Chronicles* published articles on a wide array of issues. In addition to the obligatory whaling articles, Patrick Moore wrote a piece on the negative environmental impact of nuclear power plants; Fred Easton contributed an article on the impending construction of the Trident nuclear submarine base at Bangor, Washington; and Walrus detailed the sufferings of the Ojibwa people in northwest Ontario, many of whom were exhibiting symptoms of Minimata disease as a result of the mercury discharge from a local paper factory. An English animal rights organization known as the Hunt Saboteurs outlined their campaign against fox hunting, claiming that they were the only group who used Greenpeace-style nonviolent direct action in an effort to publicize and "sabotage" the hunt. This eclectic (and occasionally conflicting) range of issues—animal rights, deep ecology, social justice, and antinuclear protest—reflected most of the major strands of modern environmentalism, indicating that, at this stage, Greenpeace was prepared to apply its direct-action approach to a wide variety of causes without necessarily being fully aware of the potential conflicts and contradictions between them.[13]

The first edition of the *Chronicles* was also important for announcing Greenpeace's plans to run an antisealing campaign in the winter of 1976. Every year in February and March, hundreds of thousands (in earlier times, millions) of female harp seals haul themselves onto the ice floes of eastern Canada in order to whelp. The fishermen of Newfoundland, many forced to remain idle throughout the winter, supplemented their generally meager incomes by "hunting" the doe-eyed, fluffy, white harp seal pups. To many middle-class city dwellers, who had become insulated and disconnected from the harsh vicissitudes of rural life, the annual slaughter of the harp seal pups seemed a vestige of a brutal and barbaric past. A "swiler," as they are known in Newfoundland, simply walks up to a defenseless pup and smashes it on the head with a spiked club known as a "hakapik," a Norwegian word that amply evokes the club's function in numerous languages.[14] The pup is quickly skinned, and the swiler moves on to the next one, often leaving behind a mother seal pathetically nudging the pup's bloody corpse. Various animal welfare groups had been protesting against the hunt for over a decade, and although they had forced some improvement in the management of the hunt, hundreds of thousands of seal pups were still slaughtered each winter.

It was Paul Watson and Walrus Oakenbough who first suggested that Greenpeace employ its confrontational direct-action approach on the ice floes off the coast of Newfoundland, Labrador, and Quebec. The harp seal, Watson insisted, was on the road to extinction as a result of the hunt.[15] From the start, however, Watson made it abundantly clear that, from his perspective, the seal campaign was not going to be about population biology and maximum sustainable yield. It would, instead, be about brutality, blood, and death. The sealers, he emoted, "greet the mothers and their babies with club and knife," turning the ice from a "peaceful nursery to [a] bloody carnage," bringing "a horrible death to the seals and international shame to Canadians." Swilers, he insisted, were "colder than the ice upon which they trod." They dispatched "baby after baby, clubbing them, kicking them and in some cases removing the skin while the baby still lives and struggles." Watson also made it clear that, as with Greenpeace's stand on whaling, mere conservation would not be good enough: "The seal hunt," he insisted, "must end completely."[16]

In addition to holistic ecology, the other branch of nonanthropocentric thought that influenced Greenpeace's turn to the antisealing campaign was animal rights. Animal welfare—preventing cruelty to animals—and animal rights—the extension of natural rights to animals—have a long history in Anglo-American philosophy. The eighteenth-century English philosopher Jeremy Bentham was a major figure in this tradition, predicting "the day may come when the rest of the animal creation may acquire those rights which never could have been withholden from them but by the hand of tyranny." Bentham's ideas were an extension of his philosophical doctrine of utilitarianism, which stated, in short, that the rightness or wrongness of an act was directly related to the degree of pain or pleasure it caused. Unlike Descartes, who insisted that animals felt no pain, Bentham argued that animals, even if they could not reason or talk, could still suffer. Therefore, for humans to behave in an ethical manner, they should strive to minimize the suffering they cause to animals.[17]

Perhaps the most well-known animal rights author and activist of the postwar era was Cleveland Amory, a popular American journalist from Boston who founded the Fund for Animals. His 1974 publication *Man Kind?* was a caustic polemic against the American hunting tradition and the government-sponsored wildlife management polices that had been developed to serve it. Animals, wrote Amory, "must be legally realigned." They did not "belong" to anyone, and it was "past high time" for people to band together and work "for the most oppressed minority of them all."[18]

While Amory's frustrated and at times misanthropic polemic outlined the emotional contours of animal rights, it was Peter Singer, the controversial Australian philosopher, who laid out its intellectual foundations. In 1973, the same year that Arne Naess published his influential deep ecology article, Singer wrote a review entitled "Animal Liberation" for the *New York Review of Books*. The opening paragraph was emblematic of the no-nonsense, confrontational style that has marked Singer's work throughout his controversial career: "The tyranny of human over nonhuman animals," he proclaimed, "is a struggle as important as any of the moral and social issues that have been fought over in recent years." With that single declarative sentence, Singer unrepentantly situated the struggle for animal rights in the same category as the civil rights movement, feminism, the antiwar movement, and environmentalism. Singer strongly supported all these struggles, but felt that they did not go far enough. To achieve a truly liberated world, he insisted, humans would have to expand their moral horizons so that "practices previously regarded as natural and inevitable [would be] seen as intolerable." Despite his family's Central European Jewish background, Singer anchored himself firmly in the British analytical and empirical tradition, and he was particularly influenced by Bentham's utilitarianism.[19] Thus, Singer's moral calculus was based on the fact that both humans and nonhumans had the capacity to experience pleasure and pain. If it was morally wrong to inflict pain and death on a human, then it was wrong to do so on any sentient creature.[20]

Naturally, Singer's work provoked a storm of controversy. Where did he draw the line as to which creatures to embrace in his ethical circle and which to leave out? "Somewhere between a shrimp and oyster" was his vague response.[21] How could he presume to "liberate" species that showed no sign of wanting to "liberate" themselves? What of the millions of rabbits, foxes, and feral cats that were destroying the native wildlife and habitat of Singer's homeland? As Nash points out, however, Singer, whatever his flaws, "deserves credit for helping liberate moral philosophy from its two-thousand-year fixation on human beings." Animal liberation represented, in one sense, the extreme branch of Anglo-American natural rights liberalism and in keeping with that tradition, sought to protect individuals from oppressive aggregates such as states or nations.[22]

Broadly speaking, animal rights and holistic ecology share a common history. Both emerged from schools of thought that sought to overcome a centuries-old heritage, both theological and philosophical, which saw humans and the rest of nature as separate entities. Both aimed to extend the circle of ethics to embrace the nonhuman world. Despite these important similarities, however, there was a deep underlying tension between the two positions. The ethical fulcrum of animal rights was the pain and suffering of the individual animal. For ecologists, however, the welfare of individual creatures was superseded by the health of the ecosystem as a whole. From an ecological perspective, an individual organism, like a cell in a human body, cannot exist outside its ecological context. Holistic ecology, therefore, blunted the individualism that was inherent in animal rights.[23] In later years, such tensions would manifest in bitter philosophical debates between Singer-inspired animal rights supporters and deep ecologists. Tom Regan, an American philosopher who wholeheartedly embraced Singer's work, characterized deep ecology as "environmental fascism," since, in his opinion, it demanded the sacrifice of individual interests and lives in favor of the ecosystem as a whole. John Rodman, a political theorist at the Claremont Graduate School, countered that animal liberation was "patronizing and perverse," and he mocked the notion that "*we* can liberate *them*." The idea of extending human rights to nonhumans, Rodman argued, ensured that nonhumans would simply be given the status of "inferior human beings" or "legal incompetents" that needed human guardianship, as well as employing a "conventional hierarchy of moral worth" based on intelligence, consciousness, and sentience. Rather than giving animals rights within the existing order, Rodman urged, it was better to work for the overthrow and replacement of the current system with a society based on the principles of deep ecology.[24]

From the rarified atmosphere of academic philosophy, the inherent tension between holistic ecology and animal rights seems obvious. But to those working on the ground, things often look decidedly different. For a group such as Greenpeace, which used direct action to attract media attention to the plight of the whales and seals, such tensions may not have been immediately apparent. Even when they were aware of the contradictions, the benefits of appealing to both ecological and natural rights arguments frequently outweighed whatever costs such an approach entailed. By using both modes of thought, Greenpeace was attempting to appeal both to people's environmental sensibilities and to their feelings for individual animals. To a group of people passionately engaged in a campaign to save as many whales and seals as possible, any logical inconsistencies in such an approach could be dismissed as mere hair-splitting. For seals, however, the difference was anything but academic. Like so many of the world's creatures, their lives could change radically depending on the thought processes occurring in the brains of one particular species of primate. Thus for the seals, just as for the whales, ideas mattered.[25]

Harp seals, which gained their name from the harp-shaped outline that appears on the backs of mature animals, are truly creatures of the cold. Born on the ice floes of Newfoundland and Quebec—the southernmost extent of their range—they spend much of their lives migrating north to the waters and pack ice of Canada's eastern Arctic and the west coast of Greenland. For millions of years, they have ridden to and fro on the Labrador Current, feeding primarily on the abundant schools of capelin—a small, smelt-like fish—that share the same seas. Expert swimmers, adult seals are able to dive to an astonishing 600 feet and can remain underwater for up to thirty minutes at a time. For several millennia up until the eighteenth century, small numbers of seals were hunted by the people who settled the northeast portion of Canada and, more recently, by Europeans who sailed across the Atlantic to exploit the region's rich marine resources. Their population in 1800 has been variously estimated as being somewhere between four and ten million.[26]

Throughout the fall and early winter, the seals feed heavily before making their way down to the ice floes around Newfoundland and the Gulf of St. Lawrence in February. Gregarious creatures, harp seals tend to assemble in large patches whose positions vary from year to year according to the climatic conditions. In a particularly cold winter, a herd may congregate almost within sight of St. John's, Newfoundland's capital, while in other years swilers may have to travel over 300 miles north of the province to find them. On the ice floes, the females give birth to the chubby, pure white, doe-eyed pups whose fur was the main object of the hunt.[27] The "whitecoats" undergo a short but intensive period of nursing. For approximately nine days, they drink copious quantities of seal milk, whose fat content is 40 percent. This would be equivalent to a human chugging several mugs of liquefied brie every day.[28]

Within three weeks, the pups balloon from their fifteen-pound birth weight to an astonishing one hundred pounds. At the same time, they begin to shed their white coat, growing in its place a black-spotted, light gray fur. Now known as "beaters," they are of less value to the fur industry. They soon enter the water and begin to feed on the small crustaceans which mass along the floes. The entire process from birth to independence takes place in just three short weeks, an adaptation made necessary by the dangers of living on the thin pack ice, which is constantly cracking, buckling, and splitting apart. As soon as possible, the females abandon the pups and slip back into the water, eager to mate with the males who have been hovering around the edge of the floes. At this point, their eleven-and-a-half month pregnancy begins, and they set out on the long journey northward to repeat their cycle once more.

The seal hunt took up only a small portion of the swilers' working lives—a mere three or four weeks toward the end of the winter. Most spent the warmer months on fishing boats or in processing plants and the winter working in the forest industry. Despite the short amount of time sealers devoted to it, the income earned from sealing could be quite substantial in certain communities, frequently constituting the difference between extreme hardship and getting by. In earlier times, as one scholar noted, the hunt was essential for many communities. However, since the mid-twentieth century its economic value can best be described as supplemental.[29]

Sealing constituted something of a rite of passage for many of the province's workingmen, and the region's folklore is filled with stories of the hardships and triumphs of the swilers.[30] Since the act of killing a seal is not particularly skillful or dangerous, the swilers' stories emphasized their battles with the elements and with unscrupulous capitalists. They tended to view themselves as poor, humble, and hardy folk living on the margins of civilization, battling the awesome forces of nature, and exploited by ruthless shipowners and fur traders who cared little for their safety or economic well-being. Newfoundland history is replete with tragic stories of swilers succumbing to sudden snowstorms and plunging through thin ice into the frigid waters of the Gulf of St. Lawrence. One of the more notable examples is the Newfoundland Disaster of 1914, which was vividly retold by Cassie Brown in her book *Death on the Ice*. For two days, a group of 120 sealers were trapped in a blinding snowstorm on the shifting ice floes, unable to find their ship. Eighty swilers froze to death on the ice, a tragedy compounded by the fact that the ship's captain, eager to make his quota, had forced the sealers off the boat despite the horrendous conditions.[31] Shipowners typically took two-thirds of a voyage's proceeds. The remainder was divided up among the crew, with the captain receiving the lion's share. Adding salt to the wounds was the fact that the swiler, as one indignant Newfoundlander put it, "is obliged to pay hospital dues and is taxed by the merchant to pay not only for the tools and materials used in the fishery, but a further sum of three pounds, ten shillings for the privilege of being allowed to hazard his life to secure a fortune for the merchant."[32] Historically, therefore, the life of a swiler bore a similar degree of hardship and indignity to that of a coal miner in an early twentieth-century Appalachian company town.

As with whaling, the Norwegians were the major technological innovators in the sealing industry. By the late 1960s, most of the swilers were working for Norwegian firms, some of which were registered in Canada, while the few independent Newfoundland sealing operations could only sell their furs to Europeans through Norwegian middlemen. Swilers received an average of $2–$3 for a single pelt, which increased in value to $100–$125 by the time it was part of a luxury fur coat. Only a very small percentage of this money found its way back to Canada. According to historian Briton Cooper Busch, by 1971, the sealing industry in Newfoundland "may safely be said to have come into Norwegian hands."[33]

Between 1949 and 1961, swilers killed an average of 300,000 harp seal pups per year. According to David Sergeant, a biologist working for the Canadian Fisheries Research Board, this was far too many to sustain the hunt in the long-term and had already more than halved the birthrate from 750,000 in 1950 to a mere 350,000 in 1961. The entire harp seal population was estimated to be less than one-and-a-half million.[34] Until the early 1960s, neither the sealers nor the Canadian government had taken an active interest in conservation, the sole measure being a vague gentleman's agreement between Canadian and Norwegian sealers that seals should only be taken from the beginning of March through early May.[35] This hands-off policy was in stark contrast to the more statist approach that generally characterized Canadian wildlife management practices throughout the twentieth century. In fact, the historian Tina Loo argues that Canadian wildlife managers have long engaged in a "normative project of social, economic, and political change," one that was often at odds with the needs of local people and particularly indigenous Canadians.[36] The situation in Newfoundland was different, no doubt in part due to the fact that Newfoundland did not become part of Canada until 1949. Furthermore, since much of the hunt took place more than twelve miles offshore, it was outside the government's jurisdiction. However, from the 1960s onward, the federal government became a staunch advocate and facilitator of the hunt. The swilers did not always appreciate the efforts of professional wildlife biologists. Nevertheless, no matter how much swiling tarnished Canada's image abroad, the Canadian government remained steadfast in its support of the industry.

By the mid-1960s, conservation-minded scientists were calling on the Canadian government to regulate the industry. In addition, animal welfare activists, incensed by what they saw as the unmitigated cruelty of the hunt, began to publicize the annual slaughter in an effort to alert the metropolitan populations of North America and Europe to what they perceived as the barbaric practices of Newfoundlanders. Ecological and conservation issues were only of passing interest to these activists, many of whom belonged to the various Canadian branches of the Society for the Protection of Cruelty to Animals (SPCA). Rather, it was the cruelty of the hunt that raised their ire. In addition to causing great pain and suffering to the pups—and to the mothers who lost them—the hunt, it was felt by many, brutalized the swilers, turning them into cold-hearted savages. For middle-class SPCA members in the suburbs of Toronto or Vancouver, it was difficult to imagine the mindset that enabled a man to pitilessly club to death one of nature's most beautiful creatures, and then proceed to skin its warm and potentially still-living body.[37]

Swilers did not fit easily into any recognizable social category. Indigenous communities, after all, also hunted seals, and few nonaboriginal Canadians would begrudge them this right. However, where native hunting practices were ennobled by thousand-year traditions and deep cultural and mythological connections to their prey, the swilers were just poor white folk who clubbed seals for the international fur trade. In reality, indigenous sealing and swiling were much more similar than many would care to admit. The methods of killing the seals were virtually identical, and by the mid-twentieth century, both groups were intimately bound to the international market economy. Perhaps the only significant difference was that the indigenous hunters included a substantial amount of seal meat in their diets, while Newfoundlanders would only dine on the occasional flipper. Nevertheless, the indigenous hunters were sheltered under the protective blanket of sacred tradition, leaving the profane Newfoundlanders to bear the brunt of middle-class opprobrium.

In 1964, the harp seal issue came to the attention of a wider audience when a Montreal production company, Artek Films Limited, signed a contract with the CBC to produce a series of thirty-minute features on fishing and hunting in Quebec. After filming several episodes on topics such as duck shooting and moose hunting, the film crew, with little foreknowledge of the harp seal hunt, flew to the Magdalen Islands off the coast of Quebec, where they hoped to film the harps in their natural habitat and perhaps witness one or two being hunted. Instead, they arrived in the middle of swiling season and were shocked by scenes of swilers clubbing and skinning seals with apparently gay abandon. This was not like skilled hunters picking off the odd caribou with a telescopic rifle; the swilers raced from seal to seal, bashing and skinning as quickly as they could to maximize their productivity. The documentary, titled *Les Phoques de la Banquise,* provoked outrage throughout Canada and beyond. In a film not short on gruesome scenes, one in particular stood out and, for antisealing protestors, came to symbolize all that was brutal and degrading about the hunt. The infamous scene shows a swiler taking his knife to a bucking and kicking harp seal—one clearly very much alive—and beginning to skin it. It ends with the flayed creature racing madly across the ice, screaming deliriously and leaving a trail of blood behind it.[38]

Both the swilers and the Canadian government, which by now strongly supported the annual hunt, contested the film's authenticity, claiming that the film crew had staged the skinning and paid one of the swilers to do something he would not have done in the normal course of his job.[39] But such charges did little to diminish the film's visceral impact, which led *Canadian Audubon* magazine to declare that, whether exaggerated or not, the film "served to focus general attention on a practice which, whether limited or general, had to be outlawed."[40]

Prior to Greenpeace's involvement in the seal hunt, the most high-profile protestor was the indefatigable Brian Davies. In 1964, while Davies was the executive secretary of the New Brunswick SPCA, he attended a gathering of government officials and seal industry representatives who were drafting the regulations for the 1965 hunt. The meeting led Davies to the conclusion that "an over-capitalized sealing industry was intent on killing the last seal pup in order to get a return on its equipment, [while] those who profited from the seals gave not one thought to their suffering." Under pressure from various humane societies, the Canadian fisheries minister offered to take three representatives to the 1965 hunt to prove that it was as humane as possible. One of the three was Davies. It was a decision that the minister and his successors would come to regret, as Davies became virtually a permanent presence on the March ice floes for the next three decades.[41]

While the SPCA was ostensibly concerned with making the hunt more humane, rather than ending it altogether, it soon became clear that Davies and various like-minded members wanted to go further. By 1966, Davies's New Brunswick chapter of the SPCA decided to pursue a policy of abolition. This put them out of step with the organization's national office, which issued an official statement supporting the hunt and the methods employed by swilers. Although they felt that the use of clubs and gaffs was cruel, it represented, they argued, the best available method for the time being. For Davies, merely putting a halt to the cruelty was not enough: "For my part," he wrote in his book *Savage Luxury*:

> I believe my efforts have reduced the cruelty of the hunt, but that was not my primary goal. The question I pose for the moral discrimination of Canada does not concern itself so much with how these animals are killed, but rather whether they should be killed at all. I see the seal issue as representing a showdown for wildlife. These animals are symbolic, and if they can't be saved it is probably not ever going to be possible to save any substantial population of wild creatures. The world will gradually fill with filth and one day, empty of all but man, this planet will become the loneliest place in the universe. *Perhaps in saving the seals, man may save himself.*[42]

To Davies and his supporters, therefore, the harp seals occupied the same role as whales did for groups such as Greenpeace and Project Jonah. They were, in essence, a "gateway species": by exploiting their innate appeal, advocates hoped that people would develop a deeper appreciation for wildlife in general, which would make it easier to save other, less glamorous species. Nevertheless, for Davies, as for Greenpeace and the whales, it was frequently difficult to separate this broader goal from the fervid desire to abolish swiling entirely. Once again, this is an example of ecology coming into conflict with animal rights.[43]

For Davies, the SPCA's grudging stamp of approval for the hunt was simply unconscionable. If even those who were supposedly defending the seals were supporting the hunt, it was clear that change could only occur if the international community pressured the Canadian government. With this in mind, Davies embarked on a decade-long media campaign designed to shame the Canadian government into ending the hunt. Although he never employed Greenpeace-style direct-action tactics in his protests, Davies nonetheless proved to be a skilled media manipulator who would not have been out of place in the company of Ben Metcalfe or Bob Hunter. In 1966, Davies convinced *Weekend* magazine, a weekly supplement that appeared in dozens of major Canadian newspapers, to publish an article that clinically but dispassionately described the swilers going about their business. In response, Davies received almost 5,000 letters, the vast majority supporting his position, as well as donations to his Save the Seals fund.[44] He then trekked off to England where he set himself the goal of convincing London's *Daily Mirror*—Britain's largest circulation newspaper with a readership of over fifteen million—to publish a lurid exposé of the seal hunt. The *Mirror* duly sent a reporter and a photographer to accompany Davies onto the ice in 1968. Davies's reward came in the form of a front-page article accompanied by a picture of a bloodstained sealer, his club raised and poised to strike another doe-eyed pup. This, the *Mirror*'s headline blared, was "The Price of a Sealskin Coat."[45]

The next year, Davies set off for France, where he recruited the high-profile magazine *Paris Match* to his cause. Like the *Mirror* the previous year, *Match* sent a reporter to the ice and used materials supplied by Davies to publish a series of antisealing articles. The behavior of swilers at the 1969 hunt did little to help their cause. Strong northerly winds pushed the ice floes much further south than usual, allowing unlicensed amateurs from all over Newfoundland to roar about the ice on their snowmobiles, clobbering thousands of seals and ignoring government demands to halt the slaughter. To outsiders, who were already predisposed to think of the swilers as little more than savages, such behavior could only be interpreted as bloodlust. Even a prosealing Newfoundlander, the schoolteacher and writer Calvin Coish, admitted that the incident "reflect[ed] badly on the seal hunt and those associated with it."[46] *Paris Match* and other publications throughout

Europe had a field day, while Canadian embassies were inundated with letters calling for an end to the hunt. The Brussels Embassy, for example, received a petition signed by over 400,000 schoolchildren. In the United States, popular publications such as *Life* magazine also criticized the hunt, and protestors picketed the Canadian embassies in Washington and New York.[47] In fact, such protests plus the evidence Davies presented at congressional hearings were significant factors in helping pass the U.S. Marine Mammal Protection Act of 1972.[48]

By this time, Davies had severed his ties with the SPCA and formed the International Fund for Animal Welfare (IFAW), an organization whose primary goal was to abolish the harp seal hunt. Davies began to advocate replacing the hunt with tourism, insisting that people from all over the world would pay a handsome price to come to the Newfoundland ice floes to look at harp seal pups, employing former swilers as their guides. To prove his point, he organized a group of a dozen "tourists" to accompany him on the ice in 1970. The visitors, including a woman with a damaged hip who walked with the aid of a cane, saw harp seals by the thousands, photographed them, and even touched the plush white coats. "Later," Davies reported, "more than one was to tell me their hours among the seals had been the most moving experience in their life." In addition to saving the seals, Davies insisted, tourism would be more economically beneficial than swiling. According to Davies's calculations, it made more sense for the Canadian government to subsidize a few thousand swilers for a few weeks every year, rather than spend millions of dollars protecting the industry and worrying about the black eye it gave to Canada's image abroad. The swilers themselves, Davies insisted, had told him time and again, "If there was something else I could do at this time of year I wouldn't kill seals."[49]

In the early 1970s, the media interest in the hunt began to wane. Protestors such as Davies had little to offer other than more images of pups being slaughtered. Despite the fact that these were as compelling as ever, to the media, the issue was becoming somewhat repetitive. In addition, the federal government succeeded in pulling the rug from beneath the antisealing advocates by announcing a series of conservation measures which, they argued, would lead to the elimination of the hunt by 1975. In 1972, the quota was reduced from 200,000 to 120,000 harps, with successive reductions forecast over the next several years. The government also passed a measure banning large sealing ships from the Gulf of St. Lawrence during the breeding season. Predictions of an end to the hunt, however, were premature. By 1975, the Norwegians and the Canadians had agreed to a quota of 150,000–160,000 each to the Norwegian and Canadian fleets operating off the coast of Newfoundland, and 30,000 to the "landsmen," or coastal residents along the Gulf of St. Lawrence.[50]

Davies's IFAW continued to run media and political campaigns throughout the early 1970s. As it became clear that the hunt was not going to end any time soon, other concerned conservationists and animal rights activists joined them. In 1974, Harold Horwood, a conservationist and long-time opponent of the hunt, wrote an article in the popular *Reader's Digest* magazine entitled "Let's End the Seal Hunt for Good." In January 1976, *National Geographic* once more raised the alarm about the hunt in an article that appeared only a few weeks before Greenpeace began their first campaign on the ice. The author of the article was David Lavigne, a zoologist at the University of Western Ontario. Lavigne and his team had used a new ultraviolet photographic technique to count the whitecoats that were otherwise invisible from the air, allowing them to cover a wide area with a great degree of accuracy. Lavigne counted 200,000 pups in 1975, a figure that supported the view that the birthrate had declined despite the conservation measures of the past several years. According to Lavigne, the blame for this rested squarely on the large factory ships rather than the landsmen who hunted from small boats or from the shore. The only way to return the hunt to a sustainable level, he argued, was to ban the factory ships entirely and allow only a greatly reduced, strictly regulated hunt by the landsmen. This distinction between the landsmen and the factory ships was one that Bob Hunter would try to exploit, but one that also caused considerable friction within Greenpeace.[51]

When Watson and Walrus presented their idea for an antisealing campaign to other members of Greenpeace, Hunter recalls that he "jumped at it" straight away: "I looked at a picture of a seal and said, 'Wow!' Not 'Wow!, this is a great media thing,' but 'Wow!, these are beautiful!'"[52] Watson, however, recalls that many in the organization, including Hunter, were skeptical. Some felt the issue was too controversial. Others, such as Spong, did not want to see precious resources directed away from the whale campaign. In this instance, Watson's version appears to be the more accurate one. While Hunter, Moore, Spong, and others devoted their efforts to organization building and preparing for the 1976 whale campaign, Watson and Walrus were largely left on their own to raise funds and make the arrangements for the harp seal protests.[53]

Among the crew they chose to accompany them onto the ice floes was Dan Willens, a music teacher from Indiana whose new local Greenpeace group had raised considerable funds for the campaign. Watson's girlfriend, Marilyn Kaga, a twenty-six-year-old Canadian of Japanese heritage, was also included, as well as Eileen Chivers, who was Patrick Moore's partner, and Henrietta Neilson, a young Norwegian woman who deplored her country's attitude toward marine mammals. Another important new Greenpeace member, and one who bolstered the animal rights brigade within the organization, was Al "Jet" Johnson, a dashing pilot who was originally from Vancouver but who was now based in San Francisco. Johnson, who was in his early forties, had a deep affinity for wildlife, particularly wolves and coyotes, and he initially approached Greenpeace to suggest that they expand their activities to embrace terrestrial mammals. Johnson was just the kind of man that Watson was looking for—mature, tough, and passionate about protecting wildlife. Furthermore, he spoke Norwegian fluently (his parents were from Norway), had Arctic survival experience, and had flying skills that could come in handy in getting Greenpeace onto the ice floes.[54]

Greenpeace's initial plan was to arrive on the floes ahead of the swilers and spray the whitecoats with a harmless permanent dye which, they hoped, would ruin the commercial value of the pelts. In addition, they stated in their press release, "Greenpeace people will place themselves bodily between the sealer's club and its intended victim," as well as "harassing the sealers constantly in an effort to slow the progress of the hunt." Such actions, they hoped, would directly save the lives of many seals and "focus world attention on the fact that harp seals are being slaughtered to the brink of commercial and biological extinction."[55] From the beginning, the campaign was based on an uneasy blend of ecology and moral outrage. Hunter was certainly aware of the pitfalls and contradictions inherent in such an approach: The hunt, he recognized,

> was an issue that brought out the worst forms of anthropomorphism and yet at the same time the highest forms of compassion. We knew we would have to walk a tightrope between a balanced "scientific" analysis that the hunt itself was simply bad for the ecology of the ocean . . . and the depths of emotion that the killing of "babies" generated in the breasts of millions of urban people, who, otherwise, with their cars and swimming pools and electric gadgets, were the worst environmental destroyers of all.[56]

Drawing on Lavigne's study and similar work conducted by scientists at the University of Guelph, Greenpeace insisted that the hunt was not only "demonstrably hideous and savage," but that there was also incontrovertible scientific evidence that if it continued for much longer, "the whole harp seal species [would] vanish forever, leaving nothing more than blood stains on the ice floes." Like Brian Davies, Greenpeace saw the harp seals as a gateway species: "Any heightened public sense of responsibility toward one species—such as the whale or the seal—inevitably has the same effect on people's thinking about lynx, beaver, polar bears or birds." Greenpeace, therefore, was working "toward an overall change in attitude toward all living creatures," a statement that once again reflected Hunter's belief in consciousness revolution. The alternative was "the slow death of human moral consciousness, the ultimate and inevitable death of an ecological system which can in time reach out to damage mankind itself."[57] This rhetorical blend of wildlife conservation and moral outrage, combined with photos of the whitecoats before, during, and after their slaughter, had worked successfully in the past for groups such as IFAW. When combined with the possibility of direct confrontation between swilers and protestors on the stark ice floes, it proved to be an irresistible formula for grabbing media attention. Within a day of announcing their plans, Newfoundland's newspapers printed headlines such as "War with Greenpeace—we must win," while Newfoundland politicians were quoted as saying that the first Greenpeace protestors to step onto the ice would be "mobbed" by irate locals.[58]

On the eve of the seal campaign in the winter of 1976, the general frame of mind among Greenpeace's chief activists could be described as one of confidence occasionally bordering on hubris. Hunter, who was blessed with the gift of being able to summarize the vast sweep of history in a language that is both succinct and evocative, described this mood in his usual vivid fashion:

> We had said it so often we were beginning to believe it ourselves: a small group of people, acting imaginatively and nonviolently, can affect the course of events in the "global village." In a planet where some four billion human beings live in gigantic ant heaps, each of them apparently powerless because of the sheer density of the masses packed around them, no more audacious idea could be conceived. Most of the thinkers on earth hold firmly to the belief that history is shaped by swarms, by the colossal momentum of numbers and bulk. The individual, even the rare genius who appears once every few centuries, can do little more than deflect the course of events for a few years before the great current closes in and sweeps onward. How can any rational person convince himself that there is a ghost of a chance of having even the most minute impact on world events? And who but a megalomaniac can dream of actually changing the consciousness of humanity?[59]

Hunter, for one, still held such dreams. He remained convinced that the revolution in international communications technology held the promise of enabling small groups of people, with no access to the traditional sources of political and economic power, to influence events on a grand scale. Hunter embraced McLuhan wholeheartedly and uncritically. Who could blame him? After all, a mimeographed press release from his cluttered little office in Vancouver's hippie ghetto would be transformed, literally overnight, into newspaper headlines across Canada, causing perceptible levels of anxiety among politicians and industrialists. The seal campaign, however, would be quite different from the whaling protests. This time, instead of coming up against an impersonal and monolithic whaling factory ship from an enemy nation, Greenpeace would be confronting working-class Canadian men at close quarters and on their own turf. It was a dynamic that would call for a greater degree of compromise than many people, on both sides, were willing to accept.

11

Blood and Death and Sex

By 1976, the Canadian government had succeeded in convincing many of the more traditional conservation organizations that the harp seal was not an endangered species and that the swilers' hunting methods were humane. Measures such as start and stop dates, catch limits, strictly defined quotas for the various hunters, licensing procedures, and the presence of government supervisors on the ice led organizations such as the World Wildlife Fund, Canada Audubon, and the Ontario Humane Society to withdraw their opposition to the hunt.[1] In many ways, the environmental philosophy of such groups was closer to the scientific conservation of government wildlife biologists, such as David Sergeant, than to that of Greenpeace or the IFAW. By this time, Brian Davies was unashamedly running an animal rights campaign. Greenpeace, however, was hedging its bets. Most of its public rhetoric was grounded in the discourse of holistic ecology: seals were a vital part of the marine ecosystem, and the hunters were killing them in unsustainable numbers, threatening them with extinction. But it soon became clear that many of the key activists—and many of Greenpeace's supporters—had come to take an abolitionist stance against sealing. While this was consistent with their attitude toward whales, the rationale for taking such a position on seals was neither as clear nor as compelling.

On March 2, 1976, the majority of the antisealing crew left Vancouver for the long train, ferry, and car ride to the remote fishing village of St. Anthony on the northwestern coast of Newfoundland. Awaiting them were two chartered Bell Jet Ranger II helicopters that would fly them to the pack ice. Watson, according to Hunter, had a "militaristic streak" and delighted in giving crew members specific titles, such as "squad leader," "flight assistant," and "squad quartermaster." For himself, he reserved the title "expedition leader." He was not particularly pleased that Hunter and Moore, who were above him in the Greenpeace pecking order, had decided to join him. He and Walrus, he felt, had done most of the work thus far and should be able to run the campaign as they saw fit. The relationship between Watson and Moore had not improved since the whale campaign and their enmity was unlikely to thaw on the ice floes. The fact that Moore and Hunter flew across Canada rather than joining the rest of the crew on the cheaper but eternally long train ride (though for apparently sound reasons) further strained relations within the group.[2]

During the whale campaign the previous summer, Greenpeace's arrival at various ports on the west coast had been greeted with great cheer and goodwill. None, of course, were expecting similar scenes in Newfoundland, but few were prepared for the genuine anger and sheer hatred that the locals directed toward the meddlesome mainlanders. As their vans pulled into St. Anthony, the small squadron of Greenpeace activists was greeted by a mob of enraged placard-waving, fist-shaking locals. Hunter saw about a dozen menacing young toughs in the front who were swinging ropes with nooses. The mob quickly surrounded the vans and began to rock them as though intent on pushing them over. Moore, who was driving the lead van, turned the engine off wearily and sighed, "We seem to be here." He and Watson managed to step out of the van and attempted to calm the angry locals while Hunter tried to remember "the best method of protecting my head while being stomped." Moore and Watson were both wearing orange-colored arctic survival suits, apparel that further distinguished them from the locals, whose attire was considerably more modest. Hunter, who was wearing an old navy-jacket, managed to move away from the van and blend into the crowd, while furious locals continued to scream obscenities at Moore and Watson, threatening them with the blunt end of their pickets. Just as it looked as if Moore and Watson would be lynched, Eileen Chivers stepped out of the van, her long black hair flowing over her orange parka. She smiled at the hecklers, who were taken aback by the sight of a woman among the Greenpeace contingent and temporarily backed off.[3]

Hunter, in the meantime, had climbed a snowbank by the side of the road to gain a better view of the overall situation. From there, he soon spotted the apparent leader of the group. Making use of his low profile, he wormed his way through the crowd and offered his hand to Roy Pilgrim, the coordinator of the Concerned Citizens Committee of St. Anthony Against Greenpeace, who shook it reluctantly. Hunter asked Pilgrim if they could have a meeting and was informed that one had already been arranged for 9:00 p.m. Pilgrim then pulled back the crowd that was still tormenting the other Greenpeacers, announcing that Greenpeace would present their case to the locals that evening, "and then we'll see what happens." To much cheering, an elderly man, employing the archaic local patois, took one last parting shot at the enemy: "Ain't gonna make no difference, b'ye . . . Meetin' or no meetin', these b'yes are gonna be on their way back to the mainland by midnight."[4]

When the Greenpeacers arrived at the local boarding house, the crowd had once again become unruly, and a young man seemed intent on picking a fight with Watson. At this point, Walrus, who had remained relatively anonymous, strode up to the belligerent swiler, jabbed his finger in the man's chest, and screamed: "*Listen, I haven't heard one of you guys say one word about Mother Earth! And that's what this is all about! We're here to protect the seals, the whales, the birds, everything! It's all part of Mother Earth! And you're not gonna stop us, because it's Mother Earth's will. Those seals are my seals too! So just get out of our way, get out of our way, that's all!*"[5] Such language further emphasized the cultural gulf between the Vancouver eco-hippies and the swilers. To locals such as Calvin Coish, "Greenpeace came through as a bunch of freaky know-it-alls from way over on the west coast." Many of the swilers vowed that they "wouldn't think twice about giving [Greenpeacers] a bash on the side of the head." Another typical sentiment was expressed in the following comment from a sealer: "if [Greenpeace] are out there trying to stop a man from making a living, there's going to be trouble."[6] To the swilers, the idea of "saving" a seal pup made no more sense than "saving" a fish. Certainly, it was appropriate to try to prevent a species that they saw as a resource from being unduly depleted. But the idea that an entity called "Mother Earth" believed that it was wrong to kill seals seemed patently absurd. In their minds, God had placed the seals there for their use.[7] Those who justified the hunt frequently referred to Genesis 1:26 with its famous declaration that man should "have dominion over the fish of the sea, and over the birds of the air, and over the cattle, and over all the earth, and every creeping thing that creeps upon this earth." It was a theme that was also stressed by the priests who frequently blessed the sealing ships before their departure for the ice floes.[8]

Soon after arriving at Decker's boardinghouse, Hunter was on the phone to Rod Marining, who had remained in Vancouver to act as Greenpeace's media spokesman. Marining had some important news: the Canadian government had just passed an order-in-council making it illegal to spray seals. An order-in-council is a statute or law that is passed by the federal cabinet and does not need to be approved by the Parliament. Somewhat melodramatically, Hunter despaired that Greenpeace had "underestimated the totalitarian streak that existed in Ottawa."[9] This development compounded the problems that Greenpeace faced as a result of amendments to the Seal Protection Act. These included banning everyone, apart from those associated with the seal hunt, from flying less than 2,000 feet over a seal herd or landing a helicopter within half a mile of a seal. These regulations were transparently designed to prevent groups such as IFAW and Greenpeace from protesting. The government argued that the amendments were necessary to prevent people from "disturbing" the seals during their breeding and nursing phase. However, as Hunter caustically observed, the law effectively stated "you may not disturb a seal unless you are definitely going to kill it." George Orwell himself, he continued, "could not have invented a nicer title for a piece of legislation aimed at the destruction of an animal."[10]

With the odds stacked firmly against Greenpeace, Hunter felt it was time for a quick change of tactics. Clearly, the federal government was not going to alter its position any time soon, but the local landsmen might be persuaded to drop their opposition to Greenpeace, and, if Hunter played his cards right, to support them on some issues. While the others were settling into the boardinghouse, Hunter arranged a quick private meeting with Roy Pilgrim. During his time among the crowd of angry locals, Hunter had surmised that the landsmen—the small-time swilers who were not crewmen on the large sealing ships—were particularly incensed by Greenpeace's plan to spray the seals with dye. Furthermore, it quickly became evident that they did not think particularly well of the predominantly Norwegian companies that dominated much of the sealing industry. Hunter was well aware that the landsmen took a relatively small number of seals compared to the predominantly Norwegian-owned ships. Here was an opportunity, he felt, to build a coalition with the landsmen against the wealthy and mostly foreign-owned sealing corporations.

With this in mind, he proposed the following deal to Pilgrim: Greenpeace would give up the spraying plan in exchange for a promise by Pilgrim that nobody would interfere with their helicopters. In addition, if Pilgrim would allow Hunter to try to recruit landsmen to Greenpeace's cause, he would leave the local swilers in peace and restrict the protest to the factory ships. From Hunter's perspective, this was a very favorable compromise. After all, the use of dye was now illegal and would only result in the protestors being arrested on the first day of the hunt. Furthermore, Greenpeace's intention had always been to focus their activities on the factory ships, rather than the landsmen, with the broader aim of pressuring the Canadian government to extend the nation's coastal boundary to the maximum 200-mile limit permitted under international maritime law, instead of the twelve-mile limit that was currently enforced. Such a deal, Hunter hoped, would enable Greenpeace "to outflank Ottawa and pull off a worker-conservationist alliance that would dumbfound everybody." By this time, the little fishing village was swarming with reporters from influential media outlets such as America's NBC and the *Washington Post*, Germany's *Stern*, and France's Gamma News Agency, as well as all the major Canadian newspapers and TV stations. Pilgrim, for his part, did not want to see his people once again depicted as angry Arctic rednecks intent upon beating up the noble seal savers. He and Hunter shook hands and agreed to abide by the terms Hunter laid out.[11]

The local media was not impressed by Hunter's apparent compromise. Wick Collins, a columnist for the St. John's *Evening Telegram* argued that it was "no more than a tactical withdrawal in the face of determined opposition from the strong minded people of St. Anthony." As soon as the "publicity hounds" were "back on safe ground in mainland Canada, they will resume their verbal assault on the Newfoundland sealers."[12] Paul Watson and his group of hand-picked supporters were also less than happy. Watson strongly objected to the idea of giving up the dye, but Moore and Hunter insisted that it was the only way they could mount a successful campaign. At the meeting that evening, Moore and Watson attempted to explain, in ecological and economic terms, why Greenpeace was against the hunt, eliciting heckles and boos from the rowdy audience. Hunter then strode to the microphone and played his ace: "Out of respect for the serious economic hardships experienced by the people of Newfoundland, the Greenpeace Foundation will drop its plans to spray the seal pups with green dye . . . instead, Greenpeacers will go out to the Front, to the icebreakers operating in international waters, focusing primarily on the Norwegians, and we will throw our bodies between those seal hunters and the seals." Hunter's speech was greeted with wild cheers and jubilation, with most of the locals interpreting it as a capitulation.[13]

Greenpeace would no longer have to worry about local swilers preventing their helicopters from taking off for the distant ice floes. But Hunter in particular would have to pay a price for it. The next day, newspapers across Canada were reporting Greenpeace's plan to abandon their spraying tactic and to form an alliance with local swilers.[14] In Vancouver, Marining and others were bombarded with angry phone calls from people demanding their money back and accusing Greenpeace of selling out. The newly formed Toronto Greenpeace crushed all their Greenpeace buttons under their heels and mailed them personally to Hunter in St. Anthony. Nevertheless, Hunter remained optimistic: with a little "luck, so long as we could get the choppers out to the ice, we would recover, and, if everything went perfectly, we might yet emerge with a formula for bringing the environmentalists together with the grass-roots people in Newfoundland against the government and the handful of fur barons who had so successfully exploited both seals and the sealers for centuries."[15]

Greenpeace's strategy, never too clear to begin with, was shifting constantly as the campaign progressed. From the start, Watson, Walrus, and the new recruits they had brought to Greenpeace wanted to pursue an animal-rights-based agenda of saving as many individual seals as possible and advocating the abolition of the hunt. Hunter, while sympathetic to this view, was pragmatic enough to realize that such an approach would create too many enemies. It was difficult enough to cope with the restrictions of the Seal Protection Act; if angry landsmen had tried to prevent Greenpeace's helicopters from taking off, then the campaign would be almost entirely ineffective. Although nobody within Greenpeace wanted to admit it, the landsmen of St. Anthony had forced them into a significant compromise. Instead of crying for the hunt's abolition, Greenpeace was now officially endorsing the landsmen's right to hunt while condemning the mostly Norwegian factory ships operating on the front. Animal rights and deep ecology had, temporarily at least, taken a back seat to a more pragmatic and traditional form of wildlife conservation.[16]

On the second day in St. Anthony, Greenpeace underwent the humiliating ritual of handing their converted fire extinguishers full of green dye to Roy Pilgrim and a group of swilers, all before the cameras of the assembled media. They then retreated to the boarding house to prepare their strategy. In order to give them maximum independence, they would set up a base camp on deserted Belle Island some thirty miles north of St. Anthony. Fuel drums would be stored on the island, along with tents and supplies, and a group of protestors would remain there as "reverse hostages," in case the government tried to prevent helicopters from leaving St. Anthony. Almost immediately, however, they ran into a major problem. The local oil dealer refused to sell them fuel. Canadian Fisheries Officers, he explained, had threatened him with heavy fines and possible imprisonment if he sold fuel to Greenpeace. Fortunately for Hunter and his crew, they had among their contingent a top-notch Vancouver lawyer named Marvin Storrow. Storrow was immediately on the phone to the local Royal Canadian Mounted Police announcing that Greenpeace was ready to file a formal complaint of attempted extortion against the fisheries officers. Within an hour, Greenpeace was able to purchase fuel.[17]

On March 12, the first contingent of protestors flew to Belle Island, where they spent two miserably cold days setting up the camp. The temperature fell to minus thirty-five degrees Celsius and a howling gale blew across the island. With only their tissue-thin tents for protection, the crew spent much of the time cocooned in their sleeping bags. Moustaches froze, and any exposed skin was soon burned by the freezing wind. On the morning of March 15, Moore, Watson, Walrus, and Jet Johnson climbed into a helicopter and headed for the ice floes. Within twenty minutes, they spotted two Norwegian ships breaking their way through the ice. From 2,500 feet, the Greenpeace "action squad" could see streaks of seal blood painted across the blindingly white surface. Some of the streaks were disjointed, indicating that an ice pan had flipped over, exposing its pristine underside. As they descended, they could see the swilers dragging bundles of pelts toward the ships, where cranes hauled them aboard. Jumping out of the helicopter, the Greenpeacers had their first sight of the baby harp seals. To Watson, they were "beautiful beyond expectation. Chubby little bundles of soft white fur, their round, jet-black eyes glistening with tears, they cried, sounding exactly like human infants in distress."[18] Watson, Walrus, and Jet Johnson immediately began to charge across the ice, heading for the nearest sealers, followed by Moore who was doubling as the photographer. Johnson was the first to employ Greenpeace's nonviolent, direct-action strategy on the ice. As a swiler was readying to club another whitecoat, Johnson bumped him aside and threw himself on top of the seal pup, pinning it to the ice by its flippers and shielding it from the sealer's raised hakapik. It was, according to Hunter, "the first seal in history to be so protected by the loving flesh of a human being." Walrus attached himself to a swiler, following him around the ice and blocking him every time he came across a pup. The furious and clearly confused swiler soon retreated to the ship. Moore filmed as much of the action as he could, as well as taking gruesomely pathetic shots of mother seals nudging the skinned remains of their offspring.[19]

Watson, too, threw himself on top of seals. He soon found himself being chased by two inspectors from the Fisheries Ministry, who demanded that he stop interfering with the hunt. The swiler, Watson, and the officials formed a bizarre conga line as they shuffled across the ice from seal to seal. As Watson came closer to one of the ships, he noticed that the blocks of ice churned up by the icebreaker's steel-reinforced bow were crushing many whitecoats. Spotting a pup that was in the path of the ship, Watson raced across the ice, bent down, and, with some difficulty, picked up the surprisingly heavy animal. The ungrateful seal bit Watson on the cheek, before resigning itself to its fate and allowing him to carry it away from immediate danger. For Watson, this was a critical moment: "I looked down into its ebony eyes which looked back quizzically with such innocence that I burst into tears. . . . Now I was personally involved, not simply making a protest of principle. I had saved that particular pup's life, held it in my arms, felt its warm body against mine."[20] It was at such moments in the life of a Greenpeacer that any abstract thoughts of animal rights, ecology, and conservation were expunged and completely overwhelmed by a wave of pure emotion and instinct. Such a visceral experience, and the passion it summoned forth, was something new to the environmental movement. It certainly was not one that

Paul Watson cradles a harp seal pup. Newfoundland, March 1976. Reproduced by permission of Patrick Moore/Greenpeace.

the average Sierra Club member or Friends of the Earth activist was ever likely to experience. For all his intense love of the wilderness, even John Muir never experienced placing his body between a live wild animal and a human hunter.

While Moore and Watson were hopping between ice pans, Hunter remained in St. Anthony, coordinating the campaign and dealing with the media. His most difficult job proved to be allocating spaces on the helicopters. Many of the journalists were desperate to get onto the ice to film the much-anticipated conflict between Greenpeace and the swilers, but spaces were limited and fuel was expensive. Hunter decided he would fly some of the journalists out but at their own expense. Most agreed, but an indignant NBC crew refused, claiming it would compromise their objectivity. When Hunter refused to alter his pay-your-own-way policy, the NBC producer stormed off with the parting words: "You're dead in New York, kid." Several Greenpeacers were also miffed when they were denied their turn on the ice.[21]

The next day, Newfoundland was enveloped in a blinding snowstorm—according to locals, the worst in a decade—and everything, including the seal hunt, came to a halt. For two days, a group of Greenpeace protestors, including Watson, Johnson, and Eileen Chivers, as well as Ron Precious and a French photographer, were trapped on Belle Island, forced to endure the freezing cold blizzard in their sleeping bags and tents. Due to an oversight, they had not brought enough food with them and had run out of gas for the cooking stoves. In order to keep warm and prepare what little food they had left, they resorted to using the helicopter fuel, which had been stored in thirty-six-gallon drums a quarter-mile from the camp. Somehow, despite the howling gale and blinding snow, they managed to tip fuel from one of the large drums into the small camp-stove fuel containers. Since jet engine fuel was not designed for camp-stoves, the only way they could light it was to build a small fire out of torn up newspaper, place the entire camp stove on top of it, and then stand back as it ignited with a small explosion. Several times the fire got out of control and burned parts of their tent. On one occasion, Jet Johnson's fuel-soaked survival suit caught fire. By the time they were rescued, the group was coughing up black phlegm from the fuel engine fumes and exhibiting signs of cabin fever.[22]

The *I Ching*, while not as evident on the ice floes as it had been in the Pacific, was nonetheless deployed when the situation demanded. Once the storm subsided, Hunter took his seat on one of the choppers and embarked upon his first trip to the floes. His *I Ching* reading had warned that he would be "Treading upon the tail of the tiger." Hunter had interpreted this as a reference to the sealers and the Canadian government, but upon landing on the ice, decided that the tiger "was nature itself, ready to snuff us out in a blink if we made a single false move." His first glimpse of a baby harp seal evoked a sense of awe similar to the one experienced by Watson and several of the other Greenpeacers. With its downy, snow-white pelt and bottomless black eyes, it seemed to Hunter that "God loved it more than almost any other creature." The "icescape," some eighty miles east of the Labrador coast, was also a "spectacle of the highest order."

> There were fractures between the ice pans, and between the fractures, sometimes three feet wide, you could see the dark blue tides of the Atlantic rushing. We had arrived on a giant wrecked marble table in an atmosphere as severe as an operating room. Above, a blinding painful light. Below, a white more perfect than hospital sheets. It was, I suddenly realized, a sacred place. A nursery. A horizon-to-horizon bed. There were frozen birth sacs everywhere. And from these had emerged creatures astonishingly like the "shmoos" in the Li'l Abner comic strips.[23]

With Watson barking orders, the Greenpeace contingent hopped from ice pan to ice pan, searching for sealers to confront. Despite his attempts to form a coalition with the landsmen and his efforts to understand sealing from the swiler's point of view, Hunter could not help but feel disdainful and superior toward them. Eileen Chivers's presence on the ice, Hunter observed, made her, in all likelihood, the first woman "to come to this place where, for centuries, Newfoundland males had entered their manhood by steeling themselves to kill the most beautiful infant creature they had ever seen." For Hunter, this was an event of historical significance, a moment when "a woman from the twentieth century would rise between a man armed with primitive killing tools and an animal that died in its infancy." Just as it had been problematic for Greenpeace to avoid the rhetoric of the Cold War during their confrontation with the Soviets, it proved equally difficult to steer clear of the pervasive middle-class view of the swilers as "primitives" and "savages."[24]

The apogee of the protest, and one of its most enduring images, occurred later that day when Hunter and Watson decided to blockade a sealing vessel. While Moore, Precious, and Chechik gathered around them to film and photograph the action, Hunter and Watson stood in the path of the icebreaker, their backs to its bow, forcing it to either halt its progress or run them down. It was an event marked by an odd verisimilitude: on the one hand, the danger was very real; much like driving a Zodiac in front of a whaler, it was a game of chicken in which a slight miscalculation could be fatal. On the other hand, the presence of the three cameramen shouting out questions and directions gave it the air of a film set. In a sense, it was the perfect metaphor for Greenpeace's style of protest: it was simultaneously artificial and real, spontaneous, yet scripted. Moore's photographs show Hunter standing, legs apart, staring determinedly ahead, his right arm extended to a slightly more nervous-looking Watson, who clasps it in a revolutionary handshake. In Chechik's film of the encounter, Hunter has the fixed air of a statue, staring resolutely forward regardless of the ice breaker's position, while Watson, somewhat belying his courageous reputation, glances back nervously across his shoulder. Hunter admitted that he was trembling vigorously but was able to allay his fear by keeping his mind "centered as much as possible on the Clear Light, a Tibetan Buddhist meditation technique." As the ship, called the *Arctic Endeavour,* made its initial run toward Hunter and Watson, a crewman warned them: "Ya better move, b'yes, the ole man ain't one ta tink twice about runnin' ya into the ice." Hunter yelled back: "Tell the old bastard to do what he wants, we're not moving." The captain took up Hunter's challenge and began to thrust the giant icebreaker forward. Hunter described the event in his usual vivid fashion:

> Everybody heard her coming. Everybody felt her coming. The vibrations of the diesel engines ruptured the air and tingled the soles of our feet, even through the thickness of arctic boots. The ice trembled, cracked, crunched, shattered. Blocks of chunky ice tumbled upward, stabbing like thick blunt broadswords. Dark lightening bolts of fracturing ice leapt outward from the steel bow as it reared over us. We stood there with our heads slightly bowed, as though before a guillotine. From far above on the deck, we heard someone yell: "Stop 'er, Cap! Stop 'er! The stupid asses ain't a-movin'!"[25]

As the ship came to a grinding halt, Chechik jumped forward and thrust a microphone in Hunter's face: "Could you tell us what you're doing, Bob?" Hunter, clearly irritated by Chechik's failure to grasp the grand symbolism of the moment, replied: "I'm standing in front of a boat, Michael. We'll talk about it later." On the ship's deck, crewmen screamed obscenities at the Greenpeacers, prompting Moore to lose his cool. "You'd kill a man, you fuckers!" he yelled angrily. "You'd kill a man, wouldn't you?" For a brief moment, as the ship drew back, it seemed that Hunter and Watson had won the confrontation. It soon became apparent, however, that the captain was merely giving himself some space in order to increase the ship's momentum. This time, the *Arctic Endeavour* charged at Watson and Hunter with the full power of its mighty engines, seemingly intent on crushing them like insects. "It's a crummy way to die," Watson commented grimly. "*Just don't look back, Paul,*" commanded Hunter. The ice began to heave and buckle under their feet, a chunk the size of an automobile bursting upward only a few feet to Hunter's left. Despite the danger, Hunter later wrote, "I had never felt stronger or more certain in my life. That ship could *not* get through us. It was possible to *will* it to stop." Just as it seemed the icebreaker had gone too far to halt its forward momentum, a cry came from the deck: "Stop 'er, Cap! *Stop 'er*! They *still* ain't a-movin'!" The captain thrust his engines into reverse, bringing the ship to a halt just a few feet behind Hunter and Watson's still-rigid bodies.

It was a classic media moment, one that was seen on televisions and in the print media throughout much of the world. It was also, as Hunter recounts, an episode that reinforced his belief that they were involved in what he liked to refer to, in an era of pre-9/11 innocence, as a jihad: "In the final few seconds before the ship had stopped, and the sound of the ice had been like a mountain-sized molar breaking apart in the middle of my head, I had felt a blaze of anger like a blowtorch. It was righteous wrath. It brought with it absolute conviction and an ecstatic, exultant feeling of strength." Later, a journalist who had been on board

Paul Watson and Bob Hunter block a sealing ship. Newfoundland, March 1976. Reproduced by permission of Patrick Moore/Greenpeace.

the *Arctic Endeavour* told Hunter that the crew had tried to convince themselves that the incident had been a victory for them: that they had succeeded in scaring off the Greenpeace publicity hounds. But such efforts, according to the journalist, came off as strained and halfhearted. In truth, he reported, the crew sensed that they had lost the initiative.[26]

During the 1976 hunt, the CBC aired a television documentary on its popular *Fifth Estate* program detailing Greenpeace's antinuclear and antiwhaling campaigns and generally lionizing Hunter and his crew. The film elicited a grudging respect for Greenpeace from a small number of locals. In the village of Raleigh, for example, Hunter claimed that seventy people signed a petition saying they were willing to accompany Greenpeace to the front to oppose the factory ships. Roy Pilgrim's younger brother, Doug, perhaps as part of some long-simmering sibling rivalry, went so far to start the local St. Anthony branch of Greenpeace, although it was not fated to last. Once Greenpeace and the media departed, according to Hunter, the younger Pilgrim was disowned by his brother, left by his wife, and beaten up by some of the local men. Doug Pilgrim and the sympathetic villagers were clearly in the minority, and a miniscule one at that, as anyone scanning the Newfoundland newspapers in March of 1976 would have quickly realized. The St. John's *Evening Telegram*, for example, called the CBC documentary "as neat a bit of propaganda as we have ever seen." The host of the program, the well-known television personality and future governor general of Canada, Adrienne Clarkson, deserved "a medal of honor for doing such a neat job of corrupting the truth." Greenpeace and Davies, the *Telegram* thundered, were engaged in a "propaganda war over the seal hunt," one that involved "lies, deceit, misrepresentation, suppression of the truth and skillful use of the half-truth."[27] This kind of invective was pervasive in the editorial and letters pages of the major Newfoundland newspapers; even the reportage was larded with sarcasm toward the protestors.[28]

Despite his efforts, Hunter's vision of a worker-conservationist coalition was not to be realized. The international media had come to see a confrontation on the ice: they were not particularly interested in a story about a grassroots campaign to form an alliance between Greenpeace and the landsmen. The successful formation of such an alliance would have required Hunter, or somebody with similar negotiating skills, to remain in the region for much longer than a few weeks. The opposition of key figures such as Watson, who continued to insist that all sealers were Greenpeace's enemies, also doomed the alliance. Furthermore, Hunter had not fully realized just how deep the cultural roots of sealing ran.[29] However distasteful the practice might appear to the outside world, for the "Newfies," swiling was a way of life—a valued part of their culture that they were not prepared to give up without a fierce struggle. To the outside observer, it appeared that the swilers were naïvely engaged in maintaining their status as an exploited underclass—a cheap, brutalized labor force that could be easily manipulated by European fur barons. In academic jargon, they were victims of "false consciousness" or a "culture of poverty." Such explanations, however, are themselves predicated on certain cultural assumptions about people's motivations. They provide little genuine insight into why Newfoundlanders tenaciously defended a brutal, bloody, and heavily subsidized industry that only provided them with a moderate, short-term income—one that could easily be replaced by other forms of employment. The swilers still await their Clifford Geertz—an anthropologist who can offer a "thick description" of their culture to the outside world. Whatever the deeper reasons for their attachment to seal hunting, however, it was not a trait that was going to be overturned by a mere fortnight of mind bombing.[30]

Greenpeace's first save the seals campaign received only a modest degree of press coverage in the United States.[31] Perhaps the NBC producer's claim that they would be "dead in New York" had been more than an idle threat. The European media, however, showed considerable interest in the issue, while the Canadian press deemed it worthy of front-page headlines. However, although Greenpeace had no trouble gaining media coverage in their home country, they could not control how their protests were depicted. Despite Hunter's efforts to downplay some of the more sentimental, emotional, and anthropomorphic aspects of the campaign, the media tended to portray Greenpeace as a narrowly focused animal rights organization that took an extremist position on the hunt.[32] This may have represented the views of many within Greenpeace, but it was not the image of scientific credibility and willingness to compromise with the landsmen that Hunter hoped to convey. For his part, Watson remained disillusioned by Hunter and Moore's compromise: "Basically, they betrayed us. I wanted nothing to do with giving up the dye and compromising with sealers. . . . I just didn't understand where it was going to get us meeting with these people. In the end, it didn't get us anywhere."[33] Watson further argued that Greenpeace should have kept their dye plan a secret, rather than announcing it to the entire world and giving the government the opportunity to counter it. In spite of his uncompromising stance, however, Watson admitted that the selective use of conservationist rhetoric could be useful, since fishermen and sealers "usually respond well to it." However, his own rhetoric remained highly emotive, with its constant reference to harp seal pups as "babies" who were "indistinguishable from [their] human counterpart[s]."[34]

Despite much lurid and sensational press coverage at home and abroad, the Canadian government remained resolute in its support for the hunt. In a letter to Greenpeace, Prime Minister Pierre Trudeau insisted that a carefully controlled hunt was "necessary for the proper management of the seal herds and of the Atlantic fishery resources." Since the seals fed heavily on fish, he continued, an unrestricted growth in their population would severely deplete the fish stocks that were the bedrock of Canada's maritime economy.[35] Such logic tended to ignore the fact that, historically, both the seal *and* the fish populations had been much higher, which suggests that seals were being used as scapegoats for poorly regulated fisheries. Furthermore, it hid the fact that Canada had recently signed an agreement with Norway that would allow Norwegian fishermen to take "surplus stocks" of capelin, the harp seal's major source of food, from Canadian waters.[36] Clearly, despite Hunter's and Watson's efforts, both the commercial and local harp seal hunts were a long way from being abolished. Greenpeace, it seemed, was destined to spend many more years on the ice floes.

By the middle of 1976, Bobbi Hunter was beginning to worry that her new husband was cracking under the strain of running Greenpeace. It was at that point that a call came from David McTaggart, asking if Hunter would be able to join him in the United Kingdom for a few months to help him write a book about his showdown with the French. Initially, Hunter balked at the idea, not wanting others to think that he was running away from the debts the organization was facing. However, with McTaggart's persuasion and Bobbi's urging, Hunter retreated to a small village in the Welsh countryside for the last months of 1976, leaving Paul Spong to take on the job of acting president. It was a break he absolutely needed, but one that he would come to rue.[37]

During Hunter's absence, Watson began to plan for a 1977 antisealing campaign that would be far more radical and uncompromising than the previous year's effort. Still seething over what he felt was Hunter and Moore's hijacking of his campaign, he insisted that Greenpeace would not reveal their tactics in advance this time. Nor would they enter into an alliance with the landsmen. The landsmen, Watson felt, were largely responsible for the fact that the number of seals killed in 1976 had been some 40,000 more than the government quota. An alliance between Greenpeace and the landsmen, he insisted, was morally repugnant and tactically naïve. It would allow the swilers to be portrayed as victims of exploitation, rather than the perpetrators of the slaughter. Any association with them would taint Greenpeace's purity.[38] He outlined his more strident approach in an article in the *Greenpeace Chronicles*:

> As conservationists, as environmentalists, we cannot allow this hunt to continue. The entire hunt must be stopped immediately and totally by both commercial and landsmen interests. The seals cannot take any more abuse. . . . The Greenpeace position . . . is that we are totally opposed to the killing of all seals by Canadians, Norwegians, Danes, and others. We are opposed to the Norwegian-Canadian commercial sealing industry. We are opposed to the killing of seals by Canadian

> Landsmen. We are opposed to the Canadian Department of Fisheries and the Minister of Fisheries for not protecting the lives and future survival of the harp and hood seal species. We are advocating a world free of all sealing and whaling activities.[39]

To emphasize just how serious he was about this, Watson told the press: "I promise a significant number of seal pups will be saved or else I and my fellow crew members will die trying."[40]

In order to carry out such a campaign, Watson knew he would have to undermine Hunter's authority and persuade Greenpeace to abandon its more pragmatic and conciliatory tactics from the year before. He traveled to the various Greenpeace offices throughout Canada and the United States, recruiting new members to his cause and persuading them to adopt his uncompromising tactics. He also heightened tension between Vancouver and the various branch offices, implying that the local groups' autonomy was under threat from Vancouver's imperial plans to establish an international organization, with itself at the head. The new Greenpeace group in Hawaii was particularly sensitive about such issues, and Watson exploited their suspicions, or, as he saw it, "sided with Greenpeace Hawaii's self-determination." Although Watson rarely admitted it, it was clear to those around him that he hoped to displace Hunter and take over as Greenpeace's president, a move that would enable him to lead the organization down a more radical and uncompromising path.[41]

Patrick Moore, in the meantime, was also beginning to question if Hunter was still the right man to lead Greenpeace. It was obvious to Moore that Hunter "was a complete nervous wreck. He had nervous breakdowns, was vomiting uncontrollably all day long and was incompetent, basically because he had burnt himself out. He had nothing left of energy or sanity. He was really needing a year off."[42] Hunter admitted that this was mostly true: "I drank and smoked too much, and was too stressed out and couldn't keep food down. I was burned out, no doubt about it." He also admitted that, in his zeal to further Greenpeace's cause, he would occasionally make injudicious statements such as: "I don't give a fuck who I take money from! I'll take it from child molesters or the CIA. I'm just trying to save whales." Such remarks were carefully noted by Watson and used against Hunter at the next board meeting. Hunter also later admitted that his leadership style "might have been appropriate in the context of a small band of borderline outlaws, but it was completely unsuited to a miniature multinational corporation struggling to bring itself into existence."[43] Not surprisingly, therefore, when Hunter returned from his two-month sabbatical in the Welsh countryside, his not-so-loyal lieutenants were in open rebellion against him. At the board meeting in January 1977, Moore decided to run against Hunter "for the good of the organization." He was backed by Watson, who, while still contemptuous of Moore, saw him as a stepping-stone toward his own ascent to power. Hunter, however, was not prepared to give up his position without a fight. He forced the issue to a vote and emerged as the victor by the slimmest of margins—one vote. He now had the dubious honor of leading the organization at a time when it was more openly divided and fractious than at any other time in its short history.[44]

Also in January 1977, the Canadian government passed an act that Greenpeace advocated for the previous two years: they extended Canada's territorial limit from twelve miles to 200 miles from the Canadian coast. At the same time, however, they increased the harp seal quota from 127,000 to 170,000. If the Canadian government thought that the protest had peaked during the previous year and that the media were no longer interested in the issue, they were sorely mistaken. Greenpeace's 1976 campaign had attracted the attention of a wealthy Swiss animal rights activist named Franz Weber, who decided to devote his prodigious PR skills and considerable wealth to the antisealing cause. He and his wife had designed and mass-produced a toy baby harp seal that was selling extremely well throughout Europe, raising considerable funds, as well as awareness and sympathy for the seals. In an effort to end the hunt, Weber made an audacious proposal to the Canadian government and to the people of Newfoundland. In exchange for giving up swiling, Weber promised, he would provide the swilers with $400,000 to compensate for their loss of income for that season. In addition, he would set up a fake fur coat factory in the area that would offer alternative employment and promote a tourism industry that would feature dogsled trips to the ice so that people could view the seals. Not surprisingly, the fisheries minister, Roméo LeBlanc, rejected the offer. After meeting with Weber, LeBlanc concluded that, although the millionaire activist appeared to be acting in good faith, he was an "obviously ill-informed man." The real flow-on value of the hunt, the minister insisted, was far higher than Weber's paltry compensation package. This prompted Weber to announce that he would spend one million dollars to fly 600 reporters from all over the world to the hunt.[45]

In addition to worrying about the prospect of a more radical and unpredictable Greenpeace protest, as well as Weber's threat to flood the ice with journalists, the government was bombarded with protests from the United States. In the first three months of 1977, the Canadian Embassy in Washington received 53,000 letters from irate Americans who were opposed to the hunt. To make matters worse, Congressman Leo Ryan, a California Democrat, introduced a House resolution "appealing to the conscience of the Canadian government" to end the "barbarous" practice.[46] Both the U.S. House and Senate passed the motion unanimously. The Canadian Parliament, however, remained undaunted, dismissing the resolution as "ill-informed."[47] To complete the Canadian government's misery, the American press also joined the antisealing chorus. For example, the *Christian Science Monitor*, in an editorial that may as well have been a Greenpeace press release, castigated the government for "ignoring the harp seal's role in the eco-system in order to court votes in the Maritime provinces." There was, the *Monitor* continued, "something supremely unnatural about taking the female harp seal's offspring only ten days after birth and using the pelt for absolutely non-essential items. . . . We agree with Greenpeace . . . that the Canadian government should at least declare a six year moratorium."[48]

To avoid the wrath of angry Newfoundlanders, Watson decided to base Greenpeace's 1977 protest in the Quebec fishing village of Blanc Sablon. He also brought with him a considerably larger crew, along with extra equipment and supplies. The cast of mostly young activists reflected Greenpeace's continuing international expansion. In addition to relative veterans such as Jet Johnson and Gary Zimmerman, there was Margaret Tilbury, the driving force behind the Portland, Oregon, Greenpeace office, and Ingrid Lustig from the new Seattle branch. Allan Thornton, originally from Vancouver, and Susi Newborn, a young Friends of the Earth activist, arrived from London, where they, along with David McTaggart, were busy establishing a UK Greenpeace office. There were three women from Norway, as well as several Canadians from the numerous Greenpeace offices that sprang up throughout the land. Rounding out the group was a film crew; a lawyer; various technical personnel; and several reporters, including Bob Cummings from the original Amchitka campaign, who was taking Hunter's place as the media coordinator. In order to try to restrain Watson's impulsive tendencies, Hunter saddled him with Patrick Moore, a move that is hard to interpret as anything other than an act of revenge for the role both men had played in the attempted coup.[49]

If the Mounties had been unprepared for the previous year's protests, they were certainly not taking any chances in 1977. Correspondence between the local RCMP and various offices in Ottawa indicates that the police knew exactly how many Greenpeace protestors to expect, where they were staying, and what kind of equipment they had hired. Watson and others, however, may have felt somewhat chagrined if they had known that the RCMP considered Greenpeace to be less of a threat than the IFAW. "Unlike Greenpeace," one officer wrote, Davies "thrives on confrontations with law enforcement agencies while at the same time puts it all together on film to show his financial supporters in the United States." Greenpeace, on the other hand, "won't put up any resistance" and "are interested in publicity alone."[50] Watson would soon give the RCMP plenty of reasons to revise this opinion.

The increased scale of Greenpeace's 1977 harp seal campaign seemed to provoke a commensurate growth in the degree of confusion and mishaps. Soon after Watson and his crew arrived in Blanc Sablon, the tiny village was flooded with some forty-five European journalists who had flown to Canada courtesy of Franz

Weber. This was far below the 600 reporters that Weber had threatened to bring with him, a fact that all and sundry were grateful for given the dearth of accommodations and the fact that Weber was only able to secure three helicopters. Disgruntled journalists, who had been expecting their own suites, found themselves sharing single rooms with half a dozen of their colleagues. Meanwhile, Brian Davies had arrived in St. Anthony, where it took some eighty RCMP officers to protect him from the furious mob of Newfoundlanders who greeted him. The locals were incensed by the expense and manpower that was devoted to "protecting" the protestors. Where, the *Western Star* cried, "is the protection for the hunters and the ship owners who are trying to carry out a legal pursuit?" "If Newfoundlanders went into British Columbia . . . to disrupt an industry, it wouldn't be long before they'd all be in jail."[51] It soon became clear that Watson's decision to base Greenpeace's campaign on the Quebec side of the Gulf of St. Lawrence was something of a logistical mixed blessing. Certainly, it was much easier for Greenpeace to go about their business without the constant protests of angry Newfoundlanders. However, most of the U.S. and Canadian media had gone to St. Anthony. Therefore, while Greenpeace received plenty of press coverage in Europe, they did not fare so well in North America, where most of their members resided.[52]

Greenpeace's policy on the hunt remained confused and disjointed. Watson's hard-line, abolitionist stance—which the media and the public generally understood to be Greenpeace's position—differed little from that of Brian Davies's animal-rights-based approach. Despite apparently sharing the same goals, however, the two groups found cooperation difficult. Moira Farrow, a *Vancouver Sun* reporter who was generally sympathetic to Greenpeace, noted that there was a tension between the groups in both their styles and operating procedures. Davies preferred to operate as a loner, generating media attention merely by his presence and by performing rather tawdry stunts, such as parading photogenic airline stewardesses on the ice. Deliberately or not, such an approach put IFAW in direct competition with Greenpeace, both in terms of media coverage and in shaping the overall message about the hunt. The result was that the media frequently confused the two organizations.[53] Such confusion was exacerbated by the efforts of some Greenpeacers to distance themselves from Davies. For example, two Winnipeg Greenpeace activists insisted that the organization took pains to distinguish itself from IFAW's sentimental anthropomorphism: "We don't take a bleeding heart stand on this matter. We're more concerned about the possible extinction of the harp seal and the fact that it's a wasteful hunt."[54]

To further complicate matters, the 1977 hunt was graced by the presence of the renowned French actress and "sex kitten" Brigitte Bardot. Bardot devoted much of her free time to animal rights activism in Europe, and Franz Weber had managed to persuade her to lend her considerable fame to the harp seals' cause.

As Hunter noted, Bardot's presence completed the last piece of the tabloid puzzle: "Until her arrival, the seal-hunt 'story' was all blood and death, but now it was blood and death and *sex*."[55] The jaundiced and mostly male media contingent in Blanc Sablon treated Bardot with a mixture of adulation and mockery, alternately fawning over her and nudging each other in the ribs and laughing in her face. At one point, someone in a crowded press conference asked her if she would like to show the journalists an example of a freshly killed seal pup, before producing a seal carcass in a plastic bag. "They ask me if I will attend the massacre as if they were asking me about a premiere at the Lido," Bardot recorded in her journal. She did not help her cause by telling the assembled throng, which now included many Canadian journalists, that "in Europe you are called Canadian Assassins." Due to various logistical difficulties, Bardot and her entourage were having trouble securing a helicopter to fly Brigitte to the ice, where they would take the all-important photograph of the beautiful actress embracing a seal pup and staring into its eyes with unadulterated maternal love. Patrick Moore, who freely confessed to being smitten by Bardot's charm, suggested that Greenpeace offer her a ride on one of their helicopters, thereby associating themselves with the actress and taking advantage of the media spotlight that followed her every move.[56]

Moore's suggestion provoked a storm of controversy within Greenpeace. Bob Cummings argued that such a cheap tactic would be beneath Greenpeace's dignity. Watson agreed. Many of the women in the crew were also furious, feeling that Bardot's presence would sully their own noble and serious protest, as well as being a slap in the face for feminism. They made snide comments to the media about how Bardot was afraid of seeing a dead seal and seemed more concerned about where she would go to the toilet on the ice than about the seals.[57] With Hunter's support, however, Moore prevailed, and Greenpeace flew Bardot to Belle Island in one of their helicopters. The next day, the *Vancouver Sun*'s front page was emblazoned with the headline: "'Brave' Greenpeace warms frigid Brigitte," and similar stories appeared throughout the world.[58] Despite the somewhat contemptuous attitude of Watson and the female crewmembers, Bardot had nothing but praise for Greenpeace after her brief trip to the ice. "I admire them. What courage and devotion.... They are fabulous people. *Vivent* Greenpeace!"[59] Although Greenpeace finally received some press coverage in North America, as well as in Europe, the stunt made a mockery of any high-minded attempt to focus on ecological rather than emotional issues. It also destroyed any lingering hope that Hunter held of building a coalition between Greenpeace and the landsmen. As Calvin Coish noted: "To the people of this corner of the world, the idea of Brigitte Bardot at the seal hunt was unthinkable, an insult, almost laughable."[60]

On the ice floes, conditions proved more difficult and dangerous than usual for both the swilers and the protestors. Warmer weather and a heaving sea had fragmented the pack ice into a giant ill-fitting jigsaw puzzle. Small, thinner-than-usual ice pans split apart and were pushed back together by the waves below, grinding against one another like giant misshapen molars. Those on the ice had to hop from pan to pan and risked falling into the frigid water or getting crushed between two blocks of floating ice. Much like the previous year, instead of remaining together and forming a well-disciplined "protest army," the Greenpeacers fanned out among the ice pans, each one attaching themselves to a swiler like an annoying jester. Perhaps inevitably, it was Watson who took the protest to a new level of confrontation. As he and Greenpeace's lawyer, Peter Ballem, approached a sealer in the process of skinning a freshly killed pup, Watson sneaked up behind him, picked up his hakapik, and threw it into the water. The hakapik, it turned out, could float, and while the angry swiler was retrieving it, Watson picked up the pelt and threw that into the water. Ballem warned Watson that such actions could be construed as theft under the Criminal Code. For Watson, however, such narrow legalistic concerns seemed absurd compared with the crimes of the swilers. When Ballam tried to ask the swiler if he was from Norway or Newfoundland, Watson, according to Ballem, pointed to the label that was still stuck to the man's sunglasses and interjected snidely: "Well, can't you tell he's a Newfie; only a Newfie would wear a sticker on his sunglasses."[61]

Watson then moved toward the nearest ship, the *Martin Karlsen,* which had stopped to pick up a pile of pelts. Sealers bundled pelts together and attached them to a cable, enabling them to be winched onto ships. Watson strode purposefully toward the pile, pulled out a pair of handcuffs that were fastened to his belt, and attached himself to the winch cable. This, he assumed, would prevent the winch operator from hauling any more pelts on board until the crew had found a way to remove Watson's cuffs. Like the Soviet harpooner from 1975, however, the winch operator aboard the *Martin Karlsen* was not about to let such an annoyance deter him. Here is Watson's version of what happened next:

> My heart almost stops when I feel the first tug. The wire line is taut. The two dozen shipboard sealers are cheering and urging the winch man on. I am being pulled off my feet and across the ice. My parka and pants have been ripped on the sharp ice. Suddenly, beneath me there is no longer solid ice. I am being hauled through a thick sludge of ice and water. I feel the line pull upward, and my body leaves the water and slams against the side of the steel hull. Ten feet above the water I'm hoisted. The line stops, slackens, and I fall back to the water, jerked to a stop while waist deep. Immediately the line tightens again, and once more I'm hoisted upward. Once more the line slackens and I fall, this time plunging into the water up to my neck. Again I'm hoisted, again I'm dropped. I'm a mouse on the end of a string being treated as

> a plaything. On the fourth pull, my belt breaks beneath the strain, and I find myself falling five feet into and under the frigid waters of the North Atlantic. The shock immobilizes me. I can't move my legs, my arms are numbing, my chest feels as if it's on fire.[62]

Unlike Hunter, Watson generally scorned the *I Ching* and all things mystical. To those of a New Age mindset, however, it seemed that Watson was suffering from a case of bad karma. In 1975, against all the odds, Fred Easton's camera had come to life at exactly the right time to capture the harpoon flying over a Greenpeace Zodiac and into the back of a sperm whale. This time, as Watson dangled from the winch, Easton stood cursing on the ice, frantically trying to reload the camera. Only when Watson was safely laid out on Peter Ballem's coat, dripping, shivering, and shouting at the sealers on the ship, did Easton manage to start filming. As if his painful and humiliating dunking had not been enough, Watson did not even get the satisfaction of creating a mind bomb. For the next half hour, he fell in and out of consciousness while Ballem and Allan Thornton tried to convince the ship's captain that Watson's life was in danger and that he needed dry clothes and a medical examination. An unsympathetic fisheries official arrived on the scene and insisted, despite Greenpeace's protestations, that he had no authority to order the captain to take Watson on board. Eventually, the captain relented and Watson was placed on a stretcher. Even as they were hauling him on board, the crew tried to give Watson one last dunking, and only the straps holding him to the stretcher prevented him from again falling into the icy water. Once on board, according to Watson, a crewman grabbed him by the hair and shouted into his ear: "We's a-going t'give ya a good taste of swile fat, b'y!" before pushing him face down into a mound of bloody seal blubber. While still in his stretcher, he was dragged through the blood and seal fat on the deck, with enraged, heavy-booted swilers kicking him along the way, before he was taken to an officer's cabin, where he was lambasted by the ship's captain who called him a "menace" and told him to "go straight to hell."[63]

Apart from Watson's misadventures, the protests were, in the main, similar to those of the previous year. Greenpeacers protected seals with their bodies while photographers snapped pictures of the action. At one point, two protestors tried to replicate Hunter and Watson's ship-blocking action. However, instead of standing with their backs to the ship, they stood facing it, thereby giving the captain a psychological advantage. This time the ship did not stop, forcing the protestors to jump out of the way at the last minute. Apart from the incident with Bardot, however, Greenpeace received little press coverage in North America. Instead of seeing pictures of Greenpeace activists on the nightly news, the people who had contributed funds to the antisealing campaign only saw images of angry landsmen confronting Brian Davies.[64] On the other hand, there was considerable coverage of the Greenpeace campaign throughout Europe, courtesy of the dozens of journalists flown in by Franz Weber. At a time when David McTaggart and others were trying to set up various European Greenpeace offices, such publicity proved extremely useful.

The 1977 seal campaign was the beginning of the end of Paul Watson's association with Greenpeace. In a scathing report on the protest, Moore called Watson "both a liar and a thief" and insisted that he would "never work with him again." Watson, he felt, was a maverick who could not be controlled: by sticking with him, Greenpeace would soon find itself involved in actions that transgressed their Quaker-inspired values and nonviolent style of protest. Watson had "shown over and over that he provokes violence." To me, Moore continued, "pacifism is not seeing how close you can get to violence without being the first to swing but it is rather seeing how far away you can get from violence." Moore was also critical of Watson's leadership skills: "Paul ran far ahead of the rest, leaving them at [*sic*] their own to make their way to the ships. He made no effort to include the others in the protest and did his most dramatic act when the cameras were not there." After his encounter with the crew of the sealing vessel, Watson lost all interest in Greenpeace's base camp on Belle Island and instead remained in his hotel room for several days during which he was comforted by one of the Norwegian women from his crew. "Yes, he was bruised a bit but leaders have gone back into action with far worse injuries than he suffered." It is clear from Moore's letter that he was deeply wounded by the disrespect shown to him by Watson and some of his supporters, who told him repeatedly that he was not part of the expedition and that they wished he had not been there. "If being in this organization means that you have to put up with insults from boorish kids who think they are my superiors then I will eventually opt out." History, Moore warned, showed that leaders such as Watson "do not fade away automatically but must be dealt with directly."[65]

Hunter, too, had become fed up with Watson, whose actions after the seal campaign, he felt, were irresponsible and divisive. The campaign, Hunter claimed, had cost twice as much as it should have, "and instead of staying at his post to pay off the bills, Watson skipped off to Hawaii to play the role of eco-hero and try to get on a Greenpeace boat leaving from Honolulu." Everywhere he went, according to Hunter, "he stirred up trouble against the Vancouver home office, trying to build a power base of his own from which he could have unchallenged authority."[66] Furthermore, his provocative actions on the ice had endangered Greenpeace's already precarious application for tax-exempt status in the United States.[67] Finally, on June 7, 1977, the Board of Directors, which included Hunter, Weyler, Marining, Moore, and Carlie Trueman, decided to ask Watson to resign from the Greenpeace Board, citing "his record of disregard for group discipline, divisiveness and a continuing pattern of unauthorized and unpredictable statements."

Watson refused, forcing Hunter to move for his dismissal. The motion was carried, with only Watson opposing it.[68] For his part, Watson found the entire episode absurd: "I was laughing at them. I said if I were in the same position again [regarding the sealer's hakapik], I'd do exactly the same thing."[69] Greenpeace's energy, he wrote, "seemed directed more now at working out little boxes and charts and creating an international complex of Greenpeaces that would be controlled out of a super-headquarters in Vancouver. Thus, it appeared, Greenpeace would spend so much of its time organizing and fund-raising that its efforts to conserve our planet would be dissipated."[70]

Apart from demonstrating some fundamental personality conflicts, Watson's expulsion represented Greenpeace's rejection of a more extreme form of direct action. If Watson had been able to gain control of the organization, as he clearly desired, Greenpeace would doubtless have taken a very different path. A glance at Watson's post-Greenpeace career provides a strong indication of where that path would probably have led. Soon after his expulsion, Watson, along with Walrus and Jet Johnson, formed Earthforce, "a crew of experienced environmentalists dedicated to direct action solutions to critical planetary problems." Like Greenpeace, Earthforce professed to be nonviolent. It was clear, however, that Watson's idea of nonviolence diverged somewhat from the Quaker ideal that the likes of Irving Stowe had imparted to Greenpeace: "Our non-violence is not pacifist but rather aggressive and obstructive. If violence is employed against us, we cannot morally resist in kind. If an attempt is made to harm us, to threaten us or to remove us, we have always stood our ground."[71]

Earthforce's first, and, it turned out, only campaign was directed at saving the African elephant from extinction at the hands of poachers. Watson and a small crew flew to Kenya, where they witnessed a gunfight between well-armed poachers and local game wardens that left two men dead. They documented the illegal elephant slaughter, passing their report on to interested members of the U.S. Congress. Soon thereafter, Earthforce folded due to a lack of funds. In 1978, Watson managed to convince Cleveland Amory to fund an anti-harp-sealing campaign. With funds from Amory and from the British Royal Society for the Protection of Cruelty to Animals, Watson bought an old fishing boat, renamed it the *Sea Shepherd*, and set off for the ice floes in March 1979. In a move that surprised both the sealers and the fisheries officers, Watson and his crew fanned out across the ice under the cover of darkness, spraying hundreds of baby seals with an indelible red dye. When furious fisheries officers, accompanied by RCMP officers, attempted to arrest him, Watson fled across the ice, eventually jumping onto a small floe. He continued to resist arrest by fighting the officers off with a stick. When a helicopter hovered above him, he threatened to throw his stick into its whirring rotor blades. Eventually, he was tackled, though not before dragging his assailants into the water for a freezing dip.[72]

In July of 1979, Watson sailed out into the North Atlantic in search of the *Sierra,* a notorious pirate whaler that had been flouting IWC regulations for years, killing any whale that came within its reach. After reinforcing the *Sea Shepherd*'s bow with concrete, Watson set out to find the *Sierra* with the intention of putting it permanently out of action. He finally tracked her down off the coast of Portugal, and, at full speed, rammed the *Sea Shepherd* into the pirate whaler's side, not once, but twice. The incredulous and terrified *Sierra* crew managed to sail the badly damaged vessel back to port, where it underwent extensive repairs, only to be blown up by a limpet mine several weeks later.[73] These and similar acts launched the legend of Captain Paul Watson, the fearless and much-feared eco-vigilante. It also spawned the radical environmental organization known as the Sea Shepherd Conservation Society, a deceptively benign name for an organization that, to this day, continues to engage in acts that skirt the bounds between nonviolent direct action and eco-vigilantism.[74]

In addition to Watson's expulsion, another major change in the organization that year was Hunter's resignation as president and chairman of the board. Apart from health problems and the sour taste left over from Moore and Watson's attempt to unseat him, Hunter suddenly found his passion for Greenpeace had been replaced by a strength-sapping apathy: "The energy of those around me seemed repugnant, embarrassing, silly. . . . Looking . . . at my old friends and comrades, I found that I did not really have anything to say to them anymore."[75] Hunter also found it difficult to come to grips with some of the demographic changes that were occurring in the organization, particularly as a result of its involvement in the harp seal issue:

> Now we had a lot of animal rights people, which didn't matter before when we were mostly anti-nuke, but now you had people who were fanatically passionate about it, and at the same time they put the back up of other people. They were attacking meat eaters. That meant that you had those kinds of people wanting a share of the power and bringing new ideological baggage to bear on the board. So, all the efforts you made to try to create alliances and trans-political gatherings, neither left nor right . . . well, I was prepared to try to suture all these political divisions, but I hadn't really thought about how all the hard-core animal rights people really hate hunters, fishermen, etc., and vice versa. And so that brought in another level of underlying, almost visceral anger between people. And from the point of view of a media campaign, that makes things really difficult.[76]

Greenpeace's foray onto the ice floes of the Gulf of St. Lawrence proved to be a fillip for the antisealing movement, as well as a watershed for Greenpeace itself. Although their campaign did not always go according to plan, their direct-action approach—with its promise of dramatic confrontations on the ice—once again focused attention on the hunt, forcing the Canadian government and the sealers to justify their actions and pressuring them to enact various conservation measures, albeit ones that groups such as Greenpeace found wholly inadequate. The campaign would continue for almost a decade, though in later years, its focus shifted away from the floes and onto the European fur market. The high point occurred in 1983, when the European Community announced a boycott of all harp seal products.[77]

To many Newfoundlanders, the antisealing movement appeared to be nothing more than a "protest business," one that was too lucrative to abandon despite the obvious absurdity and hypocrisy of IFAW and Greenpeace. Newspaper reports frequently portrayed Brian Davies as a jetsetter who had deviously converted public ignorance and sympathy into a steady stream of cash for his own self-enrichment. According to Jeremiah Allen, a political scientist at the University of Lethbridge in Alberta, groups such as Greenpeace were more like traditional businesses or bureaucracies whose primary motive was "the perpetuation of the firm, rather than ending the Atlantic seal hunt." In such a scenario, the "firm" is best served, not by the elimination of the hunt, which would remove the group's raison d'être, but by its continuation.[78] While organizational health was certainly a factor in Greenpeace's decision-making calculus, Allen was guilty of considerable exaggeration. By extension, one could argue that nineteenth-century abolitionists would have been best served by the perpetuation of slavery and that their actions were carried out with this in mind. Such an analysis completely discounts the evangelical fervor that marks many radical protest groups, as well as reducing human passion and commitment—with all their flaws and inconsistencies—to mere variables in economic equations. It also ignores the myriad cultural and ideological influences—from holistic thought to Quakerism, to animal rights—that shaped Greenpeace's actions.

Despite Hunter's efforts, Greenpeace was not able to defuse Newfoundlanders' enmity. The influence of Greenpeace's animal rights contingent, with its insistence that sealing should be abolished, combined with the intransigence of the sealers and the Canadian government, meant that there was little common ground between the two sides. Nevertheless, one can legitimately argue that groups such as Greenpeace, by keeping the issue in the media spotlight, contributed to the creation of one of the world's more sustainable hunts.[79] While this is scant consolation for animal rights activists, for those who take a more pragmatic and ecological approach to environmental issues, it can be seen as at least a partial victory. Certainly, hundreds of thousands of seals were still hakapiked to death every year. The species as a whole, however, did not appear in imminent danger, and earlier claims that harp seals were on the road to extinction appear, in hindsight, to have been somewhat exaggerated. Nonetheless, for people such as Bob Hunter and Paul Watson, this was hardly a cause for celebration. For them, the depressing reality was that after more than two decades of "mind bombing," there was still a substantial portion of the population for whom a harp seal remained a mere "natural resource."[80]

12

The Paradox of Power

The Birth of Greenpeace International

As a general rule, an institution's values and culture are usually reflected in its organizational structure and operating style. The rigid hierarchy and strict operational order of most armies, for example, reflect the discipline, obedience, and unquestioned respect for superiors that the military instills in its recruits. On the other hand, institutions whose members profess a strong allegiance to social justice, grassroots activism, and egalitarianism—such as various Green Parties around the world—will often adopt a consensus approach to decision making and a structure that is as democratic and nonhierarchical as possible. On this score, Greenpeace was faced with something of a dilemma: most of its members were committed to the countercultural values that had influenced so many of their generation. The institutional culture and structure of the military, a corporation, a government bureaucracy, or even a traditional conservation organization, would have grated against their values, which tended toward egalitarianism, libertarianism, and anarchism, rather than respect for authority and an appreciation of hierarchy. Could one imagine eco-hippie mystics such as Bob Hunter and Rod Marining operating within the boardrooms and the executive offices of the Audubon Society or the Sierra Club, with their adherence to the traditional protocols of corporate governance?

According to the German philosopher Jürgen Habermas, a society's ability to learn and respond to changed conditions depends on its ability to generate alternative worldviews. This is facilitated by a capacity for communicating new ideas and views to the broader culture, which leads social and political institutions to adapt to changes taking place within society and the world in general. If societies are to adequately address ecological problems, Habermas argues, it will require the development of a robust civil society—that is, the level of society above the individual and below the state—in which environmental concerns are first perceived and identified by citizens, "then accurately reflected and effectively presented by representative institutions of civil society, and then satisfactorily addressed by the political system."[1] If the institutions of civil society—primarily consisting of myriad nonprofit organizations, charities, and social clubs—are to ensure open communication between the everyday world of citizens, what Habermas calls the "lifeworld," and the public sphere, they must, according to Habermas, be democratic and open. He concludes, therefore, that if they are to properly fulfill their function as facilitators of communication between the "lifeworld" and the governmental sphere, civil society organizations must be "committed to democratization in their internal structure as well as in their relations with the state and each other."[2]

By the late 1970s, Greenpeace had already passed through several stages of organizational growth, a fact that was reflected in the group's internal structure and decision-making process at various points in its history. The consensus approach that had characterized the Don't Make a Wave Committee quickly gave way to a brief period of autarchy during Ben Metcalfe's reign. By the mid-1970s, when Hunter assumed control, Greenpeace returned to a more consensus-based approach. In this sense, it was in line with a general trend among self-consciously radical social movements throughout the decade. In the words of former SDS leader Todd Gitlin: "From the early '70s on, activists went into revolt against just about anybody's authority, even their own. Vertical authority had a foul odor: it smacked of colonialism, patriarchy, bad white men lording themselves over voiceless minions. In left-wing activist circles, establishments of all sorts were the immoral equivalents of The Establishment." Instead of organized hierarchies and factions, activists developed "affinity groups" and "working groups" as part of an almost fanatical devotion to egalitarianism.[3]

As the whaling and sealing campaigns of the late 1970s become more complex, requiring greater degrees of planning, flexibility, and secrecy, consensus was rapidly replaced by a top-down decision-making structure, with the likes of Hunter, Patrick Moore, and Paul Spong assuming a more traditional form of executive power. At the same time, however, Greenpeace was expanding throughout North America, Europe, Australia, and New Zealand. Since most of these groups were only loosely affiliated with the Vancouver "headquarters," they tended to develop their own internal cultures. Overwhelmingly, they leaned toward leftist and libertarian structures that invoke, in political scientist Herbert Kitschelt's words, "an ancient element of democratic theory that calls for an organization of collective decision-making referred to in various ways as classical, populist, grass-roots or direct democracy against a democratic practice in contemporary democracies labeled as realist, liberal, elite, republican or representative democracy."[4] Understandably, therefore, such groups would not respond particularly favorably to attempts at centralization, professionalization, and other developments likely to promote a more hierarchical structure. Yet, the direction in which Greenpeace's campaigns took them, combined with Vancouver's growing desire to establish a greater degree of control over the international organization, meant that the various Greenpeace groups found themselves facing exactly such a scenario. It was ironic, therefore, that a man who embodied corporate hierarchy and professionalization would come to represent these disaffected groups against what they perceived as the power hungry and authoritarian gang of hippies in Vancouver.

When Bob Hunter took over as president of Greenpeace in 1975, he was naturally wary of the kind of authoritarian approach that Metcalfe had employed during his brief tenure. Nevertheless, the months of unstructured meetings and consensus-based decision making that preceded the first whale campaign had taken some of the shine off the grassroots model. Although such a broad-based participatory structure had given everyone a voice, thereby encouraging goodwill and creativity, it had also led to endless and exhausting meetings and bureaucratic inefficiency. Furthermore, it tended to attract characters that were wacky even by Greenpeace's tolerant hippie standards. The 1975 save-the-whales campaign, while relying on a considerable amount of what could only be described as blessed luck, had also entailed a great deal of detailed planning and organization, as well as a level of secrecy and surreptitious research that would not have been out of place in the military. It became clear to Hunter that if Greenpeace was going to continue to carry out similar campaigns, they could no longer rely on the happy-go-lucky approach that had got them through so far. Paradoxically, therefore, the trappings of the traditional nonprofit organization—Robert's Rules of Order, an executive, a board of directors, sensible financial planning—began to appear positively liberating. In short, the demands and pressures of running an outfit such as Greenpeace, at least in the form that Hunter envisaged its development, dictated a greater degree of professionalization.[5]

By late March of 1976, when the antisealing group returned to Vancouver, Spong, Korotva, and others who had not gone to Newfoundland were frantically preparing for the 1976 antiwhaling voyage. This time, Greenpeace was determined to keep up with the Soviet fleet. Though everyone in the organization had a soft spot for John Cormack and his gutsy little halibut seiner, they also realized that they would need to find a larger and faster boat if they were going to seriously harass the whalers. After conducting an international search, Korotva found a decommissioned Canadian minesweeper in Seattle called the *James Bay*. Ironically, it was an exact replica of the boat that had rammed the *Vega* in 1972, as well as originally being part of the same fleet as the *Edgewater Fortune*, which had served as the *Greenpeace Too* back in 1971. Although structurally sound, the *James Bay* required a lot of work before it could sail on the high seas. In lieu of the $50,000 charter fee, which the cash-strapped organization could not afford, Greenpeace worked out a complicated contract with the boat's American owner that involved putting $50,000 worth of work into the craft. The *James Bay* gave Greenpeace one headache after another, the worst of which was figuring out where to register the vessel. If they registered it in the United States or Canada, they would have to pass Steamboat Inspection, a rigorous examination of all the boat's equipment. It was a test they knew they would in all likelihood fail. The only practical and affordable solution, ironically, was to do what leaky oil tankers and pirate-whaling vessels often did—register the boat in Panama.[6]

The crew for the 1976 voyage included most of the veterans from the previous year, as well as new members, many of whom came from the new Greenpeace branch offices throughout North America. In addition, Rod Marining, Paul Spong, and Bobbi Hunter were also aboard. The captain of the *James Bay* was George Korotva, with Moore as his first mate. On the way south, the ship pulled into Portland, Oregon, to refuel. The degree of sympathy and goodwill from the city, which had a reputation for being one of the most environmentally aware in the nation, was astounding. The city council waived the moorage fees. The fire department provided them with water. Tom McCall, the former Republican governor who had instituted a wide array of environmental measures during his term in office, presented Greenpeace with an Oregon flag. Co-ops donated food, musicians organized benefits, and a local pilot offered to fly

Rod Marining, David Garrick (a.k.a. Walrus Oakenbough), Melville Gregory, Bob Hunter, and Greenpeace USA member Peter Fruchtman prepare for the 1976 whale campaign aboard the *James Bay*. Reproduced by permission of Rex Weyler/Greenpeace.

reconnaissance flights over the ocean. After the travails of Newfoundland, the days in Portland were joyful and inspiring.[7]

On July 1, off the coast of Mendocino, California, the ship was greeted by a pod of Risso's dolphins. It was an opportunity to inspire those among the crew who had never interacted with cetaceans, as well as reminding those who had what their mission was all about. Within minutes, Hunter dived in among the large dolphins—some the size of small killer whales—and he and several of the others spent half an hour frolicking joyfully in the cold North Pacific waters. One of the young dolphins was an albino, which, according to Hunter, "surely had to be at least as rare as Moby Dick in his time." For Hunter, this was another in a growing list of transcendent cetacean moments: "As I floated there just a couple of feet beneath the surface, face-to-face with that white child-dolphin, already larger than me, with his two twinkling-eyed guardians, none of us having the least intention of hurting the other, merely fascinated, the possibility of heaven on earth did not seem at all like a vision or a fantasy or an ideal or a dream. It *was* reality." The meeting with the dolphins created an atmosphere of both exuberance and quiet introspection among the crew. For most of them, the encounter was direct evidence of a "mind in the waters": it confirmed the existence of beings at least as intelligent, and possibly far more aware, than humans. As Hunter emphatically declared: the dolphins "were *people*. Strange people. Different language, different world, different bodies, different minds. But *people*, damn it! PEOPLE!" The experience contributed to the overall feeling, as Hunter liked to put it, that they were involved in a jihad—a holy war against the whalers.

> We were united in a way that religions and revolutions can sometimes be united, at least until they get into a position of power. It was clean. It was something that few other people had gone through. For a day there in the North Pacific in July 1976, there was a boatload of us who understood what joy could be expected to come from interspecies contact—the *real feeling* of it. So I suppose we were pioneers, overcoming our fears and plunging ourselves completely into a whole new environment, not only a physical environment, but a psychological, perceptual, emotional, and experiential environment as well.[8]

Soon after arriving at their final port of call on the mainland—San Francisco—the mood on the *James Bay* turned sour. Initially, the boat was abuzz with eager Greenpeace volunteers from the busy new San Francisco office, as well as members of the public and the occasional celebrity. The Grateful Dead played a benefit concert in Union Square during which Mel Gregory got carried away, stripped off all his clothes, and was arrested. Since money was tight, Hunter decided not to post bail and left him in jail overnight. More than anything else, the financial situation was a point of major contention. Greenpeace simply did not have enough money to carry out the expedition and had to keep raising it along the way. This meant constantly approaching sympathetic entertainers, politicians, and well-heeled conservationists. For Hunter, it also meant making pragmatic business decisions that not all the members of the organization could tolerate. The most contentious of these involved the movie contract from the year before. Amy Ephron from Artists Entertainment Complex returned to San Francisco to try to renegotiate the contract, which again divided Greenpeace. Hunter and Moore led the faction in favor of the contract, while Watson, Walrus, and several of the new recruits insisted it was a sellout and a betrayal of Greenpeace's independence. Their opposition once again scuttled the prospect of a Greenpeace movie. This kind of squabbling went on for almost two weeks before Greenpeace managed to raise enough cash to finance the next stage of the voyage. By the end of that period, nearly everybody on the boat, and many of those in the San Francisco office, was feeling tense and depressed, while Hunter was having one of his periodic breakdowns and contemplating suicide.[9]

As was often the case, however, the open ocean was a tonic, and the mood soon improved under its restorative power. Hunter rediscovered the qualities that made him much respected and beloved among most of the crew: his wit and self-deprecating humor, and his ability to provide leadership and give orders

Patrick Moore and Amy Ephron attempt to hammer out a movie contract (onlooker unknown). San Francisco, July 1976. Reproduced by permission of Rex Weyler/ Greenpeace.

without adopting a superior air—a vital quality among a group of people who were not particularly predisposed to trust authority figures. He strode the deck with a toilet brush attached to his belt, assigning himself the least appealing and lowest status job on the boat. He also published a humorous daily newsletter full of exaggerated rumors and gossip about the goings-on aboard the ship.[10]

Apart from having a much larger and faster boat, there were several other significant factors that distinguished the 1976 campaign from its predecessor. One was that people such as Rod Marining, Bobbi Hunter, Eileen Chivers, and other members of the Greenpeace inner circle were on the boat this year, leaving new volunteers to run the Vancouver office. One of these was John Frizell, a young UBC biochemistry graduate who took Rod Marining's place as the press officer in Vancouver. When the *James Bay* came upon a dead giant squid, Hunter wrote an article suggesting that such a find was indicative of a severe reduction in the number of sperm whales, which had led to a corresponding boom in the population of squid, the whales' main source of food. Hunter speculated that if the squid population was not kept in check by the sperm whales, they would grow larger and more numerous, with potentially devastating effects on the ocean's ecology. The article was reminiscent of the previous year's story about the showdown between the giant squid and the sperm whale. This time, however, Frizell, who felt that the story lacked scientific credibility, refused to release it to the press.[11]

Another significant development was Greenpeace's ability, through a combination of pleading and subterfuge, to scan vast expanses of the ocean from the air in search of the Soviet or Japanese whaling fleets. Rex Weyler and Fred Easton, for example, posed as freelance journalists to get on the free weekly military flight from Honolulu to Midway Island. Others were able to convince several private Hawaiian pilots to fly them out on reconnaissance missions. One Greenpeacer even managed to talk some navy pilots into taking him aboard a high-altitude surveillance flight. The fact that none of these flights yielded any results was interpreted in a positive light: the fleets, aware of Greenpeace's presence, were steering clear of the Hawaiian Islands, thereby sparing the whales in that region. Adding credibility to this view were reports in the Hawaiian press that the U.S. ambassador to Japan had advised the Japanese whalers to stay away from the region if they wished to avoid a potentially embarrassing confrontation.[12]

As with the 1975 campaign, the crew aboard the *James Bay* could readily be divided into "mechanics" and "mystics." This time around, the mechanics had more powerful equipment to rely upon, while the mystics still resorted to the *I Ching*. The first sighting of the Soviet fleet, however, was one that warmed the cockles of the mystics' hearts. Before those on board the *James Bay* could see any sperm whales, they spotted a Soviet harpoon boat headed directly toward them from a distance. Only as the vessel came closer did they notice that it was pursuing a pod of whales that, as in the previous year, appeared to be heading unerringly toward their saviors. The new crewmembers had heard the story from the previous year on several occasions, though many had remained skeptical. However, the sight of a dozen sperm whales swimming frantically toward the safety of the *James Bay* was enough to convince most on board that that episode had not been exaggerated.

The next problem was to decide who would be allowed to ride in the Zodiacs. Passions on the boat were running high. Most of the crew seemed to be gripped by a fervent desire to place themselves between the harpoon and the whales. Hunter had to pull rank to ensure that he was given a Zodiac, taking Bobbi and Spong with him. Within minutes, he was reenacting his protest from the previous year, the Zodiac zipping deftly between the harpoon boat and the whale pod. This time, however, unlike the previous year, the harpooner refused to fire. So long as the Zodiacs remained in position between the harpoon and the fleeing pod, the whales appeared to be safe. The decision not to fire, Hunter later speculated, was probably taken months before, "perhaps at the level of the Kremlin." Whatever the case, the Greenpeacers were exultant. Their actions were having tangible results; even if it was only temporarily, they were saving actual live whales. And they were doing it not by sitting in a conference and arguing for a reduction in the kill, but with their corporeal presence, just as they had saved harp seals the previous winter. Unlike the seal pups, however, who could easily be clubbed as soon as their protector was forced to move away, the whales had a better long-term chance of escaping.[13]

Although there was no spectacular shot of a harpoon flying over a Zodiac and into a fleeing whale, the images were nonetheless dramatic, particularly since the cameramen were more experienced and were using better equipment. Perhaps the most sensational image was of Watson gunning his Zodiac onto the back of a dead whale being dragged up the *Dalniy Vostok*'s stern slipway. For a moment, the Zodiac, teetering precariously on the whale's back, looked as though it was going to be hauled into the bowels of the great factory ship, before it slipped back into the bloodstained ocean.[14]

Another highlight of the voyage, and one of its more poignant moments, occurred when a pod of sperm whales—which Spong insisted was the same group that Greenpeace had earlier saved—swam alongside the *James Bay* for an hour, before halting and allowing the boat to glide through their midst. The whales, Hunter insisted, "were saluting us. It was a conscious and deliberate action on their part." It certainly was astonishing that the whales would choose to follow a boat that, in size and shape, was very similar to the harpoon boats from which they were often forced to flee. The event was immediately followed by a rainstorm, which ended with the by-now obligatory rainbow. The behavior of the whales prompted Hunter and others on board to reflect on John Lilly's theory about why so few whales had ever turned on their human hunters. Lilly argued

The view from a Greenpeace Zodiac: sperm whales being transferred from a harpoon ship to the *Dalniy Vostok*, July 1976. Reproduced by permission of Rex Weyler/Greenpeace.

Paul Watson and Bobbi Hunter approach a harpoon ship, July 1976. Reproduced by permission of Rex Weyler/Greenpeace.

that creatures with larger brains were better able to control their lower-brain reflexes, especially those that precipitate anger and trigger the attack mechanism. The brain of the sperm whale was six times larger than the human brain and was therefore able to better control the whale's base urges. Ironically, Hunter speculated, this "superior" brain meant that whales "were helpless against a smaller-brained creature that had scarcely begun to *dream* of controlling its urge to kill, to exploit, and to dominate." It was clear, Hunter scolded, that "in the midst of wars and the innumerable calamities human beings bring down on themselves, the real horror and tragedy of the century was not to be found in man's inhumanity to man, regardless of the scale, but in humanity's blind destruction of beings so far ahead of us that we might only hope—someday—to reach the state of peacefulness they had long since achieved."[15]

From a media perspective, the 1976 campaign was once again a major success. Footage from the protest was widely shown on television throughout North America, and stories and photos appeared in countless newspapers and magazines. Even the avowedly establishment *Wall Street Journal* gave Greenpeace a lengthy and approving front-page write-up, noting that, of all the antiwhaling organizations, they were "the most dramatic and effective in publicizing efforts to curb the $200-million-a-year whaling industry."[16] There is little doubt that Greenpeace's actions helped make whaling one of the most important environmental causes of the decade, at least in North America. Furthermore, the nature of Greenpeace's protest meant that those who participated in the voyages nearly always returned with an evangelical fervor to save more whales. Not surprisingly, it was these people who frequently worked the hardest, not only on Greenpeace's antiwhaling campaign but also in fostering the growth and spread of the organization as a whole.

Despite the success of the campaign, however, the Japanese and Soviet whaling interests were still strong enough to prevent the IWC from declaring a moratorium. While remaining outwardly positive, Hunter began to experience an inner doubt about the efficacy of Greenpeace's efforts. On the one hand, he calculated, by driving the Soviets away from some of their favored whaling grounds, Greenpeace had probably saved over a thousand whales. On the other hand, the IWC quotas remained as high as before. "Maybe passivism," Hunter ruminated bleakly,

> was a crock of shit. Maybe it was time to buy some explosives and sink the bastards. This business of "bearing witness" must have an end point in it somewhere. I had borne witness to the point of nausea and soul-sickness, yet nothing in the world had really changed. Gandhi's approach may have been appropriate for India, but hadn't India gone on to detonate her own nuclear bomb, and didn't that make a mockery of

> Satyagraha, the philosophy that does not contain such a word as enemy? A hell-pit of doubt and confusion fell upon us all. . . . This business of stopping a whale ship or a seal ship for a couple of minutes here, a couple of minutes there, was so far from the solution we wanted that it seemed, in that incredibly bummed-out period after the return of the *James Bay*, that everything we'd done had been nothing much more than a diversion. Maybe it was time to take up guns.[17]

Hunter's words were more prophetic than he imagined, though it would be Watson who would soon give up in frustration and resort to violence. Not surprisingly, in light of his newfound despair regarding Greenpeace's ability to have an impact on the world, Hunter resigned his position as president at the end of the 1977 seal campaign in April. Although he continued to devote much of his time to Greenpeace and remained part of the Vancouver group's inner circle, his influence over the organization had effectively peaked.

There was never any doubt that Patrick Moore would succeed Hunter as president. This was an accession that most of those in Greenpeace's inner circle accepted with a sense of weary inevitability rather than excitement. Along with Hunter and Marining, Moore was the only member leftover from the Don't Make a Wave Committee days, and it was clear that he was above Marining in the pecking order. Knowledgeable, articulate, and highly intelligent, Moore had, on paper, all the right credentials for the job. Furthermore, his ecology PhD gave him extra cachet, at least among the media and the general public. However, there was something about his demeanor that grated on people, particularly the younger and more self-consciously radical members. Many felt that Moore had a certain air of superiority that frequently came across as patronizing and, on occasion, autocratic. In a more traditional organization, such as a business corporation or a government bureaucracy, such traits may not have counted against him. Indeed, they may have been considered virtues. But Greenpeace was not such an organization. By the late 1970s, Greenpeace was, in Hunter's colorful description, "a wild, wild blend of old folks and young folks, tough, practical technicians and visionary animal freaks, dropouts, conservationists, action junkies, traditional Quaker types, and a few flinty-eyed businessmen." The majority of these people were from what Hunter referred to as the "Post-LSD Generation"; those who had emerged "out of the apparent wreckage of the psychedelic revolution." They were not people who were likely to respond well to Moore's rather traditional and, compared to Hunter, somewhat uncharismatic authority.[18]

By the time Greenpeace embarked on their 1977 whaling campaign, the organization had entered an entirely new geopolitical realm. Much of this was due to the influence of Robert O. Taunt III, a new member of the San Francisco office. Taunt, who was from a well-to-do northern California family, had worked in the state legislature in Sacramento throughout the early 1970s, where he concentrated primarily on educational issues. He had a background in the antiwar movement but was never a hippie or a full-time student radical. In his own words, he was "formed by the sixties, but not seduced by them." In 1975 he unsuccessfully ran for office in the state legislature as a liberal Democrat. Handsome, well-spoken, and always impeccably dressed, Taunt proved to be an excellent media spokesperson. Furthermore, his political connections provided Greenpeace with assistance from unexpected, sometimes problematic quarters. Among his friends, Taunt counted U.S. Congressman Leo Ryan, a California Democrat who chaired the House Environment, Energy, and Natural Resources Subcommittee and who had played a major role in helping to pass the Marine Mammal Protection Act. It was Ryan who, with Bob Taunt's help, had instigated the congressional resolution against the harp seal hunt in early 1977, and through him Greenpeace gained entrée to people and information that would otherwise have been denied them. Most notably, and again through Ryan's efforts, the U.S. military provided Greenpeace with regular updates on the position of the Soviet whaling fleet. Convinced that the *Dalniy Vostok* was doubling as a Soviet surveillance vessel that was tracking the U.S. submarine fleet in the Pacific, the Pentagon was only too happy to see its activities disrupted by Greenpeace. The Soviets, in turn, suspected Greenpeace of being a front for the U.S. Navy or the CIA, and their submarines were keeping careful track of Greenpeace's ships. Taunt and many others within Greenpeace were desperate to confront the Japanese fleet, but the military refused to supply them with details of the Japanese position. Only once, apparently by mistake, did Greenpeace receive the coordinates for the Japanese whalers.[19] Hunter, Moore, and others within the organization had always suspected they were being used as pawns in the Cold War. Indeed, they were not above using Cold War rhetoric if it meant they could get more publicity for the whales. Now they suddenly found themselves in the awkward position of being allied with the entity whose Cold War tactics—namely, nuclear testing—had prompted the protest that had brought Greenpeace into existence in the first place.

Taunt was also largely responsible for negotiating a deal between Greenpeace and America's ABC television network that would net the organization $60,000 and valuable prime-time publicity. In early 1977, the producers of the program *American Sportsman* approached Greenpeace's San Francisco office and outlined their plan to make a sympathetic documentary about the group's whale-saving exploits. They offered to send along an eight-man crew, equipped with their own helicopter, as well as contribute money to help cover Greenpeace's expenses. This time, there was little opposition to the idea, particularly since the $60,000 would make the difference between a bare-bones, ill-prepared voyage and a well-equipped campaign. In Honolulu, Korotva, Spong, and a group of some three dozen mostly American volunteers were hard at work converting a rusty, old World War II submarine chaser into the latest Greenpeace protest vessel. They had spent much of the previous year organizing several successful fund-raising ventures and, with ABC's help, were hopeful of having the vessel, rechristened the *Ohana Kai*, in good enough shape to keep up with the whaling fleets over a long distance. However, by late May, the *Ohana Kai* was still short of passing inspection, and those in the Vancouver office worried that the ship might not be ready in time for the coming whaling season. Moore and Hunter decided that they would not wait around to see if Korotva and Spong would complete their tasks. They recruited John Cormack to captain the *James Bay*, which was lying idle in its berth on the Fraser River, while Bill Gannon once again used his financial savvy to convince the bank to extend Greenpeace's line of credit, despite the organization's considerable debts. Within three weeks, a crew had been selected and the boat had been fitted for another Pacific voyage in pursuit of the Soviet fleet.[20]

The *James Bay* left Vancouver on July 17, 1977. Two weeks later, the *Ohana Kai* departed from Honolulu. For the first time, Greenpeace had two boats that were large enough and fast enough to keep up with the whaling fleet on the open ocean. Once again, the swift and maneuverable Zodiacs buzzed along behind the fleeing sperm whales, daring the harpoon gunners to risk human lives in order to shoot their quarry. This year, one of them took up the challenge, firing a harpoon narrowly over Rex Weyler's head but missing his main target. Despite the faster ships, superior equipment, and the assistance of the U.S. military, however, there was little about the 1977 campaign that was different from its low-tech 1975 counterpart. Bigger, it seemed, was not necessarily better. A particularly soul-destroying moment occurred during a Zodiac crew change. While the inflatables were bobbing by the side of the *James Bay*, the harpoon boats surrounded the pod that Greenpeace had been protecting, killing every single whale and dragging them back to the factory ship. Furthermore, the *Ohana Kai* became a victim of its own success. Its speed and fuel storage capacity allowed it to keep pace with the whaling fleet over a long distance. When the former subchaser managed to track down the *Vostok*, however, the Soviets, realizing they could not outrun their adversaries, simply curtailed their whaling activities. Instead, they sailed in a straight line for several days, knowing that sooner or later, the *Ohana Kai* would have to head back to Hawaii to refuel. The only direct encounter between the two groups occurred when Spong and a small Greenpeace delegation were allowed to board the *Vostok*, plead with the captain to stop whaling, and leave the crew with assorted Greenpeace paraphernalia. It was not exactly the stuff of mind bombs.[21]

In addition, the deal with ABC proved to be more trouble than it was worth. The original contract had been signed with the Greenpeace Foundation of America, as the San Francisco branch was officially called, and gave ABC exclusive rights to any film that showed an encounter between Greenpeace and the whalers. ABC executives were none too pleased, therefore, when rival networks began to air film showing exactly such encounters. Nor were they thrilled to find that the film was being distributed by another group called the Greenpeace Foundation, this time based in Vancouver, which was a different entity from the San Francisco group, though both considered themselves part of the same organization. Initially, the ABC lawyers were a little confused as to whom they should sue. Nevertheless, they made it eminently clear to Hunter and Taunt, who were coordinating the campaign from the San Francisco office, that *somebody* was going to face a mammoth lawsuit. In the end, Taunt managed to strike a deal that involved ABC releasing three-minute clips immediately after they had shown them on their network news, while maintaining rights to all the footage until it was aired on *American Sportsman*, probably in eight months time.[22] The entire fiasco was another in a growing list of episodes that highlighted the inadequacies of Greenpeace's organizational structure. Clearly, if the organization was going to continue to expand internationally, and if the various offices were going to cooperate on campaigns, then they would need to develop some sort of coordinating structure. The dilemma, of course, was that such a structure would almost invariably lead to a concentration of power in a few select places. It was a question that the majority within Greenpeace, who mostly wanted to spend all their energy saving whales and seals, did not particularly wish to confront.

Given the resources and work that had been put into it, the 1977 campaign was, at best, a qualified success. Most of the best footage had come from the last-minute *James Bay* voyage operated by the experienced crew from Vancouver, rather than from the predominantly American group aboard the *Ohana Kai*. Nevertheless, there was some encouraging news. While the two Greenpeace ships were at sea, Bob Taunt received a phone call from President Jimmy Carter. Carter, it turned out, opposed commercial whaling and decided to call Greenpeace after seeing footage of their exploits on television. While Taunt reflexively stood at attention, Carter commended Greenpeace for their efforts and asked if they could send him documentation that proved that the Soviets were killing undersized whales. If the president of the United States was now on their side, it seemed to Greenpeace, then surely they were winning the battle to save the whales.[23]

Back in Newfoundland, meanwhile, the provincial government, with Ottawa's aid and approval, had launched a major public relations offensive in the hope of arousing sympathy for seal hunters. In January 1978, at a cost of some $200,000, a committee of scientists and bureaucrats, led by Newfoundland Premier Frank Moores held a series of talks and press conferences throughout United States. The hunt, Moores's committee insisted, was worth almost $6 million annually and employed over 4,000 people. Apart from its economic importance, Moores contended, swiling also had important social benefits: "The hope and expectancy of the seal hunt provides an accentuated momentum in daily life, contributing to an escape from the monotony of a long, harsh winter in certain communities where employment falls below 10 percent of the workforce during the winter months."[24]

Much to the committee's chagrin, however, they found that most Americans had made up their minds about the hunt, which they found a brutal and senseless exercise. Despite efforts to focus discussion on the plight of the sealers, reporters preferred to concentrate on the seals themselves. Why, they asked, did "baby" seals appear to be crying? Was it true that seals were often skinned alive? Those posing such questions, as one reporter noted, did "not respond well to hard facts," despite the committee's best efforts to assure them that the hunt was far less cruel than the modern industrialized slaughter of livestock.[25] Nevertheless, the committee did score some important victories, the most notable of which was a *New York Times* editorial that cautiously approved of the hunt.[26] The government of Newfoundland also began to encourage the consumption of seal meat. By early 1978, recipes for adventurous dishes such as "canned seal pie" began to appear in local newspapers and magazines.[27] The Moores's committee blundered, however, when it brought two cans of seal meat to a press conference attended by Bob Taunt. Taunt charged that the Canadians had violated the Marine Mammal Protection Act, which banned the importation of seal and whale products into the United States, and called for their immediate arrest. Soon thereafter, the two cans were winged back to Ottawa in a diplomatic pouch.[28]

The Canadian and Newfoundland governments could nonetheless take heart from a public opinion poll that indicated that, although a majority of the Canadian population (57 percent) still opposed the hunt, a sizeable 24 percent approved of it, while 18 percent remained undecided. In the Atlantic provinces, the level of support was, understandably, even higher, with 56 percent approving of the practice. Such figures belied Greenpeace's claims that an overwhelming majority of Canadians were opposed to harp sealing, bolstering the government's confidence in its hard-line position.[29]

Fed up with the annual spring antisealing ritual, the Newfoundland RCMP was determined make protest as difficult possible. In a report written at the end of the 1977 season, one Corner Brook officer warned: "There is no doubt that if DAVIES and the Green Peace people return to the hunt next year that it will not end as peacefully as it did this year. There has to be legislation passed under the Fishery Act that will bar anyone, aircraft, or vessel from going to the Front unless they are engaged in the Hunt or have a legal right to be there."[30] Ottawa duly obliged, passing an order-in-council that made it illegal for anyone except a sealer to so much as set foot on the ice within the vicinity of the hunt. The RCMP also began preparing the grounds for a conspiracy charge against Greenpeace and IFAW. If they could collect enough evidence to prove that the protestors were conspiring to break the law at the 1978 hunt, they hoped to arrest them as soon as they set foot in Newfoundland.[31]

Greenpeace, it appeared, was losing the initiative in Canada. To help it bypass the new regulations, the organization called upon its friends in high places. Just as it felt it had closed off every avenue of protest, the Canadian government suddenly found itself under pressure to grant permits to Congressmen Leo Ryan and Jim Jeffords, a Republican from Vermont, who wished to inspect the slaughter "at the invitation of Greenpeace." Naturally, diplomatic protocol demanded that they accede to the congressmen's wishes, a decision that guaranteed not only a renewed Greenpeace presence, but also the unwelcome attention of the American media. It was, Hunter admitted, a shameless and opportunistic attempt to exploit the moral superiority of the United States. Just as Greenpeace was prepared to cooperate with the U.S. military in order to track down the whalers, they were equally prepared to use the American government to pressure their own. It mattered little that, at that very moment, Walrus was leading a Greenpeace campaign against the construction of a Trident nuclear submarine base in Bangor, Washington. It was one of Greenpeace's more convoluted political exercises, Hunter admitted, "to oppose U.S. imperialism on the West Coast, yet bring it to bear against Canada on the other side of the continent."[32]

Since Watson was no longer with Greenpeace, Moore could handpick his own crew for the 1978 campaign. Dispensing with the hardcore animal rights activists of the previous year, he chose a familiar group of friends and colleagues, including his wife, Eileen; Hunter; Weyler; Taunt; and Peter Ballem, the lawyer who had saved Watson the previous year. Before Moore's group arrived in St. Anthony, Greenpeace groups from Toronto and Montreal had gathered in Halifax, Nova Scotia, in an attempt to blockade the sealing vessels departing from the port. John Bennett of the Toronto group was almost killed as a ship ignored the blockade and backed right over his Zodiac, flipping him into the water near the ship's propeller. Other protestors were pelted with bottles, rocks, and even a folding card-table. On the second day in St. Anthony, Moore led the two congressmen and Hollywood actress, Pamela Sue Martin, onto the ice. At the press conference later that day, Congressman Ryan gasped: "I'm in a state of shock. I just want to say enough . . . enough! Stop!" Jeffords was equally appalled: "I can't relate to the experience in what you might call a rational manner right now. The experience of witnessing the slaughter transcends all figures, arguments, or rationality."[33] To Newfoundlanders, such rhetoric smacked of hypocrisy. Any compassionate individual, they rightfully argued, should have been just as concerned for the fate of the millions of calves and lambs that were slaughtered every year in industrial abattoirs throughout the world. Why, therefore, were they being singled out as cruel and barbaric? Nevertheless, for

Greenpeace, it was an impressive public relations coup that infuriated local politicians and fisheries officers.

After the celebrities and politicians left Newfoundland, officials told Greenpeace in no uncertain terms that they would not be allowed on the ice. In response, Ballem, who possessed a copy of the latest regulations, challenged the officials to show him the law that prevented Greenpeace from flying to the floes. The fisheries officers merely replied that it was "departmental policy," forcing Ballem to phone his contacts in Ottawa to try to override local officials. Hunter released a series of press statements accusing the government of attempting to create a "Gulag Archipelago" on the ice floes. "It's not only the extinction and bludgeoning of seals that's occurring," he insisted, "but also that of Canadians' rights."[34] Eventually, the fisheries officials relented and offered Greenpeace four permits to go to the hunt for twenty-four hours, beginning at 2.00 p.m., on March 17. This last-minute offer was made in the full knowledge that a major storm was predicted for the area on that very day. Should they risk it? For Hunter and Weyler, there was only one way to find out—consult the *I Ching*. According to the ancient Chinese oracle, there would soon be an "awakening of a consciousness of the common origin of all creatures." Its judgment: "It furthers one to cross the great waters." The storm never arrived.[35]

Knowing that this might be their only chance to actively protest the hunt, Moore, Ballem, Taunt, and Weyler formulated a strategy designed to make the most of the opportunity. Rather than rushing about the ice and trying to save as many seals as possible, as Watson had done in the past, they would focus on the creation of a single and powerful mind bomb. Their plan was to invoke an age-old common law stating that anyone who captures an animal has a right to it. Once they were on the ice and had located a group of hunters, the Greenpeacers walked very deliberately to the seal pup that was next in line for the slaughter. As a swiler and a fisheries officer approached them, Moore knelt astride the pup and grabbed its flippers in his hands. The officer asked the swiler if he wanted to kill the seal. When the swiler nodded, the official demanded that Moore let the whitecoat go. Moore's reaction may have been rehearsed, but it was nonetheless genuinely moving: "I believe I have as much right to give this pup life as you do to kill it. . . . I want this one pup to live. We've seen them come through. They leave nothing. Just this one. Please. For all the people of the world who demonstrate through us their opposition to this hunt." Once more, the fisheries officer asked Moore to get off the seal. When he refused, the officer told him, in front of the camera and with a total lack of irony, that he was under arrest for breaching the Seal Protection Act. Finally, Moore was forced to stand up and move away from the seal, at which point the swiler smashed its skull with his hakapik and began to skin its carcass. Neither Hollywood nor Madison Avenue could have concocted a better antisealing moment. Unfortunately, the film of the protest was damaged and unusable (Moore suspects it was deliberately exposed), but Rex Weyler's photograph appeared in over 3,000 newspapers around the world.[36]

Despite the absence of Watson and the hardcore animal rights protestors, Greenpeace's rhetoric and imagery was still dominated by an emotional and rights-based attitude toward the seals. Moore's compelling protest serves as an excellent example. As well as employing the dramaturgical ploys of the classic tearjerker, it focused attention on one particular seal. By embracing that seal and appealing for its life, Moore was removing it from the realm of abstract ecology—where it was merely part of a "species"—and placing it within the context of a human value system, and a Western one at that, in which the individual rights of a given creature were paramount. Hunter continued this line of reasoning when he explicitly compared the seal hunt to slavery:

> There was a time when such customs as slavery were considered by vast numbers of allegedly rational and religious people to be completely acceptable. Values change. Standards change. Ethics change. Slavery ended. Likewise, values, standards and ethics are continuing to change, and change cannot forever be resisted. The seal hunt, like the whale hunt, has already become a symbol of an outmoded and dangerous attitude toward nature.[37]

Patrick Moore pleads to save the life of a harp seal pup. Labrador ice floes, March 1978. Reproduced by permission of Rex Weyler/Greenpeace.

Moore's plea is unsuccessful. He is arrested for breaking the Seal Protection Act, and the seal is killed. Reproduced by permission of Rex Weyler/Greenpeace.

From a historical perspective, this is certainly a legitimate argument. The question, however, is whether it was the best strategy. An abolitionist stance gave neither side room to maneuver. The animal rights advocates within Greenpeace and elsewhere would not be content until the hunt was eliminated, while the swilers and their supporters were simply not prepared to consider such a drastic step. The lack of space for a compromise meant that victory for one side could only be achieved at the expense of utter defeat for the other. Later, when Greenpeace's direct-action approach to the seal slaughter began to lose impact, antisealing advocates turned to the European market that largely sustained the industry. The 1983 European Community ban on the importation of harp seal products seemed like a great victory for animal rights activists and for the seals themselves, but it proved to be a Pyrrhic one. While the major market for seal products may have been suddenly eliminated, the slaughtering of seals, unlike slaveholding in the nineteenth century, was never outlawed. Therefore, the sealing industry, with the help of substantial government subsidies and the strong support of the swilers, was able to gradually establish new markets for its products. They have been successful to the point where the slaughter has reached an even higher level than when Greenpeace first entered the fray. Perhaps a campaign that had focused more on ecology and less on the rights of individual animals may have yielded a better long-term outcome. Through international and domestic pressure, antisealing groups may have forced sealers and the Canadian government to establish stricter and sharply reduced quotas, while enabling the industry to continue to operate at a reduced, though still viable, level. While the industry would doubtless have been angered by efforts to reduce their kill, such a strategy may have minimized the outrage among many Newfoundlanders and their representatives in Ottawa, making it easier to set up supplementary employment schemes that would gradually have reduced the swilers' dependence on sealing as an important source of winter income.

Greenpeace's emphasis on animal rights may also have contributed to a rather careless attitude toward science. For example, the San Francisco branch cited a Swedish study that claimed if the seal slaughter continued at its present rate, the seals could be extinct within a matter of ten years, a fact that Taunt emphasized at his press conferences before the U.S. media. The *New York Times* decided to check this figure with the Vancouver office, where an embarrassed John Frizell admitted: "I don't know of any evidence to support that statement. I'm sorry he said that."[38] The mix-up was compounded by the fact that the study was deeply flawed. Allan Thornton, who at that point was running the fledgling Greenpeace UK group in London, was angered both by the sloppy use of science and the excessively emotive rhetoric employed by the North Americans. Greenpeace, he insisted, should use the term "whitecoats" rather than "baby seals." "Our campaign must be scientifically justifiable and hence, people making statements on behalf of any Greenpeace office should know what they are talking about. . . . Greenpeace UK will not feel obliged to support statements which are not credible."[39]

Such problems were the result of the inherent tension that lay at the heart of Greenpeace's strategy. Since the basic premise of every campaign was to attract the attention of the mainstream media and appeal to as many people as possible, the campaigns were frequently based around: (a) the potential for violent conflict, and (b) a sentimental anthropomorphism designed to have an emotional impact on the average person. While it was important that the general public develop a sentimental attachment to whales and seals, it was equally important, from the standpoint of Greenpeace's credibility, that they not be seen as an organization whose primary goal was to foster such feelings. Therefore, it was necessary to employ the "objective" rhetoric of scientific ecology, with its emphasis on the species and the overall eco-system, rather than on individual animals. Combining these two strategies was already a tough balancing act, but it was made more difficult by the fact that so many Greenpeacers had developed deep emotional attachments to the animals they were trying to protect. It was not just a species thing: each *individual* harp seal was a creature of heartbreaking beauty; each *individual* whale was an intelligent and conscious being.[40] While ecology—deep or not—could certainly be employed as a strong argument to severely limit the number of whales and seals hunted each year, employing it to justify the abolition of the hunts was more problematic.

In 1978, the self-professed "warriors of the rainbow" found themselves facing another complicated dilemma—one that would put them in direct opposition to the native peoples they had always championed and from whom they had drawn their inspiration. The issue concerned the Alaskan Inuit and their insistence on continuing to hunt the bowhead whale, one of the most endangered cetaceans on the planet. There was little doubt that the bowhead hunt was an important part of Inuit cultural history, just as there was no question that Yankee whalers had overhunted the species in the nineteenth century and were largely responsible for its demise. The bowhead had become "commercially extinct" by the time of the First World War, when its population was estimated at less than 4,000. From that time until the late 1960s, the Inuit continued to take ten to twenty whales each year. Despite these modest numbers, however, the bowhead showed no signs of recovery, and by 1970, its population was still estimated to be between 3,000 and 4,000. Beginning in the early 1970s, the Inuit started to expand their hunt. Their share of Alaska's oil wealth enabled them to purchase more boats and modern equipment, allowing greater numbers to join a hunt that was a prestige event in their society. By 1977, they were killing over one hundred bowheads per year, the majority of which were struck by exploding harpoons but could not be landed by the enthusiastic but inexperienced young crews. The population was now thought to have dropped to below 3,000 whales, prompting environmentalists to call for an immediate halt to the slaughter.[41]

Under normal circumstances, an animal as threatened as the bowhead would have been completely protected under the Endangered Species Act. However, the act exempted native subsistence hunters. Japan and the other prowhaling nations saw this as the perfect opportunity to incapacitate the American drive to implement a whaling moratorium at the IWC. Thus, they cynically lined up with the antiwhaling nations and voted to ban the bowhead hunt. Under pressure from Alaskan politicians, the American IWC delegation, for so long the leading antiwhaling voice on the commission, found itself lobbying for the highest possible bowhead quota. Suddenly, the United States was no longer in a position to argue for a moratorium. To compound their dilemma, American delegates had to negotiate with the Japanese and Soviets, who insisted that they would only approve the bowhead hunt if the United States agreed to drastically increase the sperm whale quota from 763 in 1977 to 6,444 in 1978. The Blue Whale Unit, it seemed, was now complemented by the Bowhead Unit, in which each bowhead was worth approximately 200 sperm whales.[42]

Initially, Greenpeace reacted to the bowhead dilemma by attacking the U.S. government officials they felt responsible for caving in to the Japanese and the Soviets. The three nations, Greenpeace charged, had formed an "unholy alliance" that spelled the end for both the bowhead and the sperm whale.[43] Patsy Mink, a former congresswoman from Hawaii whom President Carter had appointed assistant secretary of state for Oceans and International Environment and Scientific Affairs, was particularly singled out for condemnation. Donald White, the head of Greenpeace Hawaii, could barely contain his outrage: "We at Greenpeace," he castigated Mink, "think it a shame that, sitting in your plush office, you will never hear the explosion of the harpoon cannon, hear the dying whistle-screams, or see the sightless eyes staring into the sky. But, now and forever, their blood is on your hands."[44]

Eventually, however, Greenpeace had to face the unpleasant fact that Mink and others were merely acting on behalf of the Inuit. Paul Spong was the first Greenpeacer to seriously wrestle with this dilemma. His solution was not one that was likely to please the indigenous peoples of North America. Although the preservation of human cultural values, Spong conceded, was obviously important, native cultures had already been seriously eroded by the impact of other cultures and by "the inexorable advance of the Technological Age." Whether they liked it or not, Spong insisted, native cultures the world over had to accept the fact that change was a "cultural constant." For the Inuit, he continued, "the end of the bowhead hunt is properly seen in the cultural context as one of many important changes. It will not destroy the culture, rather it will further modify it." Once the hunt ceases, Spong predicted, it would "remain an important part of native cultural history . . . to be recalled in dance and legend rather than in the act itself." Eventually, the Inuit would develop a different relationship with the bowhead, viewing it "as neighbor and friend, an object of curiosity and affection, rather than food." Greenpeace, Spong reiterated, was opposed to the killing of any cetaceans under any circumstances. "Short-term cultural priorities," he insisted, "cannot be permitted to over-ride the valid scientific and ethical grounds for calling an immediate and complete halt to the hunt."[45]

The bowhead controversy forced Greenpeace to abandon its romantic and monolithic view of native peoples and to adopt a more critical stance toward them. While appreciating that aboriginal North Americans lived in greater harmony with the natural world than most other North Americans, Greenpeace began to recognize that they were not infallible. As Rex Weyler observed, "Greenpeace was certainly friendly toward First Nations, however, we weren't naïve about them. . . . There are lots of examples, from the Navajo mining interests in New Mexico to the seal hunters and trappers in Canada, where modern native cultures show disregard for the environment. . . . If cultures are destructive to the earth, someone has to point it out. We felt we were the ones."[46] Paul Watson took a similar position, arguing that most so-called native hunts were merely fronts for commercial operations. By continuing to supply the fur trade, Watson insisted, native hunters were merely perpetuating their own colonial victimization.[47]

In the eyes of George Wenzel, a self-professed "advocacy anthropologist" who worked closely with the Inuit, the problem with this view is that it used an idealized native way of life from 500 years ago as a benchmark for "tradition." The only way Greenpeace could accept the hunting of whales and seals, Wenzel argued, was if it remained truly "traditional," that is, if the hunters employed pre-European technology and if the hunt remained part of a closed system of production and consumption. Therefore, it was Greenpeace that determined the cultural context in which native peoples should be allowed to hunt seals and whales.[48] This, Wenzel continued, allowed Greenpeace to have it both ways: since it was unlikely that any native culture would deliberately revert to the technology of the precontact era, Greenpeace could continue to take an abolitionist stance on the hunt while claiming to support an "uncontaminated" native tradition. Ultimately, the animal rights movement, in Wenzel's words, "was unable to categorize Inuit . . . hunting other than through its own ethnocentrically derived universalist perceptions of animal rights and values."[49] Of course, Wenzel, too, is guilty of "constructing" tradition. In his case, it is a more elastic definition that incorporates a host of phenomena, such as advanced technology, that many would regard as bearing the indelible stamp of industrialism and modernity. If taken too far, such an argument can justify almost any form of hunting, including Japanese and Norwegian whaling, as "traditional" and therefore above sanction.

It would be churlish, however, to dwell solely on the more problematic elements of Greenpeace's campaigns, for there can be little doubt that their actions ensured that whaling and sealing remained highly visible at a time when many governments would have been content to see them disappear from public debate. The Greenpeace view of *Cetaceus intelligentus* even penetrated the rigid instrumentalist walls of the IWC, where whales had always been officially cast as mere natural resources. In 1980, for example, the Smithsonian Institute hosted a special meeting of the IWC Scientific Committee entitled "Ethics and Intelligence," to which it invited such well-known proponents of *Cetaceus intelligentus* as John Lilly, Roger Payne, and Paul Spong. Sidney Holt, a population biologist who had spent two decades on the committee, had developed a close friendship with Spong. During the process, he underwent a complete conversion; instead of spouting the whales-as-natural-resources view, he became a strong proponent of the "mind in the waters" thesis. "If whales possess the intelligence and mental properties that we assume," he told the conference, "then whales have a culture. By continuing the current whaling practices, we are denying the future possibilities to discover and understand the culture of the whales. I consider it a great evil to destroy something we don't understand."[50] U.S. Congressman Leo Ryan embraced a similar philosophical position before he was tragically gunned down in Guyana's jungle during the Jonestown massacre of 1978. Ryan invited Spong to testify before the House Environment, Energy and Natural Resources Subcommittee. In a letter, he asked Spong to specifically "focus your testimony . . . on the intrinsic value of a whale, as well as . . . your personal efforts as a member of Greenpeace to prevent the killing of whales."[51]

Greenpeace also provided a platform for controversial scientists such as John Lilly, both by embodying his views in their own rhetoric and by publishing his work in the *Greenpeace Chronicles*. In 1976, for example, Lilly wrote an article that combined the "mind in the waters" argument with Christopher Stone's rights-of-nature thesis. The fact that the whales possessed a complex form of intelligence, Lilly speculated, meant that they had probably developed a very advanced system of philosophy and ethics, which they carried forward by educational means through sonic communication of a higher order. According to Lilly, this entitled them to be given rights under human laws, including complete freedom of the seas and the same individual rights that are accorded to humans.[52] By the early 1980s, animal rights philosophers such as Tom Regan were using exactly this argument. In fact, Regan not only supported the cessation of whaling but also denounced "non-consumptive" uses of whales, such as whale-watching tours. "Whales," Regan insisted, "do not exist as visual commodities in an aquatic free market, and the business of taking eager paying sightseers into their waters, though non-consumptive, is exploitative nonetheless, morally analogous to making a business of conducting sightseeing tours of human beings who either cannot or do not give their consent to be exploited by other people in this way."[53]

Animal rights philosophers, such as Regan, were not the only thinkers who were impressed with and influenced by Greenpeace. Deep ecologists were also drawn to the organization, particularly the work of Bob Hunter. In 1979, Hunter wrote a series of articles speculating on the direction that environmentalism would need to take in the 1980s in order to limit the damage that humans were doing to the planet. With his typical blend of holistic ecology, Hegelian and McLuhanesque philosophy, and assorted New Age ideas, Hunter once again took on some of the big questions facing mankind. In one article, he revisited his earlier idea that ecology would become the religion of the new era. "Nature," Hunter argued, "is quite obviously the physical manifestation of God's work. Within it, as part of it, viewing what is Our Self from the individual compartments of our little selves, we become aware that Nature is, in fact, *us*. The world is Our Body." It is not enough, he continued, "to feel worshipful toward Nature; one must view Nature as a manifested Godhead. Therefore, its contemplation becomes the stuff not just of worship, but of religion."[54]

In another article, Hunter argued that people needed to break down the mental barrier that divided nature and culture, particularly as represented by the dichotomy between our cities and the "natural" world. "The barrier between country and city," he wrote, "is imaginary. The city, in fact, *is* a part of nature. If it has seemed not to be, that was a passing phase."[55] He also reemphasized the need for humans to expand their ethical circle: "Humanistic value systems must be replaced by supra-human values that bring all plant and animal life into the sphere of legal, moral and ethical consideration." In the long run, he hinted darkly, "whether anyone likes to face it or not, force will eventually have to be brought to bear against those who would continue to desecrate the environment."[56] Bill Devall, a philosopher at Humbolt State University and one of deep ecology's leading thinkers, felt that Hunter had done "an excellent job of explaining the revolutionary potentials of environmentalism and why it cannot be considered 'just another interest group' in the calculus of interest group liberal politics." Hunter's work in particular made the *Greenpeace Chronicles,* in Devall's opinion, "one of the most lively, relevant, and interesting of the various environmental newsletters now being published." In fact, Devall used it regularly in his "Wilderness in World Perspective" graduate seminar, thus exposing numerous deep ecologists to Hunter's work.[57]

There can be little doubt that Greenpeace, through their direct-action, media-oriented approach, as well as their increasingly sophisticated lobbying efforts, played a major role in both the whaling moratorium, finally passed in 1982, and the European Community boycott of all harp seal products in 1983.[58] It is important to understand, however, that whaling and sealing were the "headline" campaigns of an organization that, by the late 1970s, was involved in myriad environmental actions throughout the world, many of them at the regional and local levels. The minutes from Greenpeace USA's national meeting in 1979, for example, list literally dozens of campaigns being carried out by the various offices throughout the country. The Oregon group was supporting innovative land-use planning laws and protesting the practice of trapping animals with leg-hold traps. The Denver outfit was part of a coalition protesting the Rocky Flats nuclear installation in Colorado.[59] The Chicago office was running a campaign to alert people to the deteriorating water quality in the Great Lakes. Greenpeace's Los Angeles branch was protesting offshore oil drilling, air pollution, and the destruction of coastal habitat to make way for property developments. In Hawaii, Greenpeace activists were involved in publicizing the fact that tuna fishermen were inadvertently killing thousands of dolphins.[60]

The various Canadian offices were conducting similar actions. In June 1977, for example, the Toronto group infiltrated the Douglas Point nuclear power plant in Ontario, placing Greenpeace stickers on the doorways to sensitive installations. The predawn "raid" was designed to show how vulnerable such facilities were to sabotage or a terrorist attack. In 1979, the same group organized a daring antinuclear action in Darlington, Ontario, which was the designated sight of what would be the world's largest nuclear power plant. With Jet Johnson as their pilot, five men, including John Bennett and Dan McDermott from the Toronto office, Bob Cummings from Vancouver, and a former Israeli commando, parachuted into the sealed-off site, where they managed to erect a Greenpeace banner before being arrested.[61]

In between the sealing and whaling actions, the Vancouver group also managed to mount some effective local campaigns. In the summer of 1977, for example, Patrick Moore organized a protest against the provincial government's plan to spray an area of the Fraser River Canyon with toxic chemicals to combat an outbreak of spruce budworm. The government, with strong backing from Union Carbide and Chevron, the companies that supplied the chemicals, and from the BC forest industry, insisted the spraying was necessary to prevent irreparable damage to a valuable stand of timber. Moore, whose PhD was in forest ecology, pointed out that the spraying would not only kill the budworms, but all the insects in the forests, the impact of which would eventually make its way through the entire food chain. When Greenpeace and local protestors announced that they would occupy the forest to prevent the spraying, the government hired Pinkerton security agents to guard the perimeter and insisted that anybody caught trying to enter the forest would be arrested and jailed. To get past the Pinkertons, Greenpeace contacted a local Indian tribe whose reservation abutted the forest and were given permission to use their land as a "base camp" from which to launch an "invasion." The military metaphors were appropriate, since Moore had planned to employ Second World War vintage troop carriers to land the protestors on the edge of the forest. At the last minute, the government backed down and cancelled the spraying.[62]

In London, meanwhile, the Greenpeace UK group led by Allan Thornton, Susi Newborn, and the peripatetic David McTaggart had purchased an old fishing trawler called the *Sir William Hardy*, which they renamed the *Rainbow Warrior*.[63] In 1978, McTaggart led a classic Greenpeace antiwhaling mission, replete with Zodiacs and camera crews, to Iceland to protest against the shore-based whalers operating out of Reykjavik.[64] Later that year, the same crew, which included enthusiastic young members from the fledgling Paris and Amsterdam groups, conducted a successful protest against the British government's plan to cull some 6,000 gray seals off the coast of the Orkney Islands north of Scotland.[65] They also employed the Zodiacs to protest the British government's dumping of nuclear waste in the North Atlantic. In a series of daring and spectacular stunts, the Greenpeacers maneuvered their inflatables alongside the dumping vessel, known as the *Gem*, placing themselves directly between the large barrels of radioactive waste and the sea. As the *Gem* sailed along at full speed, crewmen aimed high-powered fire hoses at the protestors. Several barrels were dropped just inches from the speeding Zodiacs, with one actually landing on the bow of one of the inflatables, flipping it over, and injuring one of the activists.[66] All over Europe and North America, the tactics developed by the small group of radical environmentalists in Vancouver were adapted to an array of environmental causes.

By the late 1970s, Greenpeace's "headline acts" had made the organization the most high-profile radical environmental outfit in the world. For many, the

European Greenpeace activists try to prevent a British ship from dumping nuclear waste in the North Atlantic, 1978. Reproduced by permission of Jean Deloffre/Greenpeace.

term "Greenpeace" became a synonym for nonviolent direct action on behalf of the environment, and there were few people in industrialized democracies who had not heard of the organization's campaigns against whaling and sealing. It was, by any measure, a prodigious achievement by the small band of activists from Vancouver. However, the growing scale of their campaigns—more people, bigger boats, better equipment—meant that it was increasingly difficult for the Vancouver group to meet its multiplying debts. While a combination of Bill Gannon's accounting skills and plain old good luck had enabled Greenpeace to scrape together the funds necessary to mount its campaigns, it became ever more difficult to see how the Vancouver group could pay off its considerable debts and achieve financial stability. In 1977, for example, Greenpeace Vancouver's total revenue was $373,000, while expenses totaled $483,000, forcing them to increase their bank loan by $75,000. This left them with a total debt of $174,000.[67] In Canada, the controversy caused by the sealing campaign had severely eroded the organization's funding base. While virtually everyone supported the whaling campaigns, a significant number of Canadians had been influenced by their government's prosealing arguments and were either opposed to or ambivalent about Greenpeace's hard-line stance against the swilers. The result was that, by 1978, the sealing campaign was almost entirely financed by the San Francisco and Portland offices, a trend that continued to mark Greenpeace's "headline acts" thereafter.[68] It was only a matter of time before the groups that were raising most of the funds would demand to play a commensurate role in the decision-making process and the overall running of the organization.

In retrospect, the problems that beset the Vancouver Greenpeace Foundation in the late 1970s had their origins in two events that occurred after the first whale campaign. The first was the result of ideology, the second a technical oversight. With their countercultural, Quaker, and New Leftist roots, Greenpeacers were naturally inclined toward an organizational model that was loosely structured and nonhierarchical and that favored a grassroots-based decision-making process. To Hunter, therefore, the best way to facilitate Greenpeace's rapid expansion—and the consciousness revolution it would hopefully provoke—was simply to encourage new groups to pop up everywhere. Hunter summarized this philosophy with the phrase: "Let a thousand Greenpeaces bloom." So long as all the new groups remained faithful to Greenpeace's brand of nonviolent direct action, they could do their own thing. They could run their own local campaigns, raise their own funds, and, if they desired, participate in the "headline campaigns" being run out of Vancouver. It was better, Hunter felt, to try to head problems off as they arose rather than develop a structure that stifled people's freedom and creativity. Hunter's inspiration for such a decentralized model had been the famous beatnik poet Allen Ginsberg. Hunter had met Ginsberg, one of his literary idols, in San Francisco after the first whaling voyage. When he asked him for advice on how one should best deal with power, Ginsberg replied: "You let it go before it freezes in your hands."[69]

Poets, however, do not necessarily make the best political strategists. By late 1976, the Vancouver group began to realize that their laissez-faire approach to expansion was causing more chaos than it was worth. In response, they changed the Foundation's bylaws in an effort to restore their power and bring about a greater sense of order. Greenpeace's branches, the new bylaws read, would have powers and responsibilities that did not exceed those of the Vancouver office; any policy statements issued by the branches would now require the approval of the Vancouver board of directors; and the major function of the branches was defined as "support[ing] the society by raising funds, circulating statements of policy and in any other manner as shall be determined by mutual agreement between any branch group and [Vancouver]."[70] As Hunter was soon to find out, however, power, once given away, is very hard to take back.

The second issue that would trouble Hunter and his Vancouver comrades was a technical detail in setting up the San Francisco office. Gary Zimmerman, an American engineer who had been on board the *Phyllis Cormack* during the first whale voyage, was chosen to run the new office, while a lawyer by the name of David Tussman was hired to ensure that the office met all the legal requirements that would give it tax-exempt status. Tussman was the son of Berkeley philosophy professor Jospeh Tussman, whose book on the role of citizens in a democracy, *Obligation and the Body Politic*, was admired by several of the Vancouver group. Frustrated by his inability to land a high-profile job and bored with his work as an assistant to a subrogation lawyer, David Tussman joined Greenpeace in an effort to bring some excitement and meaning into his life.[71] After researching the matter, Tussman decided that the best way to ensure that Greenpeace San Francisco—or the Greenpeace Foundation of America, as it was soon called—could obtain tax-exempt status was to set it up as a separate, but unincorporated, legal entity that was affiliated with, but not under the direct control of, Vancouver. While this strategy eventually succeeded in gaining Greenpeace tax-exempt status in the United States, it also meant that, from a strictly legal perspective, the San Francisco group was not under the direct control of its Canadian parent organization. This issue would result in a bitter dispute between the two Greenpeace groups.[72]

By 1976, the Vancouver Greenpeace group, in Bob Hunter's opinion, resembled a "tribe": "There were a couple of dozen of us whose lives had become so intertwined. . . . We shared each other's clothes, we shared our food, we pooled our resources for parties, and like an extended family network, we squabbled amongst ourselves, bitched about each other, counseled one another, and came to think of each other as brothers and sisters."[73] The "tribe" analogy was also appropriate for many of the other Greenpeace groups. In Seattle, for example, the dozen or so hardcore Greenpeace members shared a large house together, while those in Paris and London were also living, sleeping, traveling, and campaigning together almost constantly.[74] By 1978, Greenpeace resembled nothing so much as a loose confederation of tribes, each with its own elders, its own internal culture, and its own idea of what Greenpeace was or should become. Among those in Vancouver, there was never any doubt that their group constituted the leading tribe. They were, after all, the original "rainbow warriors." But as the power and confidence of the other tribes grew, they became less willing to see themselves as subordinates of the elders in British Columbia.

As the nominal umbrella organization representing all the Greenpeace groups in the United States, the San Francisco-based Greenpeace Foundation of America quickly became the largest and most complex of all the tribes. Unlike the other Greenpeace groups in the United States and Canada, the Vancouver group had specifically established the San Francisco office as a subsidiary branch that would remain under their control. It was split into three levels, though this was more a haphazard process of diffusion than the result of any sort of grand strategy. At the top were the "executives," a group of professionals such as Tussman, Zimmerman, and Bob Taunt, most of whom were college-educated and in their thirties. Operating out of a historic building in Fort Mason, which was part of the Golden Gate National Recreation Area, this group planned most of the organization's strategies and took all the important decisions. They considered themselves highly qualified environmental professionals and paid themselves salaries that they felt were commensurate with their qualifications and positions—a fact that drew considerable criticism from the grassroots members. The second tier of the organization was based in an office on Second Street. Composed of a few administrative staff and many volunteers, this was the section that planned and carried out most of the local fund-raising events, including running raffles, selling merchandise, organizing walkathons, and coordinating the citywide canvass. The third tier was composed of an assorted bunch of hippies, street people, and radicals who occupied or hung out at the dilapidated *Ohana Kai*, the boat that Greenpeace had used on the 1977 and 1978 whale voyages. This group conducted tours and gave ecology talks to members of the public, as well as offering communal meals to the assorted environmentalists and homeless people who regularly congregated there. Marginalized and scorned by the executives, the *Ohana Kai* group felt that it was they, rather than the well-groomed professionals at Fort Mason, who embodied the true spirit of Greenpeace.[75]

Unlike most other Greenpeace tribes, which regularly mounted local environmental campaigns in their own regions, the San Francisco group concentrated almost exclusively on raising funds and awareness for the whaling and sealing campaigns, which they did far more successfully than any other Greenpeace group in North America. Apart from the various walkathons, raffles, and canvasses that tapped into the Bay Area donor market, the San Francisco office also established a remarkably successful nationwide direct-mail program with the firm of Parker & Dodd, a fund-raising agency that specialized in progressive causes. The Greenpeace San Francisco mailing list was extensive and was considered extremely well targeted, a fact that prompted the campaign managers of California Governor Jerry Brown to offer to purchase it from them.[76] Given their financial clout, it was virtually inevitable that, at some point, the eclectic members of the San Francisco office would tire of taking orders from their cash-strapped masters in Vancouver.

Along with San Francisco, one of the most significant Greenpeace groups in the United States was in Seattle. Unlike San Francisco, the Seattle tribe quickly developed a reputation for being a classic consensus-oriented, Birkenstock-wearing, granola-eating band of grassroots activists. They mounted many campaigns in the Seattle area, including protest actions against polluting factories in Tacoma and pesticide spraying around local waterways. They also formed alliances with other organizations to protest against the Trident nuclear submarine facility in Bangor, Washington. Though eager to cooperate with Greenpeace offices throughout North America, the Seattle tribe was highly sensitive to any requests that sounded like "orders," regardless of whether they came from Vancouver or San Francisco. As one of their members explained to the San Francisco group: "We strive to maximize the feeling of participation for every member and volunteer that we have. Dictating terms of any sort without sufficient consultation can only deprive Greenpeace of possible valuable human resources." Such was their commitment to a grassroots structure that literally anybody was allowed to walk in off the street, join one of their meetings, and have a voice in the decision-making process.[77]

The other thirty or so Greenpeace tribes in North America tended to fall somewhere in between the elite executives from San Francisco and the grassroots activists from Seattle. The Boston group developed a reputation for tough pragmatism.[78] The Toronto outfit conducted some of the most daring and spectacular campaigns, such as parachuting into various nuclear facilities. The Hawaiian tribe prided itself on having made their island "the most thoroughly Greenpeace-indoctrinated place in the world today." According to its leader, Don White, the Hawaiian group had "written curricula for local schools, testified on important issues, earned seats on scientific commissions, launched demonstrations, influenced legislation, and filled the media with Greenpeace deeds and goals." In the process, they paid themselves "virtually nothing in the way of salaries" and even used their own personal savings to help fund various campaigns. Such sacrifices, White believed, made Hawaii "a *real* Greenpeace group."[79] Like the *Ohana Kai* crowd in San Francisco, the Hawaiian tribe was under the impression that abstemiousness and asceticism were necessary characteristics if one were to become a true Greenpeace activist. For the professionals in San Francisco, however, such an attitude was seen as sanctimonious nonsense. One did not have to completely sacrifice one's private life to be an effective environmental campaigner.[80]

In Europe, meanwhile, David McTaggart had helped set up Greenpeace offices in Paris and London. Although he was still miffed at what he felt was the lack of support from Greenpeace during his lengthy French court case, McTaggart was gradually coming around to the view that the core idea of Greenpeace—an international organization that relied on nonviolent action and was not attached to any political party or ideology—had considerable potential if it could be run by hard-nosed professionals rather than hippies. In other words, if *he* were at the helm, it might be possible to create a genuinely international organization that could effectively influence world opinion. Although he would not have shared the same terminology, McTaggart was clearly aiming to create an organization that could engage in what political scientist Paul Wapner refers to as "world civic politics."[81] With the help of Remi Parmentier, a student radical who was barely out of his teens, McTaggart started a Greenpeace office in Paris, recruiting various Les Amis de la Terre activists disaffected by that organization's growing ties to left-wing political parties in France. It was not long before McTaggart had convinced most of the new recruits that he had founded Greenpeace and that the Vancouver hippies were a bunch of incompetent fools destroying the organization he had fought so hard to establish.[82]

In London in 1977, a group of Friends of the Earth activists, fed up with what they perceived to be that organization's increasingly mainstream style, decided to set up a Greenpeace office in the United Kingdom. By chance, Allan Thornton, a Vancouverite who had been involved in the seal campaign, arrived in London, also with the intention of establishing a Greenpeace office. In another amazing Greenpeace coincidence, Thornton answered a "flatmate wanted" ad that had been placed in the newspaper by Susi Newborn, an FOE activist.[83] They proceeded to invite McTaggart and Parmentier to London, and the group officially set up a Greenpeace UK office. Much to their surprise, however, they soon found that there was another Greenpeace group that had been active (although barely so) in London since 1971. This was the group of anarchists associated with *Peace News* that Lyle Thurston had encountered in 1972. It called itself Greenpeace London, had links to various Marxist and anarchist organizations, and represented precisely the kind of marginalized, utopian radicalism that McTaggart could not stomach.[84] Newborn, who was more sympathetic toward the group, felt uncomfortable about McTaggart's rather imperialistic approach. Again, he claimed to be the true founder of Greenpeace and insisted that the Greenpeace London group were frauds who were simply trying to gain publicity by adopting the Greenpeace name. Before there was any time to debate the issue, McTaggart swiftly went ahead and incorporated Greenpeace UK. Even had they been so inclined, the anarchists were in no financial position to mount a legal challenge, and the two groups continue to co-exist to this day.[85]

By late 1977, the UK group had purchased the ship that would soon become the *Rainbow Warrior*, while McTaggart, Thornton, and Parmentier discussed the idea of forming Greenpeace Europe, which would be a loose confederation of autonomous Greenpeace groups throughout the continent. Soon thereafter, a Dutch Greenpeace office was set up, also with McTaggart's help, and the campaigns against Icelandic whaling, Orkney sealing, and Atlantic nuclear-waste dumping began. Within the space of a year, Greenpeace became a household name in much of Western Europe, and fledgling groups appeared in West Germany and Scandinavia.

Most of the Europeans were in their twenties and, unlike many of their counterparts in Vancouver, had little time for the beliefs and rituals associated with the counterculture, a fact that contributed to McTaggart's admiration for the European activists at the expense of his fellow Canadians. Nevertheless, from an organizational perspective, there was a great deal of similarity between the groups, and Parmentier's description of the Europeans closely resembles Hunter's depiction of the Vancouver group: "We were like a small tribe, spending all the time together, working together, traveling together on the cheapest night trains, sharing each others' sofas, music, beer cans and joints. Like an emerging rock band on a tour. Tribal life includes tribal wars; they can be bloody, sharp arguments—we did not fuck around with political correctness. We were faster and freer than others in the environmental movement; but it had also one basic cost: we were sectarians, and not very transparent and very competitive."[86]

McTaggart, however, was not the only one who had visions of a more organized, professional international outfit. By late 1977, Bob Hunter, Patrick Moore, and others within the Vancouver tribe were beginning to see the need to establish a more formal set of ties between the various organizations, as well as developing a chain of command that would facilitate a greater degree of efficiency in the decision-making process. With this in mind, Moore, who by then had succeeded Hunter as president, sent a letter titled "Greenpeace: Where are We Going?" to the various tribes scattered throughout North America. "We are faced with a problem," Moore began,

> that has baffled the best philosophers and politicians since the first federation of cave-people communities. Simply stated the problem is how can we achieve unity and cohesiveness as one organization and yet provide the individual and group autonomy necessary for creativity and initiative? Somehow we must be both centralized and decentralized at the same time. . . . Under the present structure, further growth is not possible without further confusion. There is a pressing and demanding need for organization.[87]

Moore suggested several organizational models, including General Motors, the United Nations, the Palestinian Liberation Organization, and the Sierra Club. However, he was particularly taken with the idea of a structure that was based on an ecosystem model. Diversity in ecosystems, he noted, in what many would now consider to be an outmoded ecological theory, "tends to result in stability." While this was an argument for a decentralized structure, it was also important to remember that "each species has a well-defined niche or function that it must keep to in order to maintain that stability. . . . We must stick to those functions and we must demonstrate the capacity to carry them out."[88]

In order to facilitate further discussion on such issues, and with the ultimate goal of forming a workable international structure, the Vancouver group organized a meeting to which it invited the various tribes scattered throughout the world. It was held over several days in January 1978 on the UBC campus. The Australian and New Zealand groups claimed they could not afford to send delegates, while a somewhat cautious David McTaggart—the representative from Greenpeace Europe—attended for only one day, thus effectively making it a North American, rather than an international gathering. Nonetheless, the United States and Canada represented by far the greatest number of Greenpeace groups and supporters, and the meeting took place on the assumption that its outcome would be binding on all Greenpeace offices. From the outset, it was clear that Vancouver and San Francisco would find it difficult to agree on a model for an international organization. Every time the Vancouver delegates proposed a structure that smacked even vaguely of centralization, the San Francisco delegates, and their supporters in various other branches, questioned if such a model would be appropriate for a grassroots organization such as Greenpeace. Since Bob Taunt and David Tussman—the San Francisco "elitists" who were most likely to side with Vancouver—were part of the advisory committee that had organized the meeting, they were not permitted to act as delegates. Their places were taken by the more grassroots-oriented members from the Second Avenue office, who were wary of both Vancouver and the Fort Mason crowd. The biggest clash of the meeting, however, occurred within the Vancouver delegation itself: specifically, between the old guard of Hunter, Moore, Marining, and Spong, on the one side, and the representative of Vancouver's own grassroots faction, John Frizell, on the other.[89]

Frizell had joined the Vancouver group in 1976, when he acted as a media liaison during the second whaling voyage. A bright and ambitious UBC science graduate, he rapidly came to represent the more disaffected grassroots elements in Vancouver, who elected him to the Greenpeace Foundation's board of directors. Throughout the meeting, Frizell continuously questioned the intentions of his fellow Vancouverites, implying that they were making a grab for power and trying to exclude grassroots members from the decision-making process. When, for example, Patrick Moore moved that a Greenpeace International board of directors should be created to "manage the affairs of Greenpeace International by carrying out the policies and objectives established by the international general meeting," Frizell immediately interpreted this in the most extreme fashion. It implied, he insisted, that "the international board will manage the entire affairs of the members." After three days of discussion, the delegates at last agreed that they would elect an international board of directors, but then became bogged down over the question of whether each office should be limited to one seat on the international board or two. Finally, it was decided that the international board would be composed of seven members, with each office allowed no more than two representatives, who would not be allowed to serve on any other Greenpeace board while he or she was an international director.[90]

The final act of the meeting was the attempt to elect this new seven-member international board. After eleven ballots, the delegates only managed to elect six: Moore and Frizell from Vancouver, Taunt and Tussman from San Francisco, and one delegate each from the Victoria and Hawaii offices. Despite another four ballots, however, they could not successfully elect the seventh member. In an effort to break the deadlock, Hunter offered to resign his position on the Vancouver board and to run for the final international slot as a delegate-at-large. When some delegates suggested that this would give Vancouver three positions on the international board, Moore replied that, although Frizell was from Vancouver, he was not supported by the rest of the Vancouver delegation and had been voted in by others. By electing Hunter, Moore insisted, they would be recognizing Greenpeace's most experienced and charismatic figure, as well as ensuring that the board was well represented by Greenpeace's strongest office. However, such sentiments meant little to the newer delegates who did not fully appreciate the role that Hunter had played in the organization's history. They continued to dispute Hunter's right to stand for the position and prevented a final ballot that might have seen him elected. In a last desperate move, the Vancouver old guard voted to remove John Frizell from their delegation, a move that prompted one of the Portland delegates to stand down and allow Frizell to take her place. Frizell may have been unpopular among the Vancouver old guard, but he clearly had considerable support among the newer and younger members, many of whom had little time for the hippie elders.[91]

Clearly, Vancouver was not going to succeed in getting Hunter on the international board nor in having Frizell kicked off it. Finally, Hunter could take no more. He stood before the group like an angry tribal chief and expressed his disgust with what he felt was their obstructive behavior and their general lack of respect for the Vancouver founders. "Eight years ago," he began, "Vancouver started off doing this trip. A few people have worked very hard. . . . There are five people on the Vancouver board who have been involved in Greenpeace for eight years [and] Vancouver now has good decision-making capabilities." For years, Vancouver had followed Allen Ginsberg's advice—give away power before it freezes in your hands—as an act of opposition "to the whole established corporate power structure." Now, he continued, "power has been given away to the point where [the Vancouver group] are lying on their backs with legs spread. Things have gotten to the point where others who have been in Greenpeace for one month have put Frizell onto the international board because he appeals to the disaffected." As a result, he felt, "Vancouver has no other option but to withdraw."[92] Thus, the gathering ended on the bitter and ironic note of the Vancouver delegation walking out of the meeting they had convened.

As punishment for his seditious behavior, Frizell was removed from the Vancouver board. However, he soon resurfaced in San Francisco, where the grassroots faction viewed him as something of a hero who had dared to stand up to tyrants such as Moore and Hunter. He was quickly elected to the San Francisco board while Taunt, who was seen as loyal to Vancouver and a close friend of Moore's, was increasingly marginalized. From that point on, relations between Vancouver and San Francisco continued to deteriorate. At one point, Bill Gannon—Vancouver's treasurer—phoned San Francisco and requested that they send $5,000 to Seattle to prepare for a campaign launch. When San Francisco equivocated, Gannon changed his tone to one of insistence: He was the treasurer of the Greenpeace Foundation and the San Francisco group was a subordinate office. If he demanded that San Francisco send money to Seattle, they had better follow his orders. They refused to comply, telling Gannon that they would raise the issue at their next board meeting and get back to him.[93]

Similarly, when Patrick Moore was arrested for disrupting the 1978 seal hunt, the San Francisco office refused to pay his legal expenses. This prompted Moore to fly down to San Francisco and try to reason with the increasingly obstreperous Americans. From the beginning of the meeting, Moore attempted to convey the gravity of the situation, urging everyone not to "get totally bunkered out on dope and booze as there are complex things to work out." Then, in an effort to curry favor with some of the self-styled radicals, he attempted to solidify his own radical credentials: "I have a solid commitment to see that GP is really revolutionary. I walked into the University of B.C. faculty club with Jerry Rubin and drank all their whiskey. I housed draft resisters and deserters. I'm from a radical background. I don't want creeping revisionism or liberalism in GP. We have to stay peacefully hard line." He then proceeded to articulate a carefully worked out plan to consolidate Greenpeace's various offices. The radicals, however, remained skeptical and did little to conceal their dislike of Moore and the corporate ethos they felt he represented. For his part, Moore grew increasingly frustrated with their objections, most of which consisted of platitudinous demands that Greenpeace remain "pure" and "democratic." By the end of the meeting, he was shouting with rage.[94]

In an effort to reestablish control of Greenpeace, the Vancouver group drafted a document they called the "Declaration and Charter." It was a contract that carefully outlined the responsibilities that all the branches had to the Vancouver office in exchange for the use of the Greenpeace trademark. From mid-1978 onward, all new Greenpeace branches would have to sign this document. Vancouver also tried, with varying degrees of success, to force all the existing North American groups to sign the contract. So long as San Francisco refused, however, Vancouver's ability to control the branches was always going to be partial at best. Finally, Hunter made one last effort at reconciliation, giving a speech before the assembled staff and volunteers at the Second Street headquarters. With his usual eloquence, Hunter explained how he had been with Greenpeace from the start and how much of his life he had sacrificed to help build the organization. Vancouver, he reassured them, did not wish to control San Francisco but was merely trying to create a workable international structure that would lead to greater efficiency and facilitate a more coordinated growth strategy. Hunter's words were met with blank stares. For those in the audience, most of whom had been with Greenpeace for less than a year, it was Frizell, rather than Hunter, who represented the true spirit of Greenpeace.[95]

By May 1979, the Vancouver elders had essentially given up on the possibility of bringing San Francisco back into line. After their lawyer, Peter Ballem, had made one last unsuccessful effort to convince San Francisco to sign the Declaration and Charter, they filed a trademark infringement suit in the U.S. Federal Court. San Francisco's lawyer, David Tussman, felt that it was "inevitable that [Vancouver] will wind up with ownership of the Greenpeace name in North America."[96] Nevertheless, the grassroots faction, whom Bob Taunt had always referred to as the "yahoos," controlled the San Francisco board and were not about to give in to the imperialist Canadians without a fight. Both sides hired lawyers and embarked on what would undoubtedly have been a lengthy, expensive, and ruinous court case.

When word of the lawsuit reached David McTaggart in Europe, he immediately boarded a plane and headed to San Francisco. If Vancouver won the suit, as they probably would, McTaggart had no doubt that they would turn their attentions to the budding Greenpeace groups in Europe. Given McTaggart's fractious relationship with Vancouver, he was not about to sit quietly by while they tried to gain control of the promising European offices. The Americans and Europeans, he told the San Francisco board, "must come out unanimously to fight, and must work towards a democratic Greenpeace U.S." He suggested that the Americans offer Vancouver a settlement: in exchange for San Francisco paying off Vancouver's considerable debts, Vancouver would relinquish the rights to the name "Greenpeace" outside Canada. Tussman replied that Vancouver would never relinquish the name and that his ultimate objective was reconciliation rather than a drawn-out battle. Nevertheless, McTaggart insisted that they put up a fight: "Canada cannot have rights to the name here, giving them rights to sue other offices." Although Tussman remained unconvinced, McTaggart managed to stiffen the resolve of Frizell and several other board members, who decided they would fight Vancouver to the bitter end.[97]

Having accomplished what he set out to achieve in San Francisco, McTaggart then flew to Vancouver. He immediately organized a meeting with Hunter, the only person on the Vancouver board whom he respected. Patrick Moore, McTaggart insisted, was leading Greenpeace down the path to ruin. He also reported that the wealthy San Francisco office would fight Vancouver for as long as it took to win their independence, though he failed to mention that he had played a large part in this decision. Couldn't Hunter convince Moore and the rest of the board to drop the lawsuit? Hunter replied that although he was in general agreement with Moore's position, he might be able to talk him into toning down some of his inflammatory rhetoric, thereby creating a better environment for any potential compromise. Moore, however, was in no mood for compromise. When Hunter tried to talk him into examining possible settlement options, Moore felt he was being lectured by Greenpeace's elder statesman. In a fit of alcohol-induced pique, he told Hunter that he was a "washed up" environmentalist whose days of leading Greenpeace were well and truly over. He should butt out of the matter and allow Moore to run things as he saw fit. Deeply wounded by his old comrade's outburst, Hunter began to think that perhaps McTaggart was right. Maybe Moore was power hungry and out of control. Perhaps it was time for the Vancouver tribe to remember the message they had received from the Kwakiutl all those years ago. It was the same advice that Fritz Perls would have given them: let go of their egos and allow a new generation to take Greenpeace forward into the 1980s.[98]

A few days later, McTaggart organized a meeting with the Vancouver board and their lawyers. With Hunter backing him up, McTaggart described his vision for the future of Greenpeace. Vancouver, he insisted, would have to drop the lawsuit and relinquish its rights to the Greenpeace name outside Canada. In exchange, a newly formed Greenpeace International would pay off Vancouver's debts. Once Moore realized that Hunter and several other board members were supporting McTaggart's plan, he eventually gave in. McTaggart's proposal, it was clear to Moore, was not so different from what he himself had in mind. The only difference—though it was a significant one—was that Moore would clearly not be at the helm of McTaggart's new organization. Remarkably, in just a few short days, McTaggart had not only solved what had seemed an intractable problem, but he had also succeeded in convincing Greenpeace's founders to effectively turn the organization over to him.[99]

With Vancouver's surrender notice in his hand, McTaggart flew triumphantly back to San Francisco, where he received a hero's welcome. The various American branches were so relieved and grateful that the lawsuit had been avoided that it became, in McTaggart's words, "an easy day's work to pull the twenty or so American offices together into Greenpeace USA. Somebody produces a map, and I draw nine different regions onto it. That's about it."[100] For McTaggart, the entire business was reminiscent of the kind of wheeling and dealing he would do on a weekly basis during his years in the building industry. Despite the palpable relief on both sides of the 49th parallel, some of the older, less impressionable Greenpeacers remained skeptical of McTaggart's goals and his modus operandi. Tussman, for example, "had mixed feelings about McTaggart. Not at all the crazy hippie environmentalist type, he defied categorization and made his own rules, changing reality in the process. I saw him as manipulative and dangerous." However, Tussman admitted, "when he focused his intense blue eyes on you it was impossible not to want to yield to his charm and the force of his personality. A born leader, he made people follow without feeling they were being led. He had built a tight organization in Europe focused on running environmental campaigns, avoiding the problems Vancouver faced. He was not an intellectual or a great thinker, simply a practical genius who could work miracles. Exactly the kind of leader Greenpeace craved."[101]

Several months later, McTaggart convened a meeting of Greenpeace delegates from around the world. At the meeting—held in Amsterdam—Greenpeace Europe agreed to change its name to "Greenpeace Council" and invited others to join the new organization. Greenpeace USA and Greenpeace Canada were immediately accepted as members but, in the process, had to accept the bylaws of Greenpeace Europe. All the national groups signed the Greenpeace Council accord, ceding their rights to the name "Greenpeace" in exchange for voting membership on the Council. Virtually overnight, the various Greenpeace tribes were merged together to create a European-dominated international organization with a bureaucracy, a hierarchical, centralized structure, and a headquarters based in Amsterdam. Not surprisingly, David McTaggart was voted in

David McTaggart discusses potential campaigns for the newly formed Greenpeace International. Amsterdam, November 1979. Reproduced by permission of Rex Weyler/Greenpeace.

as the first chairman of the new international Greenpeace organization. John Frizell, previously the voice of Greenpeace's grassroots members, became McTaggart's right-hand man for the next decade.[102]

In truth, McTaggart's Greenpeace was not so different from the organization that Moore, Hunter, and the various Vancouver elders had been trying to build. Dreams of a consciousness revolution had for quite some time been subordinated to the practical realities of organizational management. In addition to the personality conflicts and power struggles, however, there were also various structural factors that tended to channel Greenpeace's organizational development in the direction of professionalization, centralization, and hierarchy and away from grassroots and consensus. The direct-action style of protest relied on small groups of people performing dangerous stunts. Unlike mass movements which attempt to maximize participation through public gatherings and street protests, Greenpeace could not realistically expect to mobilize and coordinate large numbers of people on the ice floes of Newfoundland or in the whaling grounds of the North Pacific. Furthermore, the need for planning, discipline, and secrecy meant it was difficult to involve large numbers of people in the decision-making process without running the risk of being infiltrated by opponents. Not surprisingly, the "masses" who supported Greenpeace became, in effect, checkbook members. While many came to admire Greenpeace's dedication, few ever got to experience the emotional highs and lows of obstructing harpoons and hakapiks. Being a checkbook member of Greenpeace remains, at best, a vicarious thrill.

Greenpeace, therefore, never became an idealized version of a democratic, grassroots, and participatory movement. While social scientists like to see groups such as Greenpeace as vehicles for the expansion of civil society and participatory democracy, these organizations do not necessarily see this as their top priority.[103] Furthermore, as political scientists David Meyer and Sidney Tarrow have noted, professionalization does not necessarily entail "selling out." The core members of social movement organizations can be highly professional while putting forward ideologies of spontaneity and mobilizing temporary coalitions of nonprofessional supporters around their campaigns.[104] While the goal of consciousness revolution has never been achieved—at least, not to the extent that Hunter and his colleagues had hoped—there seems little doubt that McTaggart's professional and hierarchical organization has been a valuable contributor to the field of world civic politics.

Conclusion

The first meeting of Greenpeace International was attended by representatives from the United States, Canada, the United Kingdom, France, the Netherlands, New Zealand, Australia, and Denmark. The Canadian representatives—Moore, Hunter, Weyler, and Bill Gannon—watched the proceedings with a mixture of pride and regret. The organization they had built was now in the hands of David McTaggart and a group of energetic, highly motivated, and deeply committed activists from around the world. Nevertheless, to the Vancouver group, there was something about the original spirit of Greenpeace that was missing from McTaggart's organization, particularly the kind of grand vision that Hunter had initially outlined. After the meeting, the four Vancouverites gathered in a smoky Amsterdam beer house to discuss Greenpeace's fate.

For Hunter, it was a bittersweet moment: "We've been cloned. . . . The fact is, I'm totally redundant now. They don't need us and they don't want us around. . . . They've got it together better than we ever could. That's what we have to realize—we tried to do too much." Despite acknowledging that Greenpeace under McTaggart would probably be more "together" than when he had been in charge, Hunter was nonetheless sad to realize that his original vision had been erased to make way for a professional, well-managed international NGO. "I haven't seen the new movement yet. I haven't heard the next global consciousness shift articulated." Furthermore, Hunter continued, "there aren't too many ecologists sitting around the table." Naturally, the people at the meetings were committed environmentalists who understood a thing or two about ecology. But they did not seem to embrace it in the way Hunter did—as a "secular religion" whose laws and sensibilities should influence the future actions of humankind. In other words, they lacked the countercultural spirit that had animated so many of Greenpeace's actions throughout the 1970s. "How many other people ran out of those sessions to consult the *I Ching*?" Hunter asked his colleagues. "They'd think we were idiots if they knew."[1]

Nobody in Greenpeace thought Hunter was an idiot, although some certainly thought he was eccentric. Hunter's vision for Greenpeace emerged from a sixties counterculture characterized by a deep commitment to various forms of holism and a refusal to accept the reductive view that the world is merely the sum total of the physical and chemical forces acting upon it. In the process, it opposed the Cartesian categories that have split the world into *res cogitans* (thinking substance) and *res extensa* (extended matter), a dualism that classifies nature as a tool or resource for humans to manipulate and analyze as they see fit. The opponents of Cartesian dualism were many and varied: some spoke in the sober and detached register of science, while others resorted to full-blown mysticism. Most, including Hunter and his Greenpeace cohort, fell somewhere in between. Nevertheless, with his profound, at times apocalyptic emphasis on consciousness change, Hunter was part of a tradition that refused to see consciousness as the exclusive preserve of the human mind.

David McTaggart was not sympathetic to any of this. Nor were many of the younger activists taking over from the Vancouver hippies. People such as Steve Sawyer, John Frizell, and Remi Parmentier had little time for the *I Ching*. Cree mythology was fine as an expression of solidarity with indigenous peoples, but it did not imply that Greenpeace was literally the embodiment of the "warriors of the rainbow."[2] For McTaggart and his lieutenants, Greenpeace was not so much a vehicle for consciousness change as it was a tool of political persuasion. While many of the new leaders embraced the worldview of holistic ecology, they tended to steer clear of the kind of cultural holism that Hunter and his friends embraced. However, when shorn of Hunter's idealism, mind bombing could quickly descend into cynicism. This is evident in the words of Don White of Greenpeace Hawaii. "The grassroots counterculture loves us; we're an extension of their fantasies," he blithely intoned. But if this was the only constituency that Greenpeace was able to influence, its efforts would be wasted. It was far more important, White argued, to reach the "Great Undisturbed Middle Class, a sleeping giant with a working I.Q. of about 15 and rapacious appetites for status, energy, sex, food, and comfort. . . . If you poke the giant just right, he'll lash out so he can go back to sleep. And that's the business we're in; getting the public pissed off at things. In and of ourselves, we're totally impotent."[3]

Within a few months of the Amsterdam meeting, McTaggart's Greenpeace International developed a sophisticated management structure, with various legal, administrative, financial, and communications arms scattered throughout the world. It was not long before these offices were staffed by professionals with degrees in human resources, marketing, and accounting. In a short time, the organization's structure bore a remarkable similarity to the mainstream environmental organizations from which Greenpeace differentiated itself in the early 1970s. The baton of radical environmentalism was soon passed to groups such as Earth First!, Sea Shepherd, and the Rainforest Action Network. Despite its drift toward the mainstream, however, Greenpeace was able to avoid some of the pitfalls that beset other organizations. For example, unlike most major U.S. environmental outfits, Greenpeace has never accepted funding from corporations, political parties, unions, or large corporate endowments such as the Ford Foundation. Nor have the executives in the American branch been enticed into taking high-level positions at various regulatory agencies. This has ensured that Greenpeace has rarely been part of the so-called revolving door system in which prominent leaders circulate back-and-forth between government, industry, and NGOs, a system that forces environmental leaders to limit their demands to fit the needs of consensus-driven politics.[4] Nevertheless, there are other aspects of Greenpeace's modus operandi that serve to push it toward the political mainstream. For example, since its funding comes largely in the form of individual donations, it cannot afford to alienate large segments of the public by engaging in actions that might be perceived as too radical or closely connected to established political groups.

Throughout the 1980s, Greenpeace USA remained the largest branch of Greenpeace International, contributing the lion's share of the international budget. Although the organization's political power was centered in Europe, so long as the U.S. offices continued their successful fundraising, there were few attempts to limit their autonomy. As a result, Greenpeace USA became involved in the environmental justice movement, cooperating closely with civil rights organizations and labor unions, a move that would have gladdened the heart of Irving Stowe had he been alive to witness it. As a result, Greenpeace occupied the more progressive end of the environmental spectrum, a position that was in accord with the values of most of its employees and volunteers. In countries such as Germany and the Netherlands, however, Greenpeace rapidly became the dominant environmental group. Not surprisingly, in these nations the organization stuck to the political middle ground, assiduously avoiding campaigns that would overtly link it to any political parties, lobby groups, or labor unions. Understandably, therefore, some Europeans grew concerned about the direction in which their American brethren were taking the organization, feeling that the U.S. office's close ties to labor groups and civil rights organizations—many of which were vocal supporters of the Democratic Party—would taint Greenpeace International's image as an organization that was above party politics.

In the early 1990s, however, Greenpeace USA—like the American environmental movement as a whole—fell upon difficult times and was forced to accept a multimillion dollar bailout from the International office in Amsterdam. Thereafter the Europeans had the financial power to go with their political influence. Greenpeace International's pragmatic new leader, the German economist Thilo Bode, had always felt that Greenpeace USA's involvement in the environmental justice movement was inappropriate and that Greenpeace should stick to what it was famed for—campaigning on issues of global significance and carrying out spectacular direct actions—rather than involving itself in local grassroots movements. The result was that Greenpeace USA was forced to retreat toward the safe ground of mainstream environmentalism, a move that greatly disappointed many American activists.[5]

Despite such problems, Greenpeace has remained a successful organization with considerable political clout. Corporations and governments frequently have to prepare for a potential Greenpeace "bust" when planning their actions, thereby giving Greenpeace the power to "sting" them with an ecological sensibility. Nonetheless, the organization never reached the heights that Hunter envisioned: "It's big, but nowhere near big enough," he lamented. Greenpeace never became the leading apostle of a secular religion based on ecology.[6] It also did not develop into the kind of grassroots, participatory movement that Irving Stowe had hoped to build. Various aspects of Greenpeace's style and tactics—such as its inability to combine direct action with mass participation—compromised the development of such a movement. In contrast, groups such as the Clamshell Alliance and its West Coast counterpart, the Abalone Alliance, were 1970s movements that engaged in direct-action environmentalism, such as protesting outside nuclear power plants and engaging in peaceful "invasions" of nuclear facilities, while also embodying their intensely progressive politics in their organizational structure. The Clamshell Alliance, unlike Greenpeace, remained decentralized, nonhierarchical, participatory, and consensus driven. They engaged in what historian Barbara Epstein calls "prefigurative politics": an attempt to convey their vision of an ecologically sustainable and egalitarian society not just through their rhetoric and protests but also in the way they lived.[7]

Certain elements within Greenpeace may have wanted the organization to go in this direction—particularly people such as Walrus Oakenbough and Rod Marining. Their influence, however, never surpassed that of Hunter, Moore, Metcalfe, and McTaggart. To one degree or another, these men had come to accept the need for hierarchy and professionalism as a by-product of Greenpeace's modus operandi. Paul Watson, another of the grassroots advocates who subsequently became highly critical of what he felt was Greenpeace's "corporate" structure, has nonetheless failed to create a decentralized structure in his own organization, the Sea Shepherd Conservation Society. Though Sea Shepherd is certainly much smaller than Greenpeace, "Captain" Watson's charismatic leadership is so vital to its actions and fund-raising abilities that one can scarcely conceive of the organization without him. Again, however, we must beware of the false construction of purity: neither Greenpeace nor Sea Shepherd necessarily feels that it is incumbent upon them to develop organizational structures that reflect some distant, idealized future society. While the Clamshell Alliance may have gone a considerable way toward achieving this, it did so only by renouncing the kind of political influence that groups such as Greenpeace attained.

Could things have turned out differently? After all, McTaggart's takeover was an act of sheer opportunism and only occurred due to a series of specific events in which McTaggart himself played little part. Had Vancouver's lawsuit against San Francisco proceeded without McTaggart's intervention, it is highly likely that Vancouver would have won, forcing San Francisco to either bow to Patrick Moore's will or break away and form a new organization with a different name—an unlikely formula for success given that their growth had been based entirely on the Greenpeace name and the images associated with it. In such a scenario, Vancouver could have enforced its Declaration and Charter on all the Greenpeace groups throughout the world. In truth, however, the Vancouver group's trajectory in the late 1970s seemed to differ little from the direction in which McTaggart eventually took the organization. Hunter's power was waning and Moore was never as committed to countercultural idealism. Even without McTaggart's opportunism, therefore, it seems that Greenpeace would probably have evolved in much the same way, albeit based on the west coast of North America rather than in northwestern Europe.

What, then, is Greenpeace's legacy? The Canadian environmental journalist Alanna Mitchell argues that it is "The mindbomb that will not go away." And she does not mean this as a compliment.

> The mindbombs that Greenpeace lobbed were powerful to the point of being indelible. And rather than being strong images about the destruction of the environment, they came to represent the environmental movement itself. Thirty-odd years on, environmentalism is still seen as quintessentially left-wing, as hippie versus the establishment, as alpha male, adolescent, intuitive. As a suspiciously religious calling.[8]

Mitchell does not blame Hunter and company for this; they could not have foreseen the consequences of their actions and were just doing their best to end nuclear testing and save whales. Nevertheless, she continues, we now live in an era when "the environment is about empirical evidence, not mindbombs."[9] Therefore, the fact that Greenpeace's dramatic and highly publicized direct actions have come to stand as a synecdoche for environmentalism in general has limited environmentalism's broader appeal. Many people, even some of those who support the same causes, feel uneasy about Greenpeace's radical approach. For example, nobody wanted to save whales more than Project Jonah founder Joan McIntyre. Nevertheless, Greenpeace's macho and confrontational style did not sit well with her. The same undoubtedly holds true for many people who instinctively recoil from forms of radical protest with strong militaristic overtones.

Mitchell's critique is not entirely wide of the mark in terms of assessing the McLuhanesque side of the Greenpeace legacy. However, it tends to exaggerate the degree to which the general image of environmentalism has been conflated with Greenpeace's confrontational direct actions. The "gentle subversiveness" of Rachel Carson and the sober didacticism of Al Gore have also shaped the image of environmentalism.[10] Furthermore, by focusing intently on mind bombs, Mitchell ignores the fact that Greenpeace has always been more than a one-trick pony. Recall, for example, the educational aspects of the return voyage from Amchitka, the efforts to create alliances with workers or indigenous peoples, and the scientific papers that Bohlen and others commissioned. Such nonconfrontational, constructive strategies became increasingly important in subsequent decades. For example, in the early 1990s, Greenpeace Germany helped sponsor research into more environmentally friendly refrigeration technology.[11] Also, Greenpeace boats frequently dock at ports all around the world, welcoming people on board and conducting education campaigns about various environmental issues. In addition, Greenpeace continues to commission scientific reports, providing a valuable public forum for scientists whose work may be politically unpalatable to industry and governments.

Thus the mind bomb has increasingly become the weapon of last resort. In fact, the more Greenpeace's reputation for launching mind bombs grew, the less they needed to employ them. By the mid-1980s, governments and industries throughout the world knew that once Greenpeace announced that it was concerned about a particular issue, a spectacular direct-action protest was always a possibility. Thus the implicit threat of a mind bomb is almost as potent a weapon as the mind bomb itself. As a result, Greenpeace can initially pursue less confrontational forms of persuasion, turning to mind bombs—which, after all, are expensive, risky and difficult to organize—only as a last resort. So even as mind bombing remains Greenpeace's most high-profile tactic, it takes up far less of the organization's time and resources than most people would think.

A more controversial and polemical critique has been leveled at Greenpeace by, of all people, Patrick Moore. Unlike the other core members of the Vancouver group, Moore decided to wholeheartedly commit himself to McTaggart's new Greenpeace International. He thus became the director of Greenpeace Canada, as well as one of the five Greenpeace International directors. McTaggart and some of the newer activists found Moore annoyingly arrogant and pompous. However, his vast experience and scientific credentials made him a valuable asset, and he remained a pivotal member of the organization until the mid-1980s. For his part, Moore remained suspicious of McTaggart. He also became increasingly concerned that some of the newer activists were more interested in Marxist revolution than they were in environmental issues: "I remember visiting our Toronto office in 1985 and being surprised at how many of the new recruits were sporting army fatigues and red berets in support of the Sandinistas."[12]

While Moore had a grudging respect for McTaggart's political skills, he grew increasingly frustrated with his lack of scientific literacy. According to Moore, this meant that various radical and doctrinaire scientists easily swayed McTaggart. Thus Greenpeace embarked on numerous ill-advised campaigns that did not withstand scientific scrutiny, such as opposition to fish farming and taking unrealistically inflexible positions on the use of various chemicals such as chlorine. By the mid-1980s, Moore was becoming frustrated with what he perceived as Greenpeace's "slide into voodoo science."[13] His constant criticism annoyed the other directors and many of the organization's activists and scientists. As a result, his position as an international director became increasingly untenable. The bitterness surrounding his departure perhaps helps explain why for the past quarter of a century he has been among Greenpeace's fiercest and most relentless critics, a role that has been enthusiastically supported by numerous industry front groups and free-market think tanks.[14]

Moore's criticism needs to be understood as part of a carefully crafted narrative that culminates with his maturation into a "sensible environmentalist." This allows him to have it both ways: during his impetuous youth, environmentalists needed to raise the alarm about the impending ecological crisis, something that Greenpeace excelled at during the 1970s. By the 1980s, however, "a majority of the public, at least in Western democracies, agreed with us that the environment should be taken into account in all our activities. When most people agree with you it is probably time to stop beating them over the head and sit down with them to seek solutions to our environmental problems."[15] Thus Moore can maintain a sense of pride about his achievements with Greenpeace, while fiercely criticizing the organization's campaigns since his departure.[16]

Greenpeace has certainly been guilty of sloppy—as opposed to merely controversial—science on occasion, such as during the antisealing campaign in the late 1970s and from time to time since. For example, in 1995, Greenpeace opposed Shell's plans to sink a decommissioned oil platform in the North Sea. After an extensive and highly successful mind bombing campaign, Shell was forced to drag the platform back to shore and dismantle and recycle it there. In its rush to condemn Shell's actions, Greenpeace exaggerated the amount of oil remaining in the rig and ignored the warnings of nonindustry scientists who believed that sinking it was the safest course of action.[17] Nevertheless, Moore's attempt to use science as a cudgel to bludgeon Greenpeace's credibility mostly constitutes an exercise in polemics. Scientists are influenced by politics and ideology to a certain extent. At most, Greenpeace, like Moore, is guilty of choosing the scientific studies that best support its claims. In fact, by continually using the politically loaded term "junk science," Moore has aligned himself with industry-supported antienvironmentalist attack dogs such as Steve Milloy, the Fox News commentator and creator of the website junkscience.com.[18]

Ultimately, in a world system that primarily functions according to the self-interest of nation-states and corporations, perhaps Greenpeace's world civic politics—the "sting" of ecological sensibility promoted by nonviolent direct action—represents the organization's most important legacy. In addition, the promotion of this ecological sensibility, which is grounded in a deeply holistic worldview, constitutes the most realistic approximation of Hunter's "consciousness revolution." This commitment to a dynamic form of transnational activism and a holistic ecological worldview also reflects the various social and intellectual movements embodied by Greenpeace's founders: Quakerism, nonviolent action, holistic ecology, McLuhanism, and countercultural idealism. Whales are still harpooned throughout the world's oceans, but those who hunt them must always be prepared to endure the widespread public scorn that results from a successful Greenpeace mind bombing. The same can be said for numerous other activities that pose a threat to the environment. Whatever its shortcomings, Greenpeace created a new and potent method for confronting powerful institutions engaged in environmentally irresponsible activities. That is a legacy with which Irving and Dorothy Stowe, Jim and Marie Bohlen, Ben Metcalfe, David McTaggart, Bob Hunter, and some day, even Patrick Moore and Paul Watson, can rest in peace.

NOTES

Introduction

1. Hunter, *Warriors*, 131.
2. U.S. House, *Marine Mammals*, 65–66.
3. Sociologist James William Gibson writes that a powerful encounter with an animal is frequently the first step in the sacralization, or consecration, of animals, which is itself part of a broader search for a reenchanted natural world. See *A Reenchanted World*, 40.
4. In fact, the German magazine, *Der Spiegel*, once referred to Greenpeace as the "McDonald's of the environmental movement." See "McDonald's der Umweltszene." *Der Spiegel*, 9/16/1991, 87. For a list of the countries in which Greenpeace currently operates, see http://www.greenpeace.org/international/en/about/worldwide/.
5. Some of the more significant works include: Hunter, *Warriors*; Moore, *Confessions of a Greenpeace Dropout*; McTaggart with Slinger, *Shadow Warrior*; McTaggart with Hunter, *Greenpeace III*; Weyler, *Greenpeace* and *Song of the Whale*; Watson, *Ocean Warrior*; Wilkinson with Schofield, *Warrior*; Bohlen, *Making Waves*; Griefahn, *Monika Griefahn*; Newborn, *A Bonfire in My Mouth*; Tussman, *Greenpeace*.
6. Brown and May, *The Greenpeace Story*; Szabo, *Making Waves*; Paul Brown, *Greenpeace*; Warford, *Greenpeace Witness*; Greenpeace Deutschland, *Volle Kraft*; Greenpeace Deutschland, *Das Greenpeace Buch*. Hunter's *Warriors of the Rainbow* also fits into this category.
7. Dauvergne and Neville, "Mindbombs of Right and Wrong"; Doyle, "Picturing the Clima(c)tic"; Dale, *McLuhan's Children*; Cassidy, "Mind Bombs and Whale Songs"; Widdowson, "The Framing of Greenpeace"; Krüger and Müller-Henig, *Greenpeace auf dem Wahrnehmungsmarkt*; Hansen, "Greenpeace and the Press Coverage of Environmental Issues."
8. Prominent examples include: Shaiko, "Greenpeace U.S.A."; Wapner, *Environmentalism and World Civic Politics*; Heinz, Cheng, and Inuzuka, "Greenpeace Greenspeak"; Harter, "Environmental Justice for Whom?"; Strømsnes, Selle, and Grendstad, "Environmentalism Between State and Local Community"; Rucht, "Ecological Protest as Calculated Law-Breaking"; Lahusen, *The Rhetoric of Moral Protest*; DeLuca, *Image Politics*. Anyone expecting to learn anything about Greenpeace in James Eayrs's *Greenpeace and Her Enemies* will be sorely disappointed. In 350 pages of text, there is only one passing reference to the organization (42). On the other hand, there is a good deal of discussion of its "enemies," the U.S. and Canadian military establishments.
9. Streich, *Betrifft*; Altmann and Fritzler, *Greenpeace*; Knappe, *Das Geheimnis von Greenpeace*; Bölsche, *Die Macht der Mütigen*; Pelligrini, *Greenpeace, la Manipulation*; Reiss, *Greenpeace*; Luccioni, *L'Affaire Greenpeace*; Picaper and Dornier, *Greenpeace*; Sunday Times Insight Team, *Rainbow Warrior*; King, *Death of the Rainbow Warrior*; Gidley and Shears, *Rainbow Warrior Affair*; Jordan, *Shell, Greenpeace and the Brent Spar*.

10. Several of the more significant surveys of American environmentalism do not mention Greenpeace in their indexes. All, however, devote considerable space to organizations such as the Sierra Club, the Wilderness Society, and the Environmental Defense Fund. See Fox, *John Muir and His Legacy*; Hays, *Beauty, Health, and Permanence*; Rothman, *The Greening of a Nation?*; Crane and Egan, *Natural Protest*. The same can be said of Ramachandra Guha's popular text, *Environmentalism: A Global History*. In his otherwise solid survey of American environmentalism, Steven Stoll only mentions Greenpeace twice, both times erroneously (he refers to it as an "American organization" and gives the impression that it emerged as a direct result of the 1969 Santa Barbara oil spill). See *U.S. Environmentalism since 1945*, 19, 134. David Peterson del Mar does a better job in *Environmentalism*. His three-page overview is factually correct. The same can be said of Benjamin Kline's popular textbook, *First Along the River: A Brief History of the U.S. Environmental Movement*. Haq and Paul's *Environmentalism Since 1945* provides a brief summary of Greenpeace's history, but little more. Roderick Nash confines his discussion of Greenpeace to the group's more animal-rights-based activities of the mid to late 1970s. See *Rights of Nature*, 179–89. Gottlieb, *Forcing the Spring*, and Sale, *Green Revolution* do a better job of integrating Greenpeace into the history of the U.S. environmental movement. However, both fail to situate the organization within the broader history of the American peace movement. The more polemical works on radical environmentalism, such as Manes, *Green Rage* and Scarce, *Eco-Warriors*, take a similar approach, emphasizing Greenpeace's innovative direct-action tactics without examining their origins.
11. Wapner, *Environmental Activism*, 65.
12. McIntyre, *Mind in the Waters*.

Chapter 1

1. Denis Bell, "Story of the Greenpeace-Makers," *Vancouver Province*, 3/5/1977, 5.
2. Stowe's original name was Irving Strasmich. He did not change it until his family moved to New Zealand in the 1960s. For the sake of consistency, and to minimize confusion, I will refer to him and his wife, Dorothy, by their adopted name.
3. Hunter and Stowe interviews.
4. Stowe interview. For more on the history of Jewish radicalism, see Rothman and Lichter, *Roots of Radicalism*, and Leviatin, *Followers of the Trail*. For an exploration of modern Jewish intellectual life, see Mendes-Flohr, *Divided Passions*.
5. According to the Brown alumni magazine, this was not Irving's first marriage. He had married Harriet Johnston of Greenville, Tennessee, in April 1949. See *Brown Alumni Monthly*, June 1949, 31.
6. Stowe interview.
7. *Brown Alumni Monthly*, January 1950, 23. Stowe interview. For more on the growth of "One Worldism" in the postwar era, see Sluga, "UNESCO" and Wittner, *Rebels Against War*, ch. 6.
8. Bacon, *Quiet Rebels*, 182; Cooney and Michalowski, *Power of the People*, 130–31.
9. In his short but astute overview of the subject, Mark Kurlansky notes that no major language appears to have a proactive term for nonviolence. Instead, as in English, the concept is always denoted by the negation of *violence*. This is even the case with Sanskrit, in which the word *ahimsa*—to do no harm—is merely the negation of *himsa*. See *Nonviolence*, 5. Another useful overview is Zunes et al., *Nonviolent Social Movements*.
10. Sharp, *Politics of Nonviolent Action*, 64. Emphasis in original.
11. Ibid., 8.
12. Ibid., 68–69.
13. Joseph Kip Kosek, "Richard Gregg." For a study of Gandhi's broader influence on radical protest in the West—mostly in the United States and Great Britain—see Scalmer, *Gandhi in the West*.
14. Yavenditti, "The American People," 224–25; Boyer, *By the Bomb's Early Light*, 182–85. John Dower describes how the U.S. government and media dehumanized the Japanese, thus creating a receptive environment for their extermination. See *War Without Mercy*.
15. For a history of the WRL, see Bennett, *Radical Pacifism*.

16. Boyer, *By the Bomb's Early Light*, 352.
17. Wittner, *Rebels Against War*, 265; Winkler, "A 40-Year History of Civil Defense"; Fitzsimmons, "Brief History of American Civil Defense."
18. DeBenedetti, *Peace Reform*, 157–58.
19. Wittner, *Rebels Against War*, 246.
20. Quoted in Peace, *A Just and Lasting Peace*, 27.
21. Sharp, *Politics of Nonviolent Action*, 382.
22. Bigelow, *Voyage of the Golden Rule*, 24. The *Liberation* article, entitled "Why I Am Sailing into the Pacific Bomb Test Area," is reprinted on 44–47.
23. Ibid., 22–23, 36–37.
24. Ibid., 112 and passim.
25. Ibid., 117, 149, 162.
26. Reynolds, *Forbidden Voyage*, 28.
27. Ibid., 24, 65.
28. Wittner, *Rebels Against War*, 262; Cooney and Michalowski, *Power of the People*, 140–41; DeBenedetti, *Peace Reform*, 161. One of those arrested and jailed was the poet and musician, Ed Sanders, who later went on to form the satirical folk-rock band, The Fugs. Sanders was an important interlocutor between the beat and hippie generations, and his presence at the protest is an early example of the shared interests among pacifists and the counterculture. See Farber, *Chicago '68*, 24, and Sanders, *Tales of Beatnik Glory*.
29. Wittner, *Rebels Against War*, 263. In her study of radical pacifism, Marian Mollin notes that Polaris Action was an early example of the clash of values between older, "straight" pacifists and younger, more radical and "hip" activists. As we shall see, the same dynamic characterized the early years of Greenpeace. See Mollin, *Radical Pacifism*, 110.
30. There are countless studies exploring this theme. Those that tackle it more broadly include Boyer, *By the Bomb's Early Light*; Winkler, *Life Under a Cloud*; and Peace, *A Just and Lasting Peace*, 22–30.
31. See May, *Homeward Bound*.
32. Those wishing to understand the psychology of such activist urges can start with Kool, *The Psychology of Nonviolence*.
33. Dorothy Stowe was not exactly sure when Irving first began attending Quaker meetings, but she thought that it was probably in the late 1940s. Stowe interview.
34. AFSC, *Speak Truth to Power*, 8. The pamphlet emerged from a conference attended by leading Quaker intellectuals and activists at Haverford College in the summer of 1954. Participants included Bayard Rustin, A. J. Muste, and Amiya Chakravarty, a prominent Indian academic and friend of Gandhi. See Chmielewski, "Speak Truth to Power." Cited with author's permission.
35. AFSC, *Speak Truth to Power*, 9.
36. Ibid., 1–2, 28.
37. Ibid., 52, 60, 65.
38. *Peace Reform*, 146.
39. Nutter, "Jessie Lloyd O'Connor and Mary Metlay Kaufman"; O'Connor, *Revolution in Seattle*. For correspondence between the O'Connors and the Strasmiches (as they still were at the time), see box 61, folder 20 of the Jessie Lloyd O'Connor Papers, SSC.
40. From *Liberation*, IV (November, 1959), 14–16, summarized in Wittner, *Rebels Against War*, 238–39.
41. Mollin, *Radical Pacifism*.
42. *Peace Reform*, 197, 199.
43. For example, the *Saturday Evening Post*, edited by influential pacifist Norman Cousins, published an article by Steven Spencer entitled "Fallout: The Silent Killer" August, 29, 1959, 26, 89 and September 5, 1959, 86.
44. Divine, *Blowing on the Wind*, 262–80. For more on the doomsday literature of the era, see Wager, *Terminal Visions*. Shaheen, *Nuclear War Films* analyses the films of the period.
45. Stowe interview.
46. Letter to Jessie and Harvey O'Connor, 02/06/1963. Jessie Lloyd O'Connor Papers, SSC, box 61, folder 20. Stowe interview.

47. Stowe interview. For a discussion of peace movement activities in New Zealand in the 1960s, see Locke, *Peace People*, 168–80.
48. Stowe interview.
49. Ibid.
50. Wittner, *Resisting the Bomb*, 197–200; Kostash, *Long Way From Home*, xxii–xxvi; Palmer, *Canada's 1960s*, 58.
51. Kostash, *Long Way from Home*, 42–43. For more on U.S. draft resisters in Canada see Hagen, *Northern Passage*, Dickerson, *North to Canada*, and Haig-Brown, *Hell No We Won't Go*.
52. Stowe interview.
53. Bohlen interview.
54. Bohlen, *Making Waves*, 4–8.
55. Bohlen interview.
56. Ibid.
57. Ibid. The guest cabin, which the Bohlens were kind enough to let me use during my visit, was also a geodesic dome. I slept well.
58. Bohlen, *Making Waves*, 16; Bohlen interview. For more on the cultural and literary influence of Miller, see Fitzpatrick, *Doing it with the Cosmos*.
59. For more on Suzuki's influence on Buddhism in America, see Fields, *How the Swans Came to the Lake*, 60–61. For an example of Suzuki's work see his *An Introduction to Zen Buddhism*.
60. Bohlen, *Making Waves*, 17.
61. Bohlen interview.
62. Bohlen, *Making Waves*, 18.
63. Lutts, "Chemical Fallout"; Egan, *Barry Commoner and the Science of Survival*.
64. Bohlen interview.
65. Bohlen, *Making Waves*, 21.
66. Bohlen interview.
67. Ibid.
68. Ibid.
69. Kostash, *Long Way from Home*, 42.
70. Palmer, *Canada's 1960s*, 270–73.
71. Ibid., 43–49; Bohlen and Stowe interviews.

Chapter 2

1. Worster, *Nature's Economy*, 360.
2. Kingsland, *Evolution of American Ecology*, 4, 99.
3. Bocking, *Ecologists and Environmental Politics*, 3. For a discussion of this period in ecology's history—and particularly the influence of the Odum brothers in the development of ecosystem theory—see Craige, *Eugene Odum*, and Hagen, "Teaching Ecology During the Environmental Age."
4. Worster, *Nature's Economy*, 343, 359.
5. The history of holistic ideas in ecology is well described by Frank Golley, *A History of the Ecosystem Concept*. Michael Barbour demonstrates that holistic ideas were tendentious among ecologists during the mid-twentieth century. See "Ecological Fragmentation in the 1950s." For critiques of holistic ecology, see Kricher, *Balance of Nature*, and Phillips, *Truth of Ecology*.
6. For useful histories of cultural holism, see Craige, *Laying Down the Ladder*, and Wood, *A More Perfect Union*. For a detailed study of Carson's life and work, see Lear, *Rachel Carson*. For a shorter overview that situates Carson in the context of the environmental movement, see Lytle, *The Gentle Subversive*. For a fine study of Commoner's contribution to environmentalism, see Egan, *Barry Commoner and the Science of Survival*.
7. Zelko, "Challenging Modernity"; Eder, "The Rise of Counter-Cultural Movements against Modernity."
8. Bess, *Light-Green Society*, 58.

9. Ibid., 62.
10. Bohlen interview.
11. Cohen, *History of the Sierra Club*, 429.
12. Bohlen interview.
13. Quoted in the *Georgia Straight* (an underground newspaper founded in Vancouver in the mid-1960s), May 6–13, 1970, 7. The Simmons article in the *Straight* is an edited version of a longer article that appeared in the journal of the Pacific Northwest chapter of the Sierra Club. See "Poverty of Plenty."
14. Simmons correspondence.
15. Draper, "Eco-Activism," 1–3. The Stigant quote can be found in SPEC, "SPEC History," http://www.vcn.bc.ca/spec/spec/Spectrum/spring1999/beginframeset.htm. For a thorough examination of the history of water pollution in Vancouver—and SPEC's role in the story—see Keeling, "Sink or Swim."
16. *Vancouver Sun*, 6/13/1969. Hunter interview. SPEC kept its acronym (or "bacronym"), but finally settled on the name Society Promoting Environmental Conservation. See http://www.vcn.bc.ca/spec/spec/welcome.html.
17. Quote from SPEC "Timeline" at http://www.vcn.bc.ca/spec/spec/Spectrum/spring1999/beginframeset.htm.
18. See for example, Read, "Let Us Heed."
19. Simmons correspondence.
20. Bohlen, *Making Waves*, 25; Bohlen interview.
21. Bohlen, *Making Waves*, 25; Bohlen interview. Although Bohlen seems to have misremembered some of the details of the story, his version is broadly correct. For detailed coverage, see the UBC student newspaper, *Ubyssey*, 02/26/1969, 17 and 10/24/1969, 6-7. Bohlen's actions also helped ensure that UBC students and professors would have easy access to a nude beach. Visiting researchers should be aware of this fact before taking their lunchtime stroll.
22. Sharp, *Politics of Nonviolent Action*, 387. For more on the beach story, see Hemsing, "Production of Place."
23. Stowe to William E. Graham, director of planning for the City of Vancouver, 8/1/1970, vol. 2, file 7, GPF; interview with Dorothy Stowe.
24. United For Survival petition printed in the *Georgia Straight*, 3/26/1969. Emphasis in original.
25. Ohm, "Greenpeace," *Northword*, July 1971, 36.
26. *Daily Colonist*, 2/28/1971; Hunter and Bohlen interviews.
27. Stowe, "Greenpeace is Beautiful," *Georgia Straight*, 9/23–30/1970. For a history of the British Columbia New Left and how it intersected with the environmentalism of the late 1960s and early 1970s, see Isitt, *Militant Minorities*, 135–38.
28. Stowe, "Greenpeace is Beautiful," *Georgia Straight*, 7/15–22/1970 (capitals in original). Compare with the Editors of Ramparts introduction to *Ecocatastrophe*. See Leopold, *Sand County Almanac*, 262.
29. Stowe, "Greenpeace is Beautiful," *Georgia Straight*, 9/23–30/1970, 8/12–19/1970, 9/9–16/1970, 9/16–23/1970; interview with Stowe in *Georgia Straight*, 11/11–18/1971.
30. Not surprisingly, the scholarly literature on the counterculture is amorphous and diffuse. For a succinct and astute introduction and analysis, see Braunstein and Doyle, "Introduction: Historicizing the American Counterculture of the 1960s and '70s."
31. Roszak, *Making of a Counterculture*; Reich, *Greening of America*. Interestingly, Roszak lived in London during the mid-1960s, where he served as editor of the pacifist publication, *Peace News*. A few years later *Peace News* would be an important vehicle for the spread of Greenpeace's ideas in the United Kingdom. For more on Roszak's tenure at *Peace News*, see the *Peace News* archive catalog at Bradford University. Available at http://www.brad.ac.uk/library/special/documents/CwlPNcatalogue2010.pdf.
32. Hunter interview.
33. Hunter interview; Ruddock, *CND Scrapbook*, 22–30. The Aldermaston marches were the inspiration behind Bayard Rustin's decision to organize the 1963 March on Washington. The modern symbol for peace, consisting of a drooping cross inside a circle, was

another artifact of the marches. A young artist, Gerald Holtom, designed the symbol, explaining that the drooping cross contained the semaphore signals for the *n* and *d* of "nuclear disarmament" and the CND agreed to adopt it. See Wittner, *Resisting the Bomb*, 45–48.

34. Hunter, *Erebus*. Hunter actually worked in a slaughterhouse after he finished high school as part of a search for "real life experiences" he could then use as writing fodder.
35. Hunter interview.
36. Hunter interview.
37. Hunter interview; Perls, *Gestalt Therapy Verbatim*.
38. Hunter, *Enemies of Anarchy*, 96.
39. Ibid., 214.
40. Ibid., 214–15.
41. Ibid., 223.
42. Ibid., 226–27.
43. Hunter, *Storming of the Mind*, 16, 42; See also, Marcuse, *One Dimensional Man*.
44. Hunter, *Storming of the Mind*, 111–12.
45. Ibid., 124, 117.
46. Ibid., 183, 181.
47. Shepard, "Introduction: Ecology and Man."
48. *Vancouver Sun*, undated clipping in 6(2) GPF.
49. Turner, *From Counterculture to Cyberculture*, 5. Kirk, *Counterculture Green*.
50. Hunter, *Storming of the Mind*, 221; *Enemies of Anarchy*, 216–17; McLuhan put forth the same argument in his classic work, *The Medium is the Message*. For useful overviews of McLuhan's life and work, see Coupland, *Marshall McLuhan*, and Marchand, *Marshall McLuhan: The Medium and the Messenger*.
51. *McLuhan's Children*, 12.
52. Hunter, *Storming of the Mind*, 221, 216–18.

Chapter 3

1. Kostash, *Long Way from Home*, 121–23.
2. Letters of concern from the public regarding hippies in Vancouver can be found in General Correspondence, 45-C-6, "Hippies and Georgia Straight," 1969, MOF.
3. "Report from the Special Committee of Council Regarding Hippie Situation," 10/10/1967, 45-B-5 (10), MOF; Hank Vogel to Mayor, 6/29/1970, 45-E-1 (34), MOF.
4. Quoted in Farber, *Chicago '68*, 10–11, 20. Capitalization in original. See also Rubin, *Do It!*
5. Farber, *Chicago '68*, 14–15. Also see Hoffman's autobiography, *Soon to Be a Major Motion Picture*. Although his books on the consciousness revolution devote little space to Yippie "philosophy," Hunter admitted that Rubin and Hoffman had a considerable influence on both himself and other counterculture activists in Vancouver. Hunter interview.
6. For detailed coverage of the event, see *Ubyssey*, Oct. 22, 25, 29, 1968. Bob Hunter was among those present. Hunter interview; Marining interview.
7. Marining interview. Rocky Rococo was influenced by Living Theater, an experimental theater company founded in New York in 1947 by Julian Beck and Judith Malina.
8. Kostash, *Long Way from Home*, 134–35.
9. Watson interview. Published accounts of Watson's rise to the "ocean warrior" of the environmental movement include Khatchadourian, "Neptune's Navy"; Watson, *Sea Shepherd* and *Ocean Warrior*. Hunter's description is from his book *Warriors of the Rainbow*, 11. Hunter, Marining, and Bohlen interviews.
10. McLeod interview. A sampling of various *Straight* articles over the years can be found in Pauls and Campbell, *The Georgia Straight*.
11. For more on SAVE, see *Georgia Straight*, July 1–7, 1970, 14. The quote is in issue March 25–April 1, 1970, 3.
12. *Victoria Daily News*, 05/31/1971, 38. *Vancouver Sun*, 06/01/1971, 11. Marining, Watson, and Hunter interviews. Watson described his involvement in several articles he wrote for the *Georgia Straight*. See the following issues: July 20–23, 1971, 3 and August 3–6, 2.

The protest has been adopted as a milestone in the history of the new movement known as "guerrilla gardening." See Tracey, *Guerrilla Gardening*, 24.

13. *Georgia Straight*, June 1–4, 1971, 12–13 and June 8–11, 1971, 3. Marining interview. David Garrick, who soon adopted the nom de plume Walrus Oakenbough, describes the protest in the *Georgia Straight*, August 10–13, 1971, 6. For a video montage of the park protest, see http://www.youtube.com/watch?v=_ukUAo456BE. In 1984, after fourteen more years of lobbying and activism, the property gained official status as a public park named Devonian Harbour Park.
14. Metcalfe knew Roderick Haig-Brown, British Columbia's most famous mid-twentieth century conservationist writer and fly fisherman, and wrote a biography of him in the mid-1980s. See Metcalfe, *A Man of Some Importance*.
15. Marining interview. Metcalfe interview.
16. Metcalfe interview.
17. Ibid.
18. Ibid; Metcalfe's article describing his LSD experience appeared in the *Vancouver Province*, 9/2/1959. Also see Stevens, *Storming Heaven*, 175–76.
19. Metcalfe interview. A photo of one of the signs is reprinted in Weyler, *Greenpeace*, 186.
20. Metcalfe, "The Great MacBean," 1/12/1970. Transcript from Metcalfe's personal papers.
21. Quoted in Webb, "Protest in Paradise," 38. Metcalfe could not recall the actual date of this broadcast, but it was likely to have been sometime in mid-1969.
22. Moore interview. In 1960, under the direction of Ian McTaggart Cowan, UBC set up the first ecology department in Canada. Moore, therefore, was among the first generation of Canadians to graduate with degrees in ecology. Previously, Canadian students with an interest in the subject had been forced to go to American or British universities. See Dunlap, *Nature and the English Diaspora*, 152.
23. Moore interview. The quote is from Moore, *Confessions*, 43.
24. *Georgia Straight*, June 21–28, 1970, 9.
25. *Time*, "Environment: Ecology: The New Jeremiahs," 8/15/1969, http://www.time.com/time/magazine/article/0,9171,901238-2,00.html. Moore interview. Eugene Odum's son, Bill, spent 1970 as a postdoctoral fellow at UBC working on an International Biological Program site studying heterotrophy in a lake in the area. See Craige, *Eugene Odum*, 103.
26. Moore, "The Administration of Pollution Control," 65. Moore interview and *Confessions*, 43–45.
27. Kostash, *Long Way from Home*, 63–65; Palmer, *Canada's 1960s*, 272.
28. For more on the history of anti-Americanism in Canada, see Granatstein, *Yankee Go Home?*
29. Thompson and Randall, *Canada and the United States*, 224–29.
30. Ibid., 240, 248.
31. For a thorough history of U.S. nuclear testing in Alaska, see Kohlоff, *Amchitka and the Bomb*.
32. Hunter, *Vancouver Sun*, 9/24/1969.
33. Jacobs, "The Coming Atomic Blast in Alaska."
34. Vancouver *Sun*, editorial, 10/3/1969.
35. Hunter, *Warriors of the Rainbow*, 6.
36. *Ubyssey*, 10/3/1969, 1.
37. Bohlen and Stowe interviews.
38. Sierra Club executive director, David Brower, published an ad in the *New York Times* in January 1969, urging Americans to adopt "an international program, before it is too late, to preserve Earth as a 'conservation district' within the universe, a sort of . . . EARTH NATIONAL PARK." Cohen, *History of the Sierra Club*, 424.
39. Fox, *John Muir and His Legacy*, 321–22; Hunter interview. According to Bohlen, November 28, 1969 was the exact date on which the group decided to call themselves the Don't Make a Wave Committee. *Making Waves*, 27–28.
40. Hunter, *Warriors*, 7; Bohlen, *Making Waves*, 28; Bohlen interview.
41. Bohlen interview; *Vancouver Sun*, 2/9/1970.
42. Hunter, *Warriors*, 7–8; Bohlen, *Making Waves*, 30–31; Bohlen and Stowe interviews.
43. Wilcher to Stowe, 2/25/1970, vol. 2, file 7, GPF.

44. William Simmons to Terry Simmons, 2/27/1970, ibid.
45. Stowe to William Simmons, 4/10/1970, ibid. The Sierra Club cashed check, along with various other donations, can be found in 1(4) GPF; Venables to Stowe, n.d., 1(7) GPF.
46. Quoted in the *Vancouver Express*, 3/12/1970.
47. 1(7) GPF; *Vancouver Sun*, 6/23/1971; Barrett to Bohlen, n.d., 2(10) GPF. The New Democratic Party was the successor to the Cooperative Commonwealth Federation and the major left wing party in British Columbia.
48. Macpherson and Sears, "The Voice of Women."
49. Lille d'Easum, *Is Amchitka our Affair*, March 1970, 5. A copy can be found in the Special Collections department of the UBC library. Examples of Sternglass's work include: "Infant Mortality."; "The Death of All Children"; "Has Nuclear Testing Caused Infant Death?"; *Low-Level Radiation*.
50. Stowe interview.
51. For a detailed analysis of how groups such as Greenpeace and Amnesty International have made use of popular musicians, see Lahusen, *Rhetoric of Moral Protest*. A recording of the concert, including introductory remarks from Irving Stowe, is available at: http://www.amchitka-concert.com/listen.
52. Bohlen and Stowe interviews; Bohlen, *Making Waves*, 31–32; Stowe, "Greenpeace Is Beautiful," *Georgia Straight*, 12/22–29/1970, 17. Capitals in original.
53. Reynolds, "Irradiation and Human Evolution," Stowe and Bohlen interviews.
54. Reported in Victoria's *Daily Colonist*, 3/19/1970.
55. Bohlen, *Making Waves*, 32–33; Hunter, *Warriors*, 12–13.
56. Ambassador quoted in d'Easum, *Is Amchitka our Affair?* 4; Henry Lawless to Stowe, 10/6/69, 1(9) GPF; Department of National Revenue to Thorsteinsen, Mitchell, & Little, 1/3/1971, 1(7) GPF.
57. Davis to Coté, 5/19/1971, 2(10) GPF.
58. *Vancouver Sun*, 6/3/1971.
59. Davis to Coté, 6/24/1971, 2(10) GPF; Hunter, *Warriors*, 16; Bohlen, *Making Waves*, 39.
60. *Vancouver Sun*, 9/18/1971.
61. Quoted in the *Vancouver Sun*, 9/17/1971.
62. Barbara Stowe, "Amchitka: The Concert that Saved the World," http://www.folkworld.de/41/e/amchitka.html. Bohlen interview. If Bohlen had also decided not to take part, Hunter and several others would have dropped out as well. "I thought, I'm not going out while these guys sit back and send us on a fucking suicide mission." Hunter interview.
63. Hunter to Bohlen, 3/15/1971, 2(10) GPF.
64. Bohlen, Watson, and Marining interviews.

Chapter 4

1. The most prominent example of such an approach is the New Social Movement (NSM) school that dominated European sociology throughout the 1980s and is still influential today. Russell J. Dalton's *The Green Rainbow* provides a useful explanation of NSM theory, as well as demonstrating how it can be applied to the environmental movement.
2. Fine and Stoecker, "Can the Circle Be Unbroken?" 2–3, 15.
3. Hunter, *Warriors*, 18–19; Bohlen, *Making Waves*, 40–41.
4. *Vancouver Sun*, 7/21/1971.
5. *New York Times*, 8/31/1971, 15; Jacobs, "The Coming Atomic Blast in Amchitka," 35. In this instance, the suits were dismissed but appeals were held closer to the date of the test.
6. Editorial, *New York Times*, 8/2/1971.
7. *Time* (Canadian edition), 11/15/1971.
8. Quoted in the *New York Times*, 30/5/1971.
9. Bazell, "Nuclear Tests," 1219.
10. *Cong. Rec.*, "Planned Nuclear Bomb Tests in Alaska this Year," 18093. For dramatic video footage and photographs of the aftermath of the tsunami in Crescent City, see http://www.youtube.com/watch?v=8pitAGlz5x0.
11. *Cong. Rec.*, "Planned Nuclear Bomb Tests," 18091.

12. Jacobs, "The Coming Atomic Blast in Amchitka," 34.
13. Bazell, "Nuclear Tests," 1221; *Cong. Rec.*, "Planned Nuclear Bomb Tests," 18091.
14. *Cong. Rec.*, "Planned Nuclear Bomb Tests," 18086.
15. Ibid., 18091.
16. Ibid., 18090.
17. Bohlen, "Planned Nuclear Bomb Tests," 18091.
18. *Vancouver Sun*, 2/5/1970.
19. *Georgia Straight*, Nov. 11-18, 1971, 12–13.
20. *Wall Street Journal*, 6/24/1971.
21. 2(9) GPF.
22. Keziere, "A Critical Look."
23. Bohlen interview.
24. Hunter interview.
25. Metcalfe interview.
26. Moore, Metcalfe, and Hunter interviews. Fineberg correspondence.
27. Hunter, *Warriors*, 27; Metcalfe interview.
28. Kostash, *Long Way from Home*, 65–67.
29. Hunter, *Warriors*, 20.
30. Ibid., 23; Bohlen, *Making Waves*, 46.
31. Bohlen, 48.
32. *Warriors*.
33. "Territorial Governor Isaac Stevens & Chief Seattle," n.d., Nomadic Spirit website, http://www.synaptic.bc.ca/ejournal/muhisind.htm. For a comprehensive discussion of the history of Chief Seattle's speech, see Kaiser, "Chief Seattle's Speech(es)." For a broader discussion of Native Americans and environmentalism, see Krech, *Ecological Indian*. Ted Perry subsequently became a professor at Middlebury College in Vermont. He is somewhat embarrassed by the uses his speech has been put to over the years. His interest in Indians and ecology could best be described as fleeting (he is a scholar of avant garde cinema). He is not entirely sure where he drew upon the language for the speech, which is steeped in romanticized Indian spirituality and sentimental holism. Author's conversation with Perry, 11/17/2011, Middlebury, Vermont.
34. Hunter, *Red Blood*, 37.
35. Ibid., 38.
36. Hunter, *Storming of the Mind*, 93, n. 28. In his interview with me, Hunter insisted that the dulcimer-maker story was true in all its detail.
37. Leopold, *Sand County Almanac*, 138.
38. Willoya and Brown, *Warriors of the Rainbow*, 4, 14.
39. Ibid., 15, 78–79.
40. Keziere and Hunter, *Greenpeace*, 23. This book is largely a photo essay describing the journey to Amchitka. It has no page numbers, and my references are based on my pagination, with page 1 beginning with the passage: "Everyone was standing on the deck in the cold soiled-cloth light of the morning . . ." Some thirty years later, Keziere dug up the original, longer manuscript, which was subsequently published as *The Greenpeace to Amchitka*.
41. Hunter, *Warriors*, 19.
42. Keziere and Hunter, *Greenpeace*, 22.
43. Hunter, *Warriors*, 29; Hunter interview.
44. Keziere and Hunter, *Greenpeace*, 40.
45. Hunter, *Warriors*, 44.
46. Simmons correspondence; Hunter, *Warrriors*, 49.
47. Keziere and Hunter, *Greenpeace*, 41–42.
48. Bohlen, *Making Waves*, 58–59; Metcalfe interview.
49. *Time* (Canadian edition), 10/25/1971, 14; *Georgia Straight*, 11/11–18/1971, 3; Bohlen, *Making Waves*, 58; Hunter, *Warriors*, 68.
50. Hunter, *Warriors*, 97.
51. *New York Times*, 10/6/1971, 50.
52. Hunter, *Warriors*, 71; Moore, Metcalfe, and Bohlen interviews.

53. Quoted in *Warriors*, 71; Metcalfe interview.
54. Bohlen, *Making Waves*, 59; Bohlen interview.
55. Hunter, *Warriors*, 61; *Vancouver Sun*, 10/16/1971.
56. Hunter, *Warriors*, 71; Hunter interview.
57. Hunter, *Warriors*, 73–74. Italics in original. Hunter, Bohlen, and Metcalfe interviews.
58. Hunter, *Warriors*, 71.
59. Ibid., 75; Hunter and Metcalfe interviews.
60. *Vancouver Sun*, 10/23/1971.
61. Wittner, *Rebels Against War*, 263.
62. Bohlen interview.
63. Marining, Bohlen, and Hunter interviews; Bohlen, *Making Waves*, 61; Hunter, *Warriors*, 79–80.
64. Hunter, *Warriors*, 80–81; Bohlen, *Making Waves*, 60; Hunter, Marining, and Bohlen interviews.
65. *FID (Fight Ignorance: Dissent)*, vol. 1 (7). This is the very amateurish, mimeographed newsletter put out by the Concerned Servicemen's Movement. A copy of this volume can be found in 2(2) GPF.
66. *Island Times*, 10/22/1971. Clipping from 7(8) GPF. The file also contains clippings from the *Kodiak Mirror*, which devoted extensive coverage to the visit.
67. *Warriors*, 91.
68. Hunter, *Red Blood*, 40.
69. Hunter, *Warriors*, 91. Hunter decided to turn the generic Eyes of Fire into a Cree after meeting a Cree Indian in 1976 who told him that the warriors of the rainbow myth was part of their culture. See Hunter, *Red Blood*, 42–44.
70. *Vancouver Province*, 10/23/1971.
71. Stowe interview.
72. Stowe to Air Canada, 10/22/1971; Air Canada to Stowe, 11/12/1971, 3(2) GPF.
73. Hunter, *Warriors*, 97, 102.
74. *Vancouver Province*, 12/31/1971.
75. Hunter, *Warriors*, 98.
76. Bohlen, *Making Waves*, 65–66; Hunter, *Warriors*, 95.
77. Hunter, *Warriors*, 94; Hunter, Bohlen, and Metcalfe interviews.
78. Hunter, *Warriors*, 114; Hunter interview. Bohlen was reluctant to say anything too critical about Irving, probably due to his and Marie's good friendship with Dorothy Stowe. Metcalfe, however, corroborated most of Hunter's version.
79. *New York Times*, 11/5/1971, 20; 11/6/1971, 1.
80. *New York Times*, 11/7/1971, 1.
81. *New York Times*, 11/7/1971, 1; Hunter, *Warriors*, 110–11.
82. Ibid., 111.
83. *New York Times*, 11/7/1971, 46.
84. *New York Times*, 11/8/1971, 1.
85. *Chicago Tribune*, 11/7/1971, 1.
86. *Kansas City Star*, 11/7/1971, 1.
87. Simmons in the *Georgia Straight*, Dec. 2-9, 1971.
88. *Georgia Straight*, Nov. 11-18, 1971, 12–13.
89. Simmons's letter to the editor, *Georgia Straight*, Dec. 2-9, 1971.
90. Stowe's reply to Simmons, *Georgia Straight*, Dec. 2-9, 1971. A copy of the original DMWC certificate of incorporation can be found in 1(7) GPF.
91. *Vancouver Province*, 12/21/1971.
92. Hunter's column, *Vancouver Sun*, 11/5/1971.
93. Ibid.
94. Ibid.
95. Hunter's column, *Vancouver Sun*, 10/9/1971.
96. Ole Holsti, letter to the editor, *Vancouver Sun*, 12/15/1971.
97. Bohlen to Holsti, n.d., 2(10) GPF.
98. Quote from Keziere and Hunter, *Greenpeace*, 17.
99. Bohlen, Hunter, and Metcalfe interviews.

100. Bohlen speech notes, 3(7) GPF.
101. Hunter, *Vancouver Sun*, 10/25/1971.
102. Cassidy, "Mind Bombs," 81–82. Cummings' articles were syndicated through the Underground Press Syndicate to various alternative newspapers in both Canada and the United States. For example, *Good Times*, an alternative newspaper in San Francisco, reprinted Cummings's articles and followed the DMWC's exploits with some interest. See clipping from the 6/11/1971 edition, 3(6) GPF.
103. For example, see Brown and May, *Greenpeace Story*, 15. May was a long-time editorial director of Greenpeace Books, the organization's publishing arm, and this book served as the official history of the organization.
104. Hunter, *Warriors*, 113; Bohlen, *Making Waves*, 69. In my interviews with Bohlen and Hunter, both felt that this view was historically valid.
105. Quoted in the *Vancouver Sun*, 11/8/1971; Metcalfe interview.
106. *Vancouver Sun*, 5/19/1971.
107. Schrader, "Atomic Doubletalk," 43.
108. "Between Ourselves," written by Ben Metcalfe and produced by Bill Terry for CBC Radio. Cassette recording in vol. 9 GPF. Date of broadcast is unlisted, but according to Metcalfe, it was sometime in late November 1971.
109. Hunter interview.
110. Metcalfe interview.
111. Bohlen interview.
112. *Georgia Straight*, Nov. 11-18, 1971, 12.
113. Hunter, Bohlen, and Metcalfe interviews.
114. Bohlen and Hunter interviews.

Chapter 5

1. Myer and Tarrow, "A Movement Society," 19.
2. Bohlen and Hunter interviews.
3. For a full account of the *Rainbow Warrior* bombing, see Sunday Times Insight Team, *Rainbow Warrior*.
4. *The Citizen* (North and West Vancouver community newspaper), 1/19/1972; Metcalfe interview.
5. Metcalfe interview.
6. Hunter column, *Vancouver Sun*, 1/14/1972.
7. *Peace News* no. 1827, 7/9/1971; Moore interview. Much to Greenpeace International's annoyance, the Greenpeace London group still exists as a separate entity from Greenpeace UK. In the early 1990s, two of its activists were involved in a high-profile libel suit with McDonald's, which became known as the "McLibel" affair. McDonald's tried to implicate Greenpeace International in the lawsuit, but there was no legal connection between the small group of anarchists and the international NGO. A detailed overview of Greenpeace London and the McLibel affair can be found at the Greenpeace London website: http://www.mcspotlight.org/people/biogs/london_grnpeace.html.
8. Lamb, *Promising the Earth*, 34–35. Lovins has become an internationally renowned environmental advocate who heads the Rocky Mountain Institute in Colorado. Lalonde became a leading green politician in France and was made environmental minister during the Mitterand presidency. He was also an important figure in the early history of Greenpeace in Europe, and his name will pop up from time to time in our story.
9. Hunter, *Warriors*, 114.
10. Metcalfe interview.
11. Miriam Kahn examines the disconnect between the image and reality of Tahitian life in *Tahiti Beyond the Postcard*.
12. Quoted in Firth, *Nuclear Playground*, 11.
13. Henningham, *France and the South Pacific*, 165–67; Firth, *Nuclear Playground*, 109.
14. The *force de frappe* refers to France's nuclear strike force. For a history of its origins and early years, see Cogan, "From the Fall of France to the *Force de Frappe*."

15. Quoted in Danielsson and Danielsson, *Poisoned Reign*, 56.
16. Ibid., 73.
17. Ibid., 102–3.
18. Firth, *Nuclear Playground*, 97; Danielsson and Danielsson, *Poisoned Reign*, 169, 176.
19. Chafer, "Politics and the Perception of Risk," 13–14.
20. Ibid., 14–17; Bess, *Light-Green Society*, 30–31. For more on the French Left and nuclear weapons see Johnstone, "How the French Left Learned to Love the Bomb," and Touraine, *Anti-nuclear Protest*. For a superb study of how French national identity became intertwined with nuclear power, see Hecht, *The Radiance of France*. For an overview of environmental politics in France since the early 1970s, see Prendiville, *Environmental Politics in France*.
21. Metcalfe interview.
22. Ibid.
23. *Vancouver Sun*, 4/1/1972.
24. Metcalfe interview.
25. Ibid.
26. Hunter and Bohlen interviews. Despite the fact that New Zealand was to be a key element of the campaign and that any contacts there would be useful, Stowe never told Metcalfe that he and his family had lived there for five years during the early 1960s. In fact, Metcalfe claimed that he was not aware of this until I mentioned it to him during our interview.
27. Metcalfe interview.
28. Ibid.
29. *New Zealand Herald*, 4/4/1972, 1.
30. Firth, *Nuclear Playground*, 97.
31. McTaggart, *Outrage*, 2, 17.
32. Examples include Spencer with Bollwerk and Morais, "The Not So Peaceful World of Greenpeace," *Forbes*, 11/11/1991, and a rather hysterical report in the German language *Wiener Magazin* (June 1991) which all but accused McTaggart of being a CIA agent whose assignment was to insure that the emerging green activist movement remained under U.S. control. Similarly, the Danish documentary film, *The Man in the Rainbow* portrays McTaggart as a shady and manipulative character who ran Greenpeace like his own private eco-fiefdom. McTaggart claimed that the film was sponsored by the American antienvironmental lobby, known as the Wise Use movement, as a vehicle to discredit himself and Greenpeace, and he successfully sued to prevent it from airing outside Scandinavia. McTaggart could be notoriously paranoid, but there does appear to be a basis to these charges. The film was made in close consultation with Magnus Gudmundsson, a Danish journalist who has been heavily critical of Greenpeace, and relied for "expert commentary" on Ron Arnold, a notorious antienvironmentalist and leading Wise Use figure. McTaggart interview. For a critical study of the Wise Use movement, see Helvarg, *War Against the Greens*.
33. McTaggart's posthumously published autobiography, written with Vancouver journalist Helen Slinger, is *Rainbow Warrior: Ein Leben gegen alle Regeln*. For some reason, the German translation was published six months before the English version, so that is the one I read preparing this chapter. All translations are my own. Full disclosure: McTaggart considered me a candidate to ghostwrite the book (or, more likely, wanted me to think that he was considering me), thus my conversations with him were probably influenced by this fact. Nevertheless, he was fully aware of my project and talked freely and openly, as well as being a consummate host, during the two days I spent with him.
34. McTaggart and Slinger, *Rainbow Warrior*, 23–24.
35. Ibid., 33. Metcalfe recalls McTaggart regaling him with stories of how, as a teenager, he had seduced the mothers of several of his school friends. Though probably exaggerated, there is no reason to doubt the veracity of such stories given McTaggart's reputation as a self-confessed chronic womanizer. Metcalfe and McTaggart interviews.
36. McTaggart and Slinger, *Rainbow Warrior*, 36–37. Moore interview.
37. McTaggart and Slinger, *Rainbow Warrior*, 38, 43–45; http://www.badminton.ca/Parents_and_Fans/National_Champions/Senior.aspx?sflang=en

38. McTaggart and Slinger, *Rainbow Warrior*, 61; Bruce Orvis and Roma Orvis, McTaggart obituary in the *Alpine Enterprise* newspaper, 4/11/2001.
39. McTaggart and Slinger, *Rainbow Warrior*, 61–62. In his obituary for McTaggart, Orvis mistakenly writes that McTaggart was educated at the University of British Columbia. Perhaps Orvis misremembered the details of McTaggart's life, but it would not have been unlike McTaggart to mislead Orvis in this way to make a better impression. Similarly, Orvis writes that McTaggart had been the badminton world champion, which was also not the case.
40. McTaggart and Slinger, *Rainbow Warrior*, 65–66. McTaggart does not mention the loan from his mother-in-law in his autobiography. This fact was revealed by the former mother-in-law in an interview for the documentary film, *The Man in the Rainbow*, cited earlier.
41. McTaggart and Hunter, *Greenpeace III*, 74–76.
42. McTaggart and Slinger, *Rainbow Warrior*, 66.
43. Ibid., 68. Betty Huberty, the mother of McTaggart's third wife, was among those who would never see the money McTaggart had borrowed from them. See *The Man in the Rainbow*.
44. McTaggart and Slinger, *Rainbow Warrior*, 68–70.
45. McTaggart, *Outrage!* 12. *Outrage!* was a narrative of McTaggart's journey to Mururoa, mostly strung together from his diary entries, newspaper reports, and various other documents. *Greenpeace III*, published five years later in 1978 and largely written by Bob Hunter, is a racier, more detailed version of the same story and also deals with events beyond 1973. Although neither of the books contains notes, most of the important sources are documented in, or can be construed from, the text.
46. McTaggart and Slinger, *Rainbow Warrior*, 75–76.
47. Ibid., 77; McTaggart interview.
48. McTaggart, *Outrage!* 3.
49. Ibid., 2–3.
50. Ibid., 3.
51. McTaggart and Hunter, *Greenpeace III*, 8.
52. Ibid., 8–9.
53. McTaggart and Slinger, *Rainbow Warrior*, 81; McTaggart interview.
54. McTaggart, *Outrage!* 4; Szabo, *Making Waves*, 6.
55. Quoted in Locke, *Peace People*, 170–74. Locke was a prominent New Zealand communist until the mid-1950s, after which she turned her attentions to the peace movement, feminism, and children's writing. She was a much-beloved figure of the New Zealand political Left. See Murray Horton's obituary of her at http://www.converge.org.nz/abc/elsobit.htm.
56. Szabo, *Making Waves*, 13.
57. Quoted in Locke, *Peace People*, 286.
58. Szabo, *Making Waves*, 5.
59. Locke, *Peace People*, 290–91.
60. Ibid., 286.
61. These were the findings of a University of Canterbury political scientist, N. S. Roberts, as reported in the *New Zealand Herald*, 7/28/1972, 8.
62. *New Zealand Herald*, 5/29/1972, 1.
63. *New Zealand Herald*, 8/14/1972.
64. *New Zealand Herald*, 4/18/1972, 3, 6.
65. *New Zealand Herald*, 4/10/1972.
66. Caron, *Fri Alert*, xii.
67. Ibid., x–xi; Szabo, *Making Waves*, 21–22.
68. McTaggart, *Outrage!* 5; Metcalfe interview.
69. Metcalfe interview.
70. McTaggart, *Outrage!* 5–6.
71. McTaggart and Hunter, *Greenpeace III*, 29.
72. McTaggart, *Outrage!* 7; Metcalfe interview.
73. Ingram interview; McTaggart, *Outrage!* 7.
74. McTaggart, *Outrage!* 16–17.
75. Ibid., 18.

76. *New Zealand Herald*, 4/26/1972, 3; McTaggart and Hunter, *Greenpeace III*, 37.
77. McTaggart, *Outrage!* 33. As the *New Zealand Herald* noted, this was all part of the "the sudden interest shown in the *Greenpeace III* by Customs and Marine Department officials." 4/28/1972, 3.
78. McTaggart, *Outrage!* 32–33.
79. Ingram interview; McTaggart, *Outrage!* 23–24.
80. Bess, *Light-Green Society*, 267.

Chapter 6

1. McTaggart, *Outrage!* 37.
2. Metcalfe interview.
3. McTaggart and Hunter, *Greenpeace III*, 57; Ingram interview.
4. Metcalfe interview.
5. *Sunday Times* (New Zealand), 5/14/1972.
6. Ingram interview; McTaggart, *Outrage!* 56.
7. McTaggart, *Outrage!* 57; McTaggart and Hunter, *Greenpeace III*, 62; Ingram and Metcalfe interviews.
8. Letter from Grant Davidson to Metcalfe, 9/5/1972, Metcalfe's personal papers. The charge that Greenpeace was associated with communists was, at bottom, baseless, but was no doubt fueled by Irving Stowe's well-publicized trip to China that year as part of a delegation of Canadian citizens interested in cultural exchange between the two nations. Stowe's glowing reports on his return would only have cemented this impression in the eyes of conservatives such as McTaggart. Stowe, Hunter, and Bohlen interviews.
9. McTaggart and Hunter, *Greenpeace III*, 62.
10. McTaggart, *Outrage!* 61.
11. Metcalfe interview.
12. *New Zealand Herald*, 4/8/1972, 1.
13. *New Zealand Herald*, 4/16/1972; Metcalfe interview.
14. *Vancouver Sun*, 5/23/1972.
15. Metcalfe and Moore interviews.
16. Bohlen and Moore interviews.
17. Hunter interview.
18. *Vancouver Sun*, 6/20/1972; Marining interview; Hunter, *Warriors*, 116.
19. *New Zealand Herald*, 6/1/1972, 3. Metcalfe was convinced that his local member of Parliament, Jack Davis, was responsible for his arrest in Paris. Davis remembered Metcalfe's criticism from the Amchitka campaign, when he had tried to prevent the *Phyllis Cormack* from getting state-subsidized fishing insurance, as well as the Metcalfe-led disruption of the Liberal Party luncheon in West Vancouver back in January. Davis was the Canadian delegate at the Stockholm conference, and he knew that Metcalfe was planning a protest there. It is certainly not beyond the bounds of possibility that he was responsible for tipping off immigration authorities, though no concrete evidence has ever been produced.
20. Metcalfe and Moore interviews.
21. McCormick, *Reclaiming Paradise*, 88. Barbara Ward and René Dubos, *Only One Earth* was the major publication to emerge from the conference.
22. McCormick, *Reclaiming Paradise*, 97–100.
23. *New Zealand Herald*, 6/15/1972, 1; McCormick, *Reclaiming Paradise*, 100; Marining and Hunter interviews. France and China opposed the resolution, while the United States, the Soviet Union, Britain, and Canada abstained. Hunter declared that through their actions in the South Pacific and in Europe, Greenpeace could "rightly claim to have initiated the biggest protest against anti-nuclear testing in history." *Vancouver Sun*, 7/7/1972.
24. *Vancouver Sun*, 7/7/1972.
25. McTaggart wrote that Haddleton had a tropical fever and was not well enough to continue (*Outrage!* 61–62). Ingram, however, recalled that Haddleton had not fit in with the crew and was overwhelmed by McTaggart's personality. Ingram interview.

26. McTaggart and Hunter, *Greenpeace III*, 64–65.
27. McTaggart, *Outrage!* 80.
28. Ibid., 81–92.
29. *New Zealand Herald*, 5/3/1972, 3; 5/9/1972, 1.
30. *New Zealand Herald*, 5/30/1972, 1.
31. *New Zealand Herald*, 6/24/1972, 3.
32. *New Zealand Herald*, 6/24/1972, 3; McTaggart, *Outrage!* 131.
33. *New Zealand Herald*, 6/28/1972. See also, Mitcalfe, "Why I Am Sailing into the Fallout Area," 9–10.
34. The Shadbolt quote is from a letter to Mitcalfe, reprinted in Mitcalfe, *Boy Roel*, 116. Shadbolt later wrote a novel about a protest voyage to Mururoa which was a composite story of several actual protest voyages, including those of *Greenpeace III*. See *Danger Zone*.
35. The exchange between Sharp and other MPs is reproduced in McTaggart, *Outrage*! 145.
36. *New Zealand Herald*, 6/23/1972, 2. Several years later, investigations revealed that in early 1972 the Canadian, French, Australian, and South African governments, along with the mining giant Rio Tinto, had been involved in setting up a cartel to stabilize uranium prices. Furthermore, major Canadian mining and technology firms, some of which had had senior Liberal Party figures on their boards, were bidding for major contracts to supply France with uranium and processing equipment. See Stewart, "Canada's Role in the International Uranium Cartel," Edwards, "Canada's Nuclear Industry and the Myth of the Peaceful Atom," and Gray, *The Great Uranium Cartel*.
37. *Guardian* (UK), 6/29/1972; *New Zealand Herald*, 6/29/1972.
38. McTaggart, *Outrage!* 149.
39. McTaggart and Hunter, *Greenpeace III*, 176.
40. Ibid., 66; McTaggart, *Outrage!* 156.
41. McTaggart, *Outrage!* 174–78; Ingram interview.
42. McTaggart, *Outrage!* 181.
43. The story was also reported in the same way by the *New Zealand Herald*, 7/6/1972, 1.
44. McTaggart and Hunter, *Greenpeace III*, 199–200.
45. For example, see *New Zealand Herald*, 6/29/1972, 1, 7/3/1972, 1.
46. McTaggart and Hunter, *Greenpeace III*, 214–15.
47. Davidson to Metcalfe, 9/5/1972, Metcalfe's personal files.
48. See Spencer, "The Not So Peaceful World of Greenpeace," *Forbes*, 11/11/1991 and *Wiener Magazin* (June 1991). In the documentary film, *The Man in the Rainbow*, Metcalfe stated, rather melodramatically, that he felt like a "Dr. Frankenstein" who had brought McTaggart into the environmental movement and then lost control of him.
49. McTaggart and Hunter, *Greenpeace III*, 214–17; Metcalfe interview.
50. Metcalfe, Stowe, and Bohlen interviews.
51. McTaggart and Slinger, *Rainbow Warrior*, 119.
52. McTaggart and Hunter, *Greenpeace III*, 217–18.
53. Francis Auburn to McTaggart, 9/28/1972, reprinted in McTaggart, *Outrage!* 255.
54. Sharp to McTaggart, 10/19/1972, reprinted in ibid., 257. Detailed correspondence between McTaggart and the Canadian government is in appendix D of *Outrage!*
55. McTaggart and Hunter, *Greenpeace III*, 221–22; McTaggart interview.
56. McTaggart and Hunter, *Greenpeace III*, 229–33.
57. Quoted in Firth, *Nuclear Playground*, 98.
58. Quoted in Locke, *Peace People*, 298.
59. Ibid., 299.
60. *Globe and Mail* (Toronto), 6/9/1973; *Vancouver Sun*, 9/17/1973.
61. Szabo, *Making Waves*, 23–24.
62. McTaggart, *Outrage!* 215.
63. *Vancouver Sun*, 2/13/1973; Marining and Watson interviews.
64. Quoted in Brian Fortune, "Media Mellows Greenpeace," *Terminal City Express*, 2/23/1973, 5.
65. Ibid.
66. *Vancouver Sun*, 8/24/1973; Hunter interview; McTaggart interview.

67. *Vancouver Sun*, 6/10/1973.
68. Hunter, *Warriors*, 117; McTaggart and Hunter, *Greenpeace III*, 246.
69. Trudeau to McTaggart, 6/4/1973, reprinted in McTaggart, *Outrage!* 274–75 and in the *Vancouver Sun*, 6/9/1973.
70. McTaggart and Hunter, *Greenpeace III*, 247.
71. Ingram interview.
72. McTaggart and Hunter, *Greenpeace III*, 254; Ingram interview.
73. Szabo, *Making Waves*, 26–27.
74. Caron, *Fri Alert*, x–xii.
75. When a New Democratic member of the federal Parliament suggested that Canada, like New Zealand, should send a naval vessel with a cabinet minister to Mururoa, he was laughed out of Parliament by the Liberals and Conservatives. *Vancouver Sun*, 6/11/1973.
76. See Caron, *Fri Alert*, for a detailed narrative of the *Fri*'s voyage. Szabo, *Making Waves*, 28–33, contains a useful summary.
77. Szabo, *Making Waves*, 38.
78. *Vancouver Sun*, 7/11/1973.
79. *Peace News*, 6/8/1973.
80. *Times* (London), 6/2/1973.
81. *Vancouver Sun*, 7/11/1973.
82. *Guardian*, 2/3/1973.
83. McTaggart, *Outrage!* 211; Ingram interview.
84. McTaggart and Hunter, *Greenpeace III*, 253.
85. McTaggart and Slinger, *Rainbow Warrior*, 120.
86. Ibid., 121–22; Ingram never heard such rumors and was surprised when I mentioned them to him in our interview.
87. McTaggart's accounts of his response to the boarding tend to vary. In his books, he claims to have merely been trying to block the French commandos to prevent them coming on board (*Outrage!* 225; *Greenpeace III*, 280). However, in statements made to the *Vancouver Sun* (8/18/1973), McTaggart admitted that he "fought as hard as he could" against the boarding party, kicking and punching the French troops. "They overran my property in international waters, so, of course, I fought to defend it." Ingram was below deck trying to send out an SOS and did not see the incident, while the photographs are inconclusive on that score. Although the crew had discussed nonviolence and had generally agreed that it was the best method of protest, they were not experienced in the techniques of nonviolent resistance, which usually requires considerable training and practice to work effectively. Ingram interview.
88. McTaggart and Hunter, *Greenpeace III*, 281.
89. Ibid., 283.
90. Ibid., 297–98; Ingram interview.
91. *Vancouver Sun*, 8/20/1973.
92. Ingram interview.
93. 9/10/1973. The photos are reproduced in *Outrage!* 230–34, and *Greenpeace III*, 278.
94. *Vancouver Sun*, 9/26/1973.
95. McTaggart and Hunter, *Greenpeace III*, 302; Ingram interview.
96. *Vancouver Province*, 12/12/1973. The BC provincial legislature unanimously resolved to urge the federal government to demand compensation from France for the Greenpeace incident. *Vancouver Sun*, 10/12/1973.
97. As late as August 1973, the French minister responsible for the armed forces had declared that France would "never undertake to stop tests in the air." Quoted in the *Vancouver Sun*, 8/31/1973.
98. *Vancouver Sun*, 11/14/1973.
99. Wapner, *Environmental Activism*, 4.
100. Ibid., 14–15, 65. For a similar analysis, see Princen and Finger, *Environmental NGOs and World Politics*.
101. Ingram, Bohlen, and Hunter interviews.
102. McTaggart and Slinger, *Rainbow Warrior*, 135.

103. McTaggart and Hunter, *Greenpeace III*, 315–16. While such ideas may have gone through McTaggart's mind, the language here is clearly Hunter's.
104. McTaggart interview.
105. Quoted in Szabo, *Making Waves*, 38.
106. Ibid.
107. Bohlen interview.
108. Hunter interview; Hunter, *Warriors*, 124.

Chapter 7

1. Hunter interview.
2. Hunter, *Warriors*, 124.
3. Ibid.
4. *Peace News*, 5/5/1973, for example, simply announced that there would be a "Greenpeace activity" in Dundee, Scotland. Clearly, the editors felt no need to elaborate: their readers understood that this meant an antinuclear protest with an environmental focus.
5. Spong interview; Weyler, *Song of the Whale*, 6. Weyler was a major participant in the Greenpeace antiwhaling campaign and will be introduced shortly. Spong's dissertation, submitted at UCLA in 1966, was titled "Cortical Evoked Responses and Attention in Man."
6. Interview with Linda Gannon (Spong's first wife), Vancouver, 5/2/2000.
7. Spong interview.
8. Moore, "The I Ching in Time and Space"; Hook, *I Ching and Mankind*; McEvilly, "Synchronicity and the I Ching."
9. Kesey, "The I Ching."
10. Turner, *From Counterculture to Cyberculture*, 93.
11. For more on this antinomial interpretation of modernity, see Saler, "Modernity and Enchantment."
12. Spong interview; Weyler, *Song of the Whale*, 3–6.
13. Spong, "Introduction," in Hunter and Weyler, *To Save a Whale*, 5.
14. Spong and White, *Cetacean Research*, 5–8; Weyler, *Song of the Whale*, 13–19.
15. Weyler, *Song of the Whale*, 23–24.
16. Spong, "Introduction," 7.
17. Spong and White, *Cetacean Research*, 38–39.
18. Murray Newman's memoir notes, 1969, file 3, 619-A-4, MNF.
19. Weyler, *Song of the Whale*, 50–51.
20. Ibid., 52.
21. Ibid., 52–53.
22. Ibid., 53–54.
23. Lilly, *Man and Dolphin*. We will examine Lilly in more detail in the next chapter.
24. Newman's memoir notes.
25. *Vancouver Sun*, undated clipping from 1969, file 3, 619-A-4, MNF.
26. *Georgia Straight*, June 26–July 2, 1969, 9–10. Spong also invited various local musicians to play for Skana, including jazz flautist, Paul Horn, the man subsequently dubbed the "father of New Age music." Horn, like Spong, was a product of LA's counterculture. He had recently completed a stint—along with the Beatles—at the Maharishi Mahesh Yogi's ashram in India. Yogi, of course, was the founder of Transcendental Meditation. For more on the connection between whales and musicians, see Rothenberg, *Thousand Mile Song*.
27. Weyler, *Song of the Whale*, 62–77. The homepage of Spong's whale research station, Orcalab, is www.orcalab.org.
28. Weyler, *Song of the Whale*, 118.
29. Spong interview.
30. Ibid.
31. Ellis, *Men and Whales*, 428–30.
32. Tønnessen and Johnsen, *History of Modern Whaling*, 674.
33. Weyler, *Song of the Whale*, 127–29.
34. Spong interview.

35. Weyler, *Song of the Whale*, 135–36; Spong interview; Hunter interview.
36. Weyler, *Song of the Whale*, 136–37. Emphasis in original. Weyler reconstructed the conversation from interviews with Spong and Hunter and from their respective notes.
37. Ibid., 137–38.
38. Ibid., 138–39.
39. Hunter and Spong interviews; Weyler, *Song of the Whale*, 139–40.
40. Bohlen interview.
41. *Georgia Straight*, December 9–16, 1970.
42. Hunter and Marining interviews.
43. Hunter, *Warriors*, 125.
44. Ibid., 124–25.
45. Hunter and Spong interviews; Weyler, *Song of the Whale*, 141. The *I Ching* served another useful function among anti-elitist activist groups: it deflected attention away from the power of leaders. This gave people the impression that the group's fate was being submitted to some sort of ancient oracular wisdom; that they were participating in a sort of cosmic hippie democracy. See Turner, *From Counterculture to Cyberculture*, 65.
46. Weyler, *Song of the Whale*, 140.
47. Ibid., 141–42.
48. Hunter, *Warriors*, 127.
49. Spong and Hunter interviews; Hunter, *Warriors*, 127; Weyler, *Song of the Whale*, 139. Paul Watson claims that he came up with the idea of using the Zodiacs. None of the other participants, however, remembers this being the case.
50. Hunter, *Warriors*, 129.
51. Hunter, *Warriors*, 130.
52. Ibid., 131. Emphasis in original.
53. Ibid.
54. Ibid., 143; Hunter, Bohlen, Spong, and Marining interviews.
55. Hunter, *Warriors*, 137,125.
56. Spong interview.
57. Ibid., 132–36; Weyler, *Song of the Whale*, 145–47; Spong interview.
58. Bohlen and Hunter interviews; Hunter, *Warriors*, 142–43.
59. Hayes, "France Bombs Mururoa, India Bombs India."
60. Hunter, *Warriors*, 143; Weyler, *Greenpeace*, 245–47. Heimann, like McTaggart and numerous others associated with Greenpeace, wrote a book about his voyage: *Knocking on Heaven's Door*. Heimann is now a well-known cartoonist and children's book writer in Australia.
61. Bohlen interview.
62. Dorothy Stowe interview.
63. Hunter, *Warriors*, 149.

Chapter 8

1. "Killing the Killers," *Time*, October 4, 1954, available at http://www.time.com/time/magazine/article/0,9171,857557,00.html.
2. Mowat, *A Whale for the Killing*, 54.
3. Shoemaker, "Whale Meat in American History," 281–82; Ash, *Whaler's Eye*.
4. For a discussion of whale hunting as an elite sport, see Burnett, *Sounding of the Whale*, 273–79.
5. There is a vast literature on the history of whaling. I have primarily drawn on: Tønnessen and Johnsen, *History of Modern Whaling*; Ellis, *Men and Whales*; Roman, *Whale*; Davis, Gallman, and Gleiter, *In Pursuit of Leviathan;* and Burnett, *Sounding of the Whale*.
6. Ellis, *Men and Whales*, 427. For an account of Onassis's foray into pelagic whaling, see Fraser et al., *Aristotle Onassis*, 115–31. Soviet data falsification has been confirmed by post-Soviet-era research. See Yablokov, "On the Soviet Whaling Falsification, 1947–1972."
7. Burnett, *Sounding of the Whale*, ch. 2.
8. Epstein, *Power of Words*, 77–79.

9. Burnett, *Sounding of the Whale*, 668.
10. McBride, "Evidence for Echolocation by Cetaceans." For a more recent study of the subject, see Au, *Sonar of Dolphins*, 2–3.
11. McBride, "Meet Mister Porpoise," 17.
12. Davis, *Spectacular Nature*, 35. Burnett convincingly argues that dolphins (and I would add killer whales) became, in the public mind, stand-ins for cetaceans as a whole. See *Sounding of the Whale*, 622.
13. Alan Bryman uses the term "Disneyization" to describe the reification of Disney's aesthetic and moral values. See *Disneyization of Society*. To my ears at least, "Disneyfication" has slightly less euphonic dissonance.
14. Merle, *Day of the Dolphin*. For a fascinating, if occasionally speculative, cultural analysis of the intersection of cetaceans and technology in the human imagination, see Bryld and Lykke, *Cosmodolphins*.
15. Lilly recounts the isolation tank stories in numerous autobiographical works. For example, see *The Deep Self* and Lilly and Gold, *Tanks for the Memories*.
16. Burnett, *Sounding of the Whale*, 570–71; Lilly, "A Feeling of Weirdness," 71.
17. Lilly, *Man and Dolphin*, 7.
18. Burnett, *Sounding of the Whale*, 640. For an example of Turner and Norris's collaborative work, see "Discriminative Echolocation in a Porpoise." The article, as the authors acknowledge, was supported by the Office of Naval Research, NASA, and the Brain Research Institute. Jay Stevens has chronicled CIA-sponsored LSD research. See *Storming Heaven*, 79–87.
19. Burnett, *Sounding of the Whale*, 615–18.
20. Ibid., 611.
21. Ibid.
22. The quote is drawn from Howe's research notes and reproduced in Lilly, *Mind of the Dolphin*, 251. For family-friendly photographs of Howe and Peter, see Lilly, *Communication Between Man and Dolphin*.
23. Lilly, *Mind of the Dolphin*, 55–56; Burnett, *Sounding the Whale*, 622. For a summary of the critical reviews of *Man and Dolphin*, see Burnett, 591–94. For a thoughtful and more recent discussion of dolphin intelligence, see Wynne, *Do Animals Think?* ch. 8. In recent years, a number of cetacean researchers have taken up where Lilly left off, albeit somewhat shorn of Lilly's mysticism and eccentricity. Examples include Dudzinski and Frohoff, *Dolphin Mysteries*; Kelsey, *Watching Giants*; and Reiss, *The Dolphin in the Mirror*.
24. It is worth noting that prior to his appointment at UCLA, Norris was the founding curator of Marineland of the Pacific, which was the country's second oceanarium (after Florida's Marineland). He is thus an excellent example of the fluid relationship between marine theme parks, cetacean research, brain science, and Cold War bioscience. See Davis, *Spectacular Nature*, 51.
25. Burnett, *Saving the Whales*, 640. The fact that Spong had already conducted significant research into how the human brain responds to visual stimuli probably also played a role in this decision. See Spong, Haider, and Lindsley, "Selective Attentiveness and Cortical Evoked Responses to Visual and Auditory Stimuli."
26. For a fascinating discussion of Esalen's influence on the counterculture and the New Age, see Kripal, *Esalen*.
27. Burnett, *Sounding of the Whale*, 585, 621, 626.
28. McVay, "The Last of the Great Whales," 21.
29. McVay, "Can Leviathan Long Endure So Wide a Chase?" 40. The essay was also published in the new British environmental journal, *Ecologist* 1, no. 16 (Oct. 1971): 5–9.
30. McVay, "Can Leviathan Endure," 42. Also see McVay, "Stalking the Arctic Whale," and "Reflections on the Management of Whaling."
31. For a discussion of the song and of whale-related music in general, see Rothenberg, *Thousand Mile Song*, 47–49.
32. Payne and McVay, "Songs of the Humpback Whales"; Rothenberg, *Thousand Mile Song*, ch. 1. Bob Hunter recalls giving a lecture on whales during which he turned off the lights and played Payne's recordings. When he turned the lights back on, half the audience was in tears (though he also admits that half were probably stoned). Hunter interview.

33. Fichtelius and Sjölander, *Smarter than Man?* 36, 40. Emphasis in original.
34. Ibid., 144–45. Fichtelius and Sjölander were strongly influenced by the work of ethologists such as Konrad Lorenz, who used evolutionary theory to explain behavioral traits. See Lorenz, *On Aggression*, and Morris, *Human Zoo*.
35. Bunnell, "The Evolution of Cetacean Intelligence," 57–58.
36. Sutphen, "Body State Communication Among Cetaceans," in McIntyre, *Mind in the Waters*, 141–42.
37. Spong, "The Whale Show"; McVay, "One Strand in the Rope of Concern," 225; Scheffer, "The Case for a World Moratorium on Whaling," 230; Talbot, "The Great Whales and the International Whaling Commission," 236.
38. *Mind in the Waters*, 8.
39. Ibid., 94. Emphasis in original.
40. Ibid., 224. The Norwegian social anthropologist, Arne Kalland, has spent many years exploring the symbolic meaning of whales in various cultures, particularly Japan. For him, McIntyre's words exemplify the fact that "as metaphors whales have come to epitomize values and qualities that we like to see in our own species but that many of us feel we have lost." See *Unveiling the Whale*, 2. For Kalland's work on Japan, see Kalland and Moeran, *Japanese Whaling: End of an Era?*
41. Pinchot, "Whale Culture—A Proposal." Pinchot Sr., it is worth mentioning, was also interested in whales, both hunting them and studying them. In 1928, he and his son set out on what Pinchot called the South Sea expedition. They hunted numerous whales and porpoises, carefully preparing specimens according to the detailed instructions of zoologist and whale expert Remington Kellogg, who had urged Pinchot to use his high-profile voyage in the service of cetacean science. See Miller, *Gifford Pinchot*, ch. 12.
42. At least one well-known cetologist, David Gaskin, suggested that the bovine comparison was, in fact, entirely appropriate. There is little evidence, he wrote, "of behavioral or social complexity (in baleen whales) beyond that of an ungulate herd." See Gaskin, *Ecology of Whales and Dolphins*, 151. For a skeptical study of numerous claims about animal cognition, including that of dolphins, see Wynne, *Do Animals Think?* For a fascinating and complex way of approaching the issue, see Lurz, *Mindreading Animals*.
43. Lovejoy, *Revolt Against Dualism*.

Chapter 9

1. Hunter, Spong, Marining, and Moore interviews.
2. "Whole Earth Church" manifesto and certificate, Patrick Moore's personal papers. Some writers sympathetic to deep ecology have argued that New Age thought and deep ecology are fundamentally incompatible. See for example Sessions, "Deep Ecology and the New Age," 21, and Manes, *Green Rage*, 145–46.
3. Author's correspondence with Gary Snyder, Jan. 2003. Similar declarations extend at least as far back as the mid-1930s, when Secretary of Agriculture Henry Wallace used the phrase. See Worster, *Nature's Economy*, 320.
4. "The Three Laws of Ecology" and "Declaration of Interdependence" are in Moore's personal papers. For a lucid discussion of biospheric holism, see Ivakhiv, *Claiming Sacred Ground*, ch. 2.
5. Hunter, "The Politics of Evolution" (unpublished essay, 1974), 11. Emphasis in original. My copy is from Rex Weyler's personal papers.
6. Ibid., 9–10, 15. See Stone, "Should Trees Have Standing?"
7. Hunter, "Politics of Evolution," 25.
8. Hunter, *Warriors*, 146–48. Emphasis in original.
9. Ibid., 149.
10. Moore interview; Hunter, *Warriors*, 150.
11. Naess, "The Shallow and the Deep." Hunter, *Warriors*, 132. There is some confusion among Hunter, Moore, and others about when they first became aware of Naess and deep ecology. Moore claims to have been aware of Naess's work, if not actually having read it, by the time the whale campaign got underway. Hunter did not read Naess until many years later but

was familiar with Leopold and American biocentric thought. Rex Weyler, an influential American who joined Greenpeace at this time, claimed that many Greenpeace members had deep ecological outlooks before they had ever heard of the term. Weyler interview.

12. Hunter, *Warriors,* 169. Hunter and Watson remained good friends until Hunter's death in 2005, despite a few falling outs along the way. When I interviewed him, Hunter had just come back from a voyage to the Faeroe Islands aboard the *Sea Shepherd.* He returned with footage of Watson and his eco-commandos in a gunfight with Fairoese authorities, who were none too pleased with the *Sea Shepherd*'s forceful protest against the islanders' ritual whale hunt.
13. Paul Watson as told to Warren Rogers, *Sea Shepherd,* 45. For an excellent biographical essay, see Khatchadourian, "Neptune's Navy."
14. Watson interview.
15. SFU's communications department was quite innovative during the 1960s and 1970s. It was the academic home of R. Murray Schafer, who was in the process of developing the field of acoustic ecology, which focused on the way that animals interrelated with their environment through sound. See Wrightson, "An Introduction to Acoustic Ecology," 10–13.
16. Watson, *Sea Shepherd,* 47–50; Watson, Hunter, and Weyler interviews; Hunter, *Warriors,* 11. One needs to adopt a healthy dose of skepticism toward Watson's autobiographical details. As Khatchadourian tactfully put it, he is at times "astray in the labyrinth of his own illusions." See "Neptune's Navy."
17. Watson interview; Watson, *Sea Shepherd,* 59–68; Hunter, *Warriors,* 172. As usual, the details of Watson's story are disputed. Ward Churchill, for example, insists that none of the key participants at Wounded Knee could remember Watson and that he certainly would not have been given a tribal name. See the debate between Watson and Churchill at the 2001 Environmental Land, Air and Water Conference in Eugene, Oregon: http://www.youtube.com/watch?v=spHn7Twt8w8.
18. Weyler interview.
19. Ibid.
20. Hunter, *Warriors,* 150; Hunter, Weyler, and Marining interviews.
21. Hunter, *Warriors,* 150–51; Weyler, *Song of the Whale,* 150–51.
22. Hunter, *Warriors,* 151–52.
23. Ibid., 153. Emphasis in original.
24. Ibid., 158. Not all female participants would agree with Hunter's version of events. While some I interviewed felt they were treated equally and given important duties, others felt that they were merely cast as supporting actors to the macho heroes who rode in the Zodiacs and captained the ships.
25. Hunter, *Warriors,* 156–57; Hunter interview.
26. *Vancouver Sun,* 4/26/1975; Hunter, *Warriors,* 154–56. The MacMillan Bloedel donation is particularly ironic given their lengthy struggle against Greenpeace over the past decade on the issue of old growth forest clear-felling in British Columbia.
27. Spong interview; Weyler, *Song of the Whale,* 152–53; Hunter, *Warriors,* 159–60.
28. Spong interview; Weyler, *Song of the Whale,* 153–55; Hunter, *Warriors,* 160–63.
29. *Vancouver Sun,* 3/15/1975.
30. Hunter, *Warriors,* 178. Emphasis in original.
31. Weyler's personal journal, book no. 11, Weyler's personal papers; Marining, Hunter, and Weyler interviews; Hunter, *Warriors,* 165, 173.
32. Marining, Hunter, Moore, and Weyler interviews; Hunter, *Warriors,* 170–82.
33. Weyler's journal, book no. 11.
34. Hunter and Watson interviews; Hunter, *Warriors,* 175–77.
35. Hunter, "The Politics of Evolution," 26; see also, Capra, *The Tao of Physics.*
36. Hunter, Weyler, and Moore interviews.
37. Hunter, *Warriors,* 194–95.
38. Weyler's journal, book no. 11.
39. Weyler and Watson interviews.
40. Weyler's journal, book no,. 11. Emphasis in original.
41. Hunter, *Warriors,* 184.

42. Weyler's journal, book no. 11. Emphasis in original.
43. *Vancouver Sun*, 5/24/1975.
44. Hunter, *Warriors*, 187.
45. Ibid., 190, 196.
46. Ibid., 206.
47. Spong interview; Weyler, *Song of the Whale*, 157–64.
48. Weyler, Moore, and Hunter interviews; Hunter, *Warriors*, 210.
49. Hunter, *Warriors*, 211–12.
50. Ibid., 215.
51. For a description of Hunter's time in the slaughterhouse, see *Erebus*, 6–8.
52. Moore interview.
53. Quoted in Hunter, *Warriors*, 217.
54. Ibid., 217–18.
55. Ibid., 220.
56. Watson, Weyler, Hunter, and Moore interviews; Hunter, *Warriors*, 220–21.
57. Hunter and Watson interviews; Hunter, *Warriors*, 223.
58. Weyler's journal, book no. 11.
59. Hunter interview; *Warriors*, 225–26.
60. Ibid., 225; The film can be seen in the documentary *Ecology in Action*.
61. Hunter, Weyler, and Moore interviews; Hunter, *Warriors*, 226–27.
62. Hunter, *Warriors*, 229.
63. Nash expands on this comparison between abolitionists and animal rights activists. See *Rights of Nature*, 213.
64. Hunter, *Warriors*, 229. Emphasis in original.
65. Weyler's journal, book no. 11.
66. Hunter, *Warriors*, 229.
67. Spong interview; Weyler, *Song of the Whale*, 169–70.
68. Cassidy, *Mind Bombs*, 117.
69. *San Francisco Chronicle*, 7/2/1975.
70. Flowers, "Between the Harpoon and the Whale"; Weyler, *Song of the Whale*, 170; Hunter, *Warriors*, 231–32. In a letter to Hunter, Farley Mowat conveyed what was probably a widespread sense of surprise at the campaign's success: "I must frankly admit that, when you first announced your plans, I didn't give them a chance of success. Well, I was wrong. Happily wrong, I might add." Mowat to Hunter, 10/14/1975, box 1, file 1, R4377, GF.
71. Cassidy, *Mind Bombs*, 135–40.
72. Weyler, "Portfolio," 16. Weyler's article, which appeared in a special edition of the magazine subtitled "Photography as an Agent of Change," includes many of the spectacular photos he shot throughout the campaign.
73. Quoted in the *New York Times Magazine*, 8/24/1975, 14.
74. For example, see Hunter, *Storming of the Mind*, 216.
75. Weyler, *Song of the Whale*, 170; Hunter, *Warriors*, 233–34.
76. Hunter interview; *Warriors*, 232–33. Emphasis in original.
77. Ibid., 234–35.
78. For more on the concept of guerrilla theater, see Durland, "Witness: The Guerrilla Theater of Greenpeace," 30–35.
79. For example, see Hunter, "Taking on the Goliaths of Doom," *Greenpeace Chronicles* 1, no. 1 (Autumn 1975): 5; Similar comments can be seen in the footage from a press conference contained in the film *Voyage to Save the Whales*.
80. *Greenpeace Chronicles* 1, no. 1 (Autumn 1975). Capitals in original.

Chapter 10

1. 7/22/1975.
2. Strong to Greenpeace Foundation, 8/14/1975, Patrick Moore's personal papers. Hunter invited Strong to join the Greenpeace honorary board of governors. Strong, however, was about to leave his job at the UN Environmental Programme to work for Petro Canada.

Strong to Hunter, 12/09/1975. Hunter also invited Allen Ginsberg, Carl Sagan, and John Lilly to join the honorary board. Box, 1, file 1, GF.

3. *Vancouver Sun*, 7/15/1975.
4. For an intelligent and sensitive discussion of this subject, see White, "'Are You an Environmentalist or Do You Work for a Living?'"
5. Weyler interview.
6. Bill Gannon interview.
7. Hunter, *Warriors*, 244.
8. Ibid., 245–46; Hunter interview.
9. Hunter and Moore interviews.
10. Hunter, *Warriors*, 240–41.
11. *Greenpeace Chronicles* 1, no. 1 (Autumn 1975): 2.
12. Ibid. Capitals in original.
13. Ibid., 6, 3, 7.
14. N.B.: "Swiling" rhymes with "smiling."
15. Ibid., 3. While there was no dispute that the harp seal population had declined dramatically over the course of the twentieth century, to say that it was on the road to extinction was alarmist. Up until the 1960s, the hunt remained largely unregulated. At that point, Watson's comment might have been less inaccurate, but by the mid-1970s, various conservation measures had helped stabilize the population, and it appeared that, from a natural resource perspective, the hunt was well on the road to becoming sustainable. See Busch, *War against the Seals*, 248–52; Henke, *Seal Wars!* ch. 3.
16. *Greenpeace Chronicles*, 1, no.1 (Autumn 1975): 3.
17. Nash, *Rights of Nature*, 23–24. For an analysis of the changing attitude toward animals in Victorian England, see Ritvo, *Animal Estate*. For the United States, see Grier, *Pets in America*.
18. Amory, *Man Kind?* 356–57.
19. Singer describes his Viennese roots in *Pushing Time Away*.
20. Singer, "Animal Liberation," 17–18; Singer was ostensibly reviewing Stanley Godlovitch, Rosalind Godlovitch, and John Harris, eds., *Animals, Men and Morals*. For his fellow philosophers, Singer published a more academic version of his essay entitled "All Animals Are Equal." He expanded his ideas for a broader audience in his controversial classic, *Animal Liberation: A New Ethics for Our Treatment of Animals*.
21. Singer, *Animal Liberation*, 188. For critiques of Singer's work, see Pierce, "Can Animals Be Liberated?" and Lamb, "Animal Rights and Liberation Movements." Sprague, "Philosophers and Antivivisectionism" analyzes the main arguments in the debate.
22. Nash, *Rights of Nature*, 140, 160.
23. Ibid., 159. For a more detailed analysis of the philosophical incompatibility of biocentric and animal rights thought, see Sagoff, "Animal Liberation and Environmental Ethics."
24. Quotes from Nash, *Rights of Nature*, 159, 152. For an elaboration of Regan's and Rodman's arguments, see Regan, "Holism as Environmental Fascism," and Rodman, "The Liberation of Nature?"
25. Nash places Greenpeace in the same category as the Fund for Animals, Sea Shepherd, and the Animal Liberation Front. Such groups, he contends, "concentrated on the rights of individual organisms," in contrast to groups such as Earth First! which "preferred to act on the basis of a more holistic philosophy" (189). This demonstrates a limited understanding of Greenpeace's philosophy and campaigns. An animal rights approach was never a clear-cut Greenpeace policy and was always controversial within the organization, where such views came up against the more ecologically holistic ideas held by various members.
26. My description of the harp seal is based on the following sources: Lavigne, "Harp Seal: *Pagophilus groenlandicus*"; Ronald and Dougan, "The Ice Lover: Biology of the Harp Seal"; Busch, *War against the Seals*, 42–44; Henke, *Seal Wars*, ch. 2.
27. The number of pups born in a given season varies from 250,000 to 400,000. See Busch, *War against the Seals*, 43.
28. When long-time sealing protestor Brian Davies brought two orphaned whitecoats home from the ice (females give birth to one pup and will only nurse their own offspring), he

found, somewhat ironically, that the only way he could nurse them was by feeding them a mixture of cow's milk and whale oil. See Davies, *Savage Luxury*, 43.

29. Sanger interview. Sanger, a native Newfoundlander, is an emeritus professor of geography at Memorial University and has a deep knowledge of the history of sealing and whaling in the region.
30. In 1913, the Newfoundland House of Assembly reported: "There are few young men in the colony who have not been on the 'ice' and an expedition is looked on as a test of manhood." Quoted in Busch, *War against the Seals*, 57.
31. Brown, *Death on the Ice*.
32. Quoted in Mowat and Blackwood, *Wake of the Great*, 91.
33. Busch, *War against the Seals*, 246–47. For some fascinating historical photographs of Newfoundland sealers, see Ryan, *Seals and Sealers*.
34. Sergeant, "Harp Seals and the Sealing Industry," 34; Coish, *Season of the Seal*, 72.
35. At this time, Canadian control was restricted to a distance of twelve miles from the Canadian shore. The 200- mile limit that applies today was established in 1977. See Busch, *War against the Seals*, 248.
36. Loo, *States of Nature*, 6. Loo's study is the best critical scholarly history on the subject. Studies that celebrate Canadian wildlife management include: Burnett, *A Passion for Wildlife*, and Foster, *Working for Wildlife*.
37. Despite appearing brutal and bloody, most biologists seem to agree that the clubbing and bleeding method is the most humane, as well as the most efficient method of dispatching the pups. The first blow usually renders the seal unconscious, the second kills it. Its movement thereafter is simply the result of reflexive twitches that occur once it is unconscious or brain dead. For a useful overview (albeit from a prosealing perspective), see Henke, *Seal Wars!* ch. 4.
38. Coish, *Season of the Seal*, 74; Henke, *Seal Wars!* 68–72.
39. Coish, *Season of the Seal*, 74. Two years later, a sealer from the Magdalen Islands signed an affidavit declaring that the film crew had paid him to "torment the said seal and not to use a stick, but just to use a knife to carry out this operation where in normal practice a stick is used to first kill the seals before skinning them." Quoted in Coish *Season of the Seal*, 95. Despite such apparent evidence, one is left wondering why Artek, who appeared to have no intention of making a propaganda film, would need to embellish a documentary that was already studded with images of cruelty and gore.
40. *Canadian Audubon*, Nov.–Dec., 1964, 75.
41. Davies, *Savage Luxury*, 17–21.
42. 207–8. Italics in original. The passage is remarkably similar to many arguments used by whale advocates, particularly Scott McVay.
43. Many years later in a television interview, Davies claimed that he had primarily been motivated by animal rights, even if, for strategic purposes, he did not always present the issue that way. From his perspective, the issue was a moral one revolving around the question of whether the rights of animals are superior to human interests. See *All Things Bright and Beautiful*, TV program aired February 10, 1985.
44. Davies, *Savage Luxury*, 68–69.
45. Ibid., 150. The 1968 hunt was marked by the appearance of several veterinarians on the ice who were burdened with the gruesome task of examining the smashed skulls of the harp seal carcasses to determine if they had been killed "humanely." In a twelve-hour period, they examined 361 carcasses for fractures, hemorrhages, and lesions. They concluded that 97 percent of the pups had in all likelihood been rendered unconscious by the blows from the club before they were skinned. Coish, *Season of the Seal*, 106–7.
46. Davies, *Savage Luxury*, 166; Coish, *Season of the Seal*, 109.
47. Coish, *Season of the Seal*, 111–12.
48. Henke, "Canada and the WTO in the 21st Century"; Ellis, *Empty Ocean*, 203.
49. Davies, *Savage Luxury*, 201–2, 54–57. The ferocity with which many swilers defended their right to hunt, along with the enthusiasm with which they frequently went about it, somewhat belies Davies's claim. Fred Bruemmer, who wrote widely on the hunt, described it as an exhilarating release from the boredom and grinding poverty of winter, as well as representing a traditional rite of the early spring. See Bruemmer, *Life of the Harp Seal*.

50. Coish, *Season of the Seal*, 116–17; Lavigne, "Life or Death for the Harp Seal," 141–42.
51. Lavigne, "Life or Death for the Harp Seal," 129, 137–38, 142. The distinction between the "landsmen" and the large ships was not so clear-cut, since many swilers were local Newfoundlanders hired by Norwegian and Canadian firms to work on the factory ships.
52. Hunter interview.
53. Watson and Spong interviews. Watson and Walrus had first proposed the campaign in July 1975, but the Greenpeace board did not give it their approval until late November. See Hunter, *Warriors*, 249.
54. Ibid., 253–55; Johnson interview.
55. Greenpeace press release, 1976, n.d., Rex Weyler's personal papers.
56. Hunter, *Warriors*, 250.
57. Greenpeace, "Save the Seals Expedition." Direct-mail fund-raising letter from early 1976, Patrick Moore's personal papers.
58. St. John's *Evening Telegram*, March 12, 1976; Hunter, *Warriors*, 251.
59. Hunter, *Warriors*, 252.

Chapter 11

1. George Wenzel, *Animal Rights, Human Rights*, 47.
2. Watson interview; Hunter, *Warriors*, 253–55.
3. Hunter, *Warriors*, 257–59; Watson, Hunter, and Moore interviews. Had Greenpeace had access to the Newfoundland newspapers, they might not have been as surprised by the locals' viciousness. The editorial and letters pages of the St. John's *Evening Telegram* and the Corner Brook *Western Star* were filled with furious denunciations well before their arrival.
4. Hunter, *Warriors*, 258–59; Hunter interview.
5. Quoted in Hunter, *Warriors*, 260. Italics in original.
6. Coish, *Season of the Seal*, 127. Coish was a high school teacher in Grand Falls-Windsor. Although he supported the hunt, he did so with reservations. Unlike many Newfoundlanders, he could sympathize with both Davies and Greenpeace, even though he ultimately disagreed with them. Coish interview.
7. Interestingly, though perhaps not surprisingly, local religious leaders were strong backers of the hunt. The Roman Catholic archdiocese of St. John's, for example, circulated a letter of support, suggesting "it would be far more appropriate if the emotional rhetoric given to the humane harvesting of the seals were applied to the plight of the unborn and suffering children of the world." Quoted in the St. John's *Daily Mail*, 03/15/1977. Many letters in the Newfoundland newspapers voiced similar sentiments, bemoaning a society that seemed to care more for seals than human fetuses.
8. Coish, *Season of the Seal*, 179–80. In the 1981 film, *Bitter Harvest*, a priest is shown speaking exactly these words as a sealing ship departs for the floes. Busch also makes the telling point that it is *sealing*, rather than the seals, that occupies the most important role in Newfoundland's folklore. Newfoundlanders have none of the traditions respecting the spirit and power of the seals that are common in old northern European or Native American folk traditions. Perhaps because the Newfoundland seal hunt was, almost from its inception, a commercial venture, seals have always been seen as a mere resource. See *War against the Seals*, 41–42.
9. Hunter, *Warriors*, 260–61. The following year, all the parties in the Canadian House of Commons passed a unanimous resolution supporting the continuation of the seal hunt. It is unlikely, therefore, that the Trudeau government's actions represented a "totalitarian streak." See *International Herald Tribune*, 3/26/1977.
10. Hunter, *Warriors*, 286–87.
11. Ibid., 261–63.
12. *Evening Telegram*, 3/13/1976, 13.
13. Watson interview; Hunter, *Warriors*, 264.
14. For example, see the *Montreal Star*, 3/10/1976. To Newfoundlanders such as the writer Calvin Coish, "The sudden swing to the side of the landsmen seemed like a desperate attempt by Greenpeace to counter the unexpected hostility of the Newfoundland sealers." See *Season of the Seal*, 126.

15. Hunter, *Warriors*, 263.
16. Watson, Hunter, and Moore interviews.
17. Hunter, *Warriors*, 268–70.
18. Watson, *Sea Shepherd*, 92. Harp seals' eyes are constantly lubricated by tears to protect them from salt water. Unlike terrestrial mammals, they lack ducts to drain away the tears. Thus they look like they are constantly crying. See Lavigne, "Harp Seal," 543–44.
19. Hunter, *Warriors*, 279–80; Johnson, Watson, and Moore interviews. Some biologists argue that it is anthropomorphic to infer that mother seals are horrified when they discover the newly skinned corpse of their pups. After about ten days, the mother seal resolutely stops nursing her pup and effectively abandons it, preferring to leap back into the ocean and copulate with numerous males. At that point, so the argument goes, they could care less what happens to the infants, and any apparent display of maternal horror is merely a combination of curiosity and territoriality. See Henke, *Seal Wars!* 58–60.
20. Watson, *Sea Shepherd*, 93–94.
21. Hunter, *Warriors*, 273–74. Brian Davies, who had earned a pilot's license and was flying journalists to the ice in his own hired chopper, relieved some of the pressure. As an indication of how far his tabloid instincts had evolved, Davies brought with him a team of highly photogenic airline stewardesses to photograph with the seal pups. See Watson, *Sea Shepherd*, 90, and Coish, *Season of the Seal*, 135.
22. Ibid., 283–86; Johnson interview.
23. Hunter, *Warriors*, 287–88.
24. Ibid., 290.
25. Ibid., 290–91. Photographs of the event are widely available in various Greenpeace publications and on the World Wide Web. The most accessible published version is probably the one on the back cover of the paperback edition of Hunter's *Warriors of the Rainbow*. Film of the event can be seen in the documentary *Bitter Harvest*.
26. Ibid., 291–93. Italics in original.
27. Hunter interview; "Cruel Propaganda," *Evening Standard*, 03/19/1976.
28. This was my unmistakable impression after reading through all of the March 1976 issues of the St. John's *Evening Telegram* and the *Western Star*, which is based in the west Newfoundland town of Corner Brook.
29. Hunter and Watson interviews.
30. George Wenzel has done this for indigenous sealers in the Canadian Arctic. See *Animal Rights, Human Rights*.
31. Cassidy, *Mindbombs*, 158.
32. Widdowson, *Framing of Greenpeace*, 104.
33. Watson interview.
34. Watson article in *Georgia Straight*, April 15-22, 1976.
35. Trudeau to Patrick Moore, 3/29/1976, Moore's personal papers.
36. *Georgia Straight*, Jan. 16-22, 1976.
37. Ibid., 363; McTaggart, *Shadow Warrior*, 109; Hunter interview; McTaggart interview.
38. Watson interview.
39. Watson, *Greenpeace Chronicles*, Winter 1976/77.
40. Quoted in the *Vancouver Sun*, 1/10/1977. The Norwegian students who accompanied Watson reported that they were similarly willing to embrace martyrdom. Calgary *Albertan*, 03/10/1977.
41. Watson, Hunter, Moore, Weyler interviews.
42. Moore interview.
43. Hunter interview; *Warriors*, 365.
44. Ibid., 365–66; Hunter, Moore, and Watson interviews.
45. St. John's *Evening Telegram*, 03/14/77; Watson, *Sea Shepherd*, 125–27; Coish, *Season of the Seal*, 140–45.
46. St. John's *Daily News*, 03/23/77.
47. *Winnipeg Tribune*, 3/24/1977; *Western Star*, 3/26/1977; Coish, *Season of the Seal*, 2, 166.
48. 3/30/1977.
49. *Montreal Gazette*, 2/23/1977; Watson, *Sea Shepherd*, 130; Hunter, *Warriors*, 368–69.

50. Senior Sergeant G. R. Butt, Corner Brook Subdivision, February 16, 1977. RG 18, vol. 6649, file 77-HQ-102-15-4. Secured through an Access to Information request through Archives Canada (GP AIR). Copy in author's possession.
51. Editorial, *Western Star*, 03/19/77.
52. *Vancouver Sun*, 3/21/1977; Hunter, *Warriors*, 371–72.
53. *Vancouver Sun*, 3/21/1977.
54. *Winnipeg Tribune*, 3/28/1977.
55. Hunter, *Warriors*, 380–81. Emphasis in original.
56. Bardot journal quotes in ibid.; Moore interview.
57. Watson interview; *Vancouver Province*, 3/21/1977.
58. *Vancouver Sun*, 3/19/1977.
59. Quoted in Hunter, *Warriors*, 382.
60. Coish, *Season of the Seal*, 1. Remi Parmentier of the fledgling French Greenpeace outfit was also unimpressed by Bardot's antics and Greenpeace's willingness to cooperate with her. "We did not decide to create a Greenpeace group in France to get involved with Brigitte Bardot whose positions we think are *irresponsible*." Parmentier to Moore, 08/02/1977, box 5, file 15, GF.
61. Watson, *Sea Shepherd*, 143–44; Hunter, *Warriors*, 375.
62. Quoted in Hunter, *Warriors*, 376. The same quote is largely reproduced in Watson, *Sea Shepherd*, 145–46.
63. Watson, *Sea Shepherd*, 145–50. The ship's captain, Per Lyngvaer, told a different story. Watson, he insisted, had been dunked, not by any mishandling by his men, but rather, by the erratic heaving of the ship. Thereafter, he was taken on board, given dry clothes, and allowed to stay for two nights, during which he was treated as any injured man would be. The fisheries officer backed up the captain's version, and the RCMP also downplayed the incident in their reports. Peter Ballem, who witnessed the entire event and was with Watson on the ship, corroborated Watson's version. Even allowing for a certain amount of exaggeration on both sides, the two stories are irreconcilable. However, given Watson's provocative behavior and the pent up anger of the swilers, combined with the fact that fisheries officials strongly supported the sealers, one is inclined to believe Watson's version more than that of the captain. See Coish, *Season of the Seal*, 153–54; telegram from Corner Brook RCMP to Ottawa, March 16, 1977, RG 18, vol. 6649, file 77-HQ-102-15-4, GPA AIR.
64. Hunter, *Warriors*, 378, 373. In addition to the fact that most of the U.S. media was in St. Anthony, Moore felt that Bob Cummings had been a "disaster" as the Greenpeace media coordinator. If they had relied entirely on him, Moore reported to Hunter, "I don't think there would have been a single story." Letter from Moore to Hunter, 3/23/1977, Moore's personal papers.
65. Moore to Hunter, 3/23/1977, Moore's personal papers.
66. Letter to the *Vancouver Sun*, 10/28/1977.
67. Hunter, *Warriors*, 387–88.
68. Minutes of the Board of Directors Meeting, 6/7/1977, Rod Marining's personal papers. While most of the Vancouver inner circle felt that Watson was more trouble than he was worth, others were disappointed with the board's decision. The Hawaiian branch of Greenpeace, for example, reassured Watson that its members were "not in agreement with the reason for your dismissal . . . We feel you are the victim of internal political motivations." E. Ross Thornwood to Watson, 6/16/1977, SSCS.
69. Watson interview.
70. Watson, *Sea Shepherd*, 153.
71. Earthforce brochure, 1977, Al Johnson's personal papers.
72. Watson, *Sea Shepherd*, 188–206.
73. Paul Watson, "Pirate Whalers Rammed out of Business," *Greenpeace Chronicles*, no.19, Sept. 1979; Watson, *Sea Shepherd*, 207–50; Hunter, "Fighting Back." No one was ever charged with blowing up the *Sierra*. Watson claims to know who was responsible, and, while supporting their action, insists that he played no role in this particular act of "ecotage."
74. For a rundown of Watson's actions to the early 1990s, see Watson, *Ocean Warrior*. For more recent campaigns and actions, see the Sea Shepherd website at www.seashepherd.org.

For a lively, though strongly sympathetic account of the more radical elements within the environmental movement, including Sea Shepherd, see Rik Scarce's *Eco-Warriors*.

75. Hunter, *Warriors*, 386–87.
76. Hunter interview.
77. For a discussion of the European ban, see Barry, *Icy Battleground*. Wenzel analyzes the ban's impact on indigenous sealers. See *Animal Rights, Human Rights*, 1–2, 128–33. In the past two decades, the Canadian government and the sealing industry have managed, with very little publicity, to once again build up the industry to the point where the number of seals hunted today is as high as it was in the 1960s. The EC ban has largely been overcome by product diversification—seals are used to make everything from salami, various medicines, briefcases, and leather trinkets, as well as fur coats. The Canadian government has also helped open up large new markets for seal products, particularly in Russia and China. The campaigns of animal rights activists have been partially defused by a ban on hunting whitecoats. This does not protect seals—it just delays their slaughter until they are several weeks old and have lost their pure white fur. For an insightful analysis of how the Canadian government revived the industry, see Dauvergne and Neville, "Mindbombs of Right and Wrong." Also see Lavigne, "Estimating Total Kill of Northwest Atlantic Harp Seals."
78. Allen, "Anti Sealing as an Industry," 427. From another angle, John-Henry Harter argues that Greenpeace has always had an inherently anti-working-class bias. In fact, he goes so far as accusing them of "devastat[ing] two entire economies and communities: those of the Inuit and the Newfoundland sealers." Clearly, the story is far more complex. For one thing, Newfoundland provincial politics have been dominated by the Conservative, and to a lesser extent, the Liberal Party. The socialist NDP has barely gained a toehold. Harter appears reluctant to acknowledge that sometimes the interests of the working class are best served by allying themselves with capitalists in order to defeat environmentalists. See Harter, "Environmental Justice for Whom?"
79. Sanger interview. For an analysis of the effectiveness of both the antiwhaling and antisealing campaigns, particularly since the 1980s, see Guevara, "Assessing the Effectiveness of Transnational Activism."
80. Hunter and Watson interviews.

Chapter 12

1. My summary of Habermas is drawn from Brulle, *Agency, Democracy, and Nature*, 67–70. The Habermas quote appears on 68.
2. Ibid., 68–69.
3. By 1967, SDS had become "so disdainful of the formal structures of its first five years, as to abolish its own presidential and vice-presidential offices." Todd Gitlin, "The Left Declares Its Independence," *New York Times*, 10/08/2011, http://www.nytimes.com/2011/10/09/opinion/sunday/occupy-wall-street-and-the-tea-party.html?emc=eta1&pagewanted=all. Barbara Epstein's fine-grained study of the Clamshell Alliance is a useful analysis of an antinuclear organization committed to the affinity group model. See *Political Protest and Cultural Revolution*.
4. Kitschelt, "Social Movements, Political Parties, and Democratic Theory," 15.
5. Hunter interview.
6. Hunter, *Warriors*, 296–99.
7. Ibid., 316–17. For more on Tom McCall's environmental reforms in Oregon, see Walth, *Fire at Eden's Gate*.
8. Hunter, *Warriors*, 318–20. Emphasis in original.
9. Ibid., 321–23; Hunter, Watson, Moore, and Weyler interviews.
10. Weyler interview.
11. Hunter, *Warriors*, 329–30. Ironically, Hunter's prophecy proved to have some credibility. A squid's metabolism allows it to feed almost constantly, thereby enabling it to grow at a phenomenal speed. Recent studies by marine biologists indicate that the population of giant squid seems to be increasing at a substantial rate. For more information, see Johnson, "Jumbo Squid Following Low-Oxygen Zone."

12. Moore interview; Hunter, *Warriors*, 343–44.
13. Hunter, *Warriors*, 334–36; Hunter and Spong interviews.
14. Watson, *Sea Shepherd*, 116. Film of the incident can be seen in *Ecology in Action*.
15. Lilly, *Mind of the Dolphin*; Hunter, *Warriors*, 339–40.
16. *Wall Street Journal* (Pacific Coast edition) 9/24/1976. See also, Herron, "A Not-Altogether Quixotic Face-Off with Soviet Whale Killers in the Pacific."
17. Hunter, *Warriors*, 362–63.
18. Hunter, Weyler, and Marining interviews; Hunter, *Warriors*, 387.
19. Taunt interview; Moore, *Confessions*, 79.
20. Hunter, *Warriors*, 398–403.
21. Ibid., 411–15; Weyler, *Song of the Whale*, 192–93.
22. Taunt interview; Hunter, *Warriors*, 413.
23. Taunt interview.
24. Quoted in the *Vancouver Sun*, 1/12/1978.
25. Ibid. A report written for the British RSPCA, however, found that the killing of seals was "by any normally accepted civilized standards an ugly, brutal and violent business." See "The Killing of the Harp Seal Pups, 1978," RSPCA (UK). A copy can be found in vol. 12, scrapbook 7, GPF.
26. Editorial, *New York Times*, 1/17/1978. The *Washington Post*, 4/23/1978, also published an article that made a case for the continuation of the hunt.
27. Busch, *War against the Seals*, 243. Such tactics have also been used by the Japanese government in an attempt to encourage its people to eat more whale meat. Since most Japanese display little interest in the product, the government has resorted to including the meat in school-lunch programs in an effort to inculcate the nation's youth with a taste for the product, thereby allowing the whaling industry to argue that whale meat constitutes an important and traditional food source. See the BBC documentary, *Whale Hunters*, produced by Jeremy Bristow, 2002, and "Whale meat back on school lunch menus," *Japan Times*, 09/04/2010, http://www.japantimes.co.jp/text/nn20100904x2.html.
28. Taunt interview.
29. *Weekend Magazine*, 3/11/1978, 3.
30. Inspector J. McGuire, Corner Brook subdivision, April 4, 1977, 77 HQ-102-15-4, vol. 2, GPA AIR.
31. Superintendent R. C. Richards, Officer in Charge, Criminal Investigation Branch, St. John's to Corner Brook subdivision, June 14, 1977, 77HQ-102-15-4 (supplement A), GPA AIR.
32. Taunt interview; Hunter interview; *Warriors*, 439. On the Trident campaign, see Walrus Oakenbough, "Greenpeace Flagship to Spearhead Anti-Trident Mission," *Georgia Straight*, June 15–22, 1977, and Terry Glavin, "Over the Fence," *Georgia Straight*, May 26–June 2, 1978.
33. Quoted in Hunter, *Warriors*, 441.
34. The press releases were printed throughout the March 1978 issues of the *Georgia Straight*.
35. Weyler's journal, no.17; Hunter, *Warriors*, 443.
36. Moore, *Confessions*, 96–100; Hunter, *Warriors*, 444–45, and Weyler's journal, no.17.
37. Greenpeace Foundation of America Press Release, March 1978, Robert Taunt's personal papers.
38. *New York Times*, 1/10/1978. The Swedish study was by Steffan Soderberg and Lennart Almkvist, *In Prospect of the Seal Hunt in Canada 1977*.
39. Thornton to various Greenpeace offices, 4/13/1978, box 13, scrapbook 5, GPF.
40. Several years later, animal rights philosophers Dale Jamieson and Tom Regan made exactly this point in their argument for the abolition of whaling: "As with humans, so also with whales, it is individuals, not species, who have rights. We must take care not to accept that science that smothers the individual whale in numbers, graphs, charts, and so on." Jamieson and Regan, "Whales Are Not Cetacean Resources," 107.
41. *Oregonian*, 9/25/1977; Day, *Whale War*, 36–37.
42. Day, *Whale War*, 37–38. The Japanese bolstered their case by arguing that the pressure put on the sperm whale by the Japanese and Soviet harvests was causing an increase in the pregnancy rate, thus allowing them to adjust future birthrate projections upward. More hunting, happily, was good for the species. See Weyler, *Song of the Whale*, 195.

43. *Honolulu Advertiser*, 12/12/1977.
44. White to Mink, n.d., vol.12, scrapbook 5, GPF.
45. Spong, "In Search of a Bowhead Policy," *Greenpeace Chronicles*, Nov. 1978, 2.
46. Weyler interview.
47. Watson interview.
48. Wenzel, *Animal Rights, Human Rights*, 59.
49. Ibid., 41.
50. Quoted in Weyler, *Song of the Whale*, 225.
51. Ryan's November 1978 letter to Spong is reproduced in ibid., 196.
52. John C. Lilly, "The Rights of Whales," *Greenpeace Chronicles*, no. 2, Spring/Summer 1976, 10.
53. Jamieson and Regan, "Whales Are Not Cetacean Resources," 107. Paul Watson also agrees with the idea that whales should simply be left alone. Watson interview.
54. Hunter, "Ecology as Religion," *Greenpeace Chronicles*, no.18, August 1979, 3. Italics in original.
55. Hunter, "Reclaiming the Cities," *Greenpeace Chronicles*, no.16, May 1979, 2. Italics in original.
56. Hunter, "Endgame Ecology," *Greenpeace Chronicles*, no. 15, April 1979.
57. Bill Devall, *Greenpeace Chronicles*, no.19, September 1979, 2. Devall's published work includes: *Simple in Means, Rich in Ends* and, with George Sessions, *Deep Ecology*.
58. The 1982 vote was 25 to 7, which was just barely enough to achieve the three-quarter majority that IWC regulations required. The seven opposing votes came from Japan, the USSR, Norway, Peru, Iceland, Brazil, and South Korea. See Day, *Whale War*, 98. The EC bans and their impact are discussed in Wenzel, *Animal Rights, Human Rights*, 1–2, 128–33.
59. For more details, see Rex Weyler's article in *Greenpeace Chronicles*, no.7, June 1978, 6–7.
60. Greenpeace USA, Minutes from the National Meeting, October 20–22, San Francisco, 1979, Rod Marining's personal papers.
61. McDermott interview; Johnson interview; Bob Cummings, *Greenpeace Chronicles*, no.18, August 1979, 12.
62. *Vancouver Province*, 5/19/1977; *Vancouver Sun*, 5/27/1977; *North Shore News*, 6/1/1977. The left-wing New Democratic Party had recently been ousted by a rejuvenated Social Credit Party, with Bill Bennett, son of long-term former premier, W. A. C. Bennett, as the new leader.
63. *Vancouver Sun*, undated clipping, 12, scrapbook 5, GPF.
64. McTaggart, *Shadow Warrior*, 121–30. For a film about the campaign, see *Destination Iceland*.
65. McTaggart, *Shadow Warrior*, 138–41; *Time*, 9/23/1978; *Daily Mail*, 10/17/1978.
66. *Greenpeace Chronicles*, November 1978, 6–7; Wilkinson with Schofield, *Warrior*, 26–30. For footage of the *Gem* action, see *Desperate Measures*.
67. Minutes of the Greenpeace Foundation Board of Directors Meeting, Vancouver, January 12, 1978, Rod Marining's personal papers.
68. Ibid.; Hunter interview.
69. Hunter interview.
70. Extraordinary Resolutions of the Greenpeace Foundation AGM, Dec 20, 1976, personal papers of Gerhard Wallmeyer, Greenpeace Germany. Wallmeyer was Greenpeace Germany's chief fundraiser for over twenty years and an avid collector of documents. On one of his visits to Vancouver in the 1980s, he managed to salvage a number of important materials that would otherwise have been thrown out during a move. He allowed me to sift through them in his office at Greenpeace Germany's headquarters in Hamburg.
71. Tussman interview.
72. Letter from Tussman to Martin Majestic, 7/9/1979, Tussman's personal papers.
73. Hunter, *Warriors*, 246–47.
74. Interviews with Kay Treakle (an early member of Greenpeace Seattle) and Remi Parmentier (a co-founder of Greenpeace France).
75. The description of the San Francisco group is drawn from the following sources: "Greenpeace San Francisco, Greenpeace Foundation of America," a fact file outlining the group's

structure from the personal papers of David Tussman; Macy, "Greenpeace," 67–69; Tussman and Taunt interviews.

76. Macy, "Greenpeace," 68. Greenpeace refused to sell the list to Brown, feeling that it would alienate some of their supporters. After all, there were also plenty of Republicans who opposed whaling and sealing.
77. Boe interview. Boe was an early member of the Seattle organization and would go on to lead Greenpeace's sealing campaign from 1979 onward; Treakle interview. The quote is from Janus Masi, the correspondence secretary of Greenpeace Seattle in a letter to Gary Zimmerman, 5/26/1977. It was in a box labeled "Greenpeace History Dump Re: Internal Structure" which was found at the former Seattle Greenpeace office.
78. Sawyer interview.
79. Don White to various Greenpeace offices, 3/27/1979. Box labeled "1978 General Overview of Greenpeace Autonomy Bullshit from Various Perspectives" (italics in original) in the Seattle Greenpeace office. Clearly, the people in the Seattle office did not take filing particularly seriously.
80. Taunt interview.
81. McTaggart interview.
82. Parmentier interview. Numerous people I interviewed corroborated Parmentier's story of how McTaggart tried to convince people that he was the true founder of Greenpeace. Some even thought that McTaggart had convinced himself that this was the case.
83. Thornton to Frizell, 12/21/1976, box 5, file 12, GF.
84. Thornton described the London anarchists as incompetent and inefficient: "I doubt anyone would even know if they disappeared, if they ceased to function." Thornton to Frizell, 12/15/1976, box 5, file 12, GF.
85. Newborn, *A Bonfire in My Mouth*. A letter from the Greenpeace London group to Greenpeace New Zealand conveys a sense of the growing schism between the advocates of a more organized and hierarchical organization, such as McTaggart and the Vancouver group, and the grassroots anarchists.

    > We are getting an increasing number of people confused about which Gp is which . . . we have tried hard to work out our problems with the people at GP LTD, but have had very little come-back from them. The way that they work is very hierarchical, as opposed to our structure of collective working, so it does make trying to talk to them difficult. While traditionally we have had small autonomous GP groups in various parts of the country (& world), it is now a situation in which the Vancouver Foundation & GP Ltd are setting up "branches" of their own in the UK and around the world. According to what we have heard, one of the people at GP LTD thinks that our group is "too democratic" because of our way of working. That should give you some idea of what we are up against. Another thing that we have noted of late, while looking through some documents put out by the Greenpeace Foundation in Vancouver, is a claim that they are "Apolotical." This is in contrast to our own libertarian viewpoint.

    Martin Howe to Greepeace New Zealand, 5/13/78. London Greenpeace, Assorted Newsletters (1970s), CCBU.
86. Weyler's interview with Parmentier. Used with permission.
87. Moore to various Greenpeace offices, n.d. (though clearly late 1977), Moore's personal papers.
88. Ibid.
89. International Greenpeace Meeting Minutes, January 21–25 1978, Vancouver, David Tussman's personal papers.
90. Ibid., 39–53.
91. Ibid.
92. Ibid., 64–65.
93. Gannon interview.
94. Minutes of Extraordinary Meeting of Board of Directors, GP Foundation of America, San Francisco, August 29, 1978, box 2, file 2, GF. Tussman interview. Moore subsequently

viewed his San Francisco visit as a tactical error. In addition to losing his temper, he had called a press conference during which he spelled out the reasons why Vancouver was upset with its San Francisco brethren. This prompted a not-so-brotherly $1 million libel and defamation lawsuit, not against Greenpeace, but against Moore personally. See Moore, *Confessions*, 106–7. It seems that San Francisco was not the only place where Moore lost his temper. Will Anderson, of Greenpeace Alaska, complained to Moore: "You alienated the hell out of me when you said, 'You'll find yourself on the Alaskan docket if you don't honor the charter.' What do you hope to accomplish by that? You'll end up with dust in your hands." Anderson to Moore, n.d., correspondence, box 6, file 11 (Alaska), GF. The San Diego representative voiced similar sentiments: "Why don't you just come out and plainly say that GP Vancouver is filing a court suit against Gp America, namely San Francisco, because you don't like the way they do things and you want full control over the money and power of *all* Greenpeace offices." Geoffrey T. Moulton to Moore, 05/11/1979, box 6, file 12 (California), GF.

95. Tussman and Hunter interviews. Frizell did not consent to an interview.
96. Tussman to Zimmerman, 9/8/1979. Tussman's personal papers.
97. Greenpeace San Francisco, Minutes of the Meeting of the Board of Directors, 7/25/1979, Tussman's personal papers.
98. McTaggart interview; Hunter interview.
99. McTaggart and Slinger, *Shadow Warrior*, 148–49; McTaggart interview; Hunter, Weyler, and Moore interviews.
100. McTaggart and Slinger, *Shadow Warrior*, 149.
101. The quote is from Tussman's unpublished memoir. Used with author's permission. Tussman has subsequently made the memoir available online, with lots of interesting photos from his Greenpeace days. See: http://quandmeme.wordpress.com/greenpeace-2/
102. Minutes of the Greenpeace Council International Meeting, November 16–20, 1979, Amsterdam, Patrick Moore's personal papers.
103. Jordan and Maloney, *Protest Business?* 191.
104. Meyer and Tarrow, "A Movement Society," 17.

Conclusion

1. The conversation was recorded and transcribed by Rex Weyler.
2. Don White, from Greenpeace Hawaii, was among those who voiced concerns about some of the more mystical elements within Greenpeace. There are many things, he argued, that Greenpeace should leave behind. "These include the rampant use of the I Ching, Astrology, Erik von Daniken books, razor-blade-sharpening pyramids, hallucinogens, and bizarre personal appearances. Cleansing from pseudo-science is in order." White, *What Is Greenpeace?* Undated memo (probably late 1979) in box 6, file 13, GF.
3. Ibid.
4. For more on how mainstream American environmental organizations have been co-opted by governments and corporations, see Dowie, *Losing Ground*.
5. Bode and Sawyer interviews. (Sawyer is a former executive director of Greenpeace International.) A similar story played out in Canada.
6. Hunter interview.
7. Epstein, *Political Protest and Cultural Revolution*.
8. Mitchell, "The 'Mindbomb' That Was Greenpeace," 13–14.
9. Ibid., 13.
10. I derive the phrase from the title of Mark Lytle's biography of Carson, *The Gentle Subversive*.
11. Known as "Greenfreeze," the technology has been employed in over 400 million refrigeration units since it was developed in the early 1990s. However, its butane and propane cooling technology is banned in the United States, a situation Greenpeace attributes to the lobbying power of DuPont and Honeywell, who see it as a potential threat to their own Tetrafluoroethane refrigerant, R-134a. The EU has banned Tetrafluoroethane refrigerants because of their contribution to climate change. See Greenpeace Deutschland, "30 Jahre, 1980–2010."

12. Moore, *Confessions of a Greenpeace Dropout*, 5.
13. Ibid., 142.
14. For a summary of Moore's ties to various industry front groups, see "Patrick Moore," n.d., Sourcewatch.org, http://www.sourcewatch.org/index.php?title=Patrick_Moore#cite_note-1 (accessed 11/15/11). Hunter labeled Moore an "eco-Judas." However, he was never one to hold a grudge for very long, and the two reconciled and remained on friendly terms for the rest of Hunter's life. Hunter interview.
15. Moore, *Confessions*, 4. Another well-known countercultural scientist of Moore's generation, Stewart Brand, has taken a similar path. While Moore is now a self-proclaimed "sensible environmentalist," Brand styles himself an "ecopragmatist." He, too, is highly critical of some of Greenpeace's campaigns, albeit without Moore's psychological baggage. In fact, he accuses Greenpeace of directly contributing to the starvation of millions of Africans as a result of its campaigns against genetically modified crops. See Brand, *Whole Earth Discipline*, ch. 5.
16. The major exception has been Greenpeace's antiwhaling campaigns, which he continues to wholeheartedly support: "No whale or dolphin should be killed or captured anywhere, ever. This is one of my few religious beliefs. They are the only species on earth whose brains are larger than ours and it is impossible to kill or capture them humanely" (*Confessions*, 10). Thus Moore's allegiance to *Cetaceus intelligentus* remains firm. However, when a timber company clearcuts an old growth forest near Moore's childhood home on Vancouver Island, thus destroying a rich habitat for numerous species, we should merely think of it as the creation of "a temporary meadow." See his "Green Spirit Trees Are the Answer," n.d., Michigan Forest Forever website, http://mff.dsisd.net/Balance/GreenSpirit.htm (accessed 11/15/11).
17. For a full exploration of the controversy, see Jordan, *Shell, Greenpeace and the Brent Spar*. Also useful is Wöbse, "Die Brent Spar-Kampagne: Plattform für diverse Wahrheiten."
18. For an examination of how various industry front groups have attempted to dismiss environmentalists as purveyors of "junk science" see Monbiot, *Heat*, ch. 2. Thomas O. McGarity has explored how industry has worked to conflate the term "junk science" with any science employed by environmentalists. See "Our Science is Sound Science and Their Science is Junk Science."

BIBLIOGRAPHY

Note on the Sources

There is, unfortunately, no single archive that houses all of Greenpeace's historical documents. Instead, important documents remain scattered throughout the world, with many stored in the attics and basements of current and former Greenpeace members. For the 1970s, the most complete set of documents available to the public can be found in the Greenpeace Fonds at the Vancouver City Archive. Greenpeace USA has an archive at its headquarters in Washington, D.C., though the materials mostly cover the period after 1980. The International Institute for Social History in Amsterdam is the repository of much of Greenpeace International's historical documentation. Since GPI corresponds with all the national offices, this archive also contains material pertaining to those offices. For information, see http://www.iisg.nl/archives/en/files/g/10918786.php. Archives Canada in Ottawa also has an important collection of documents, particularly for the post-1980 period. (The abbreviation for an archive cited in the notes follows the entry.)

The institutions mentioned are, to varying degrees, in the process of collecting materials from various Greenpeace offices (many now defunct) and the papers of former employees. Some of the records I found in personal papers, therefore, may now be housed in one of these archives. For example, Patrick Moore's papers are now available at Archives Canada in Ottawa. If one is willing to wait for up to a year for the redactors to do their work, the police reports and similar surveillance documents available through Access to Information requests in Canada also contain useful information.

Most of the current and former Greenpeacers I approached generously allowed me to examine and photocopy their private papers. These have been cited in the notes as "personal papers." Some materials were found in ad hoc boxes in various Greenpeace offices.

Archives and Personal Papers

Commonweal Collection, Bradford University, United Kingdom. (CCBU)
Greenpeace Fonds (R4377), Archives Canada, Ottawa. (GF)
Greenpeace Foundation Fonds, City of Vancouver Archives. (GPF)
Greenpeace USA Archives, Washington, D.C.
Johnson, Al. Personal Papers.
Marining, Rod. Personal Papers.
Mayor's Office Fonds, City of Vancouver Archives. (MOF)
Metcalfe, Ben. Personal Papers.
Moore, Patrick. Personal Papers.

Murray Newman Fonds. AddMSS no. 1287. City of Vancouver Archives. (MNF)
Sawyer, Steve. Personal Papers.
Sea Shepherd Conservation Society Archives, Friday Harbor, WA. (SSCS)
Seal Hunt and Greenpeace, Access to Information Request (A-2009-00353/BAS), Library and Archives Canada, Ottawa. (GP AIR)
Sophia Smith Collection, Smith College, Northampton, MA. (SSC)
Taunt, Robert. Personal Papers.
Tussman, David. Personal Papers.
University of British Columbia Library, Special Collections, Vancouver.
Wallmeyer, Gerhard. Personal Papers.
Weyler, Rex. Personal Papers.

Interviews

Bode, Thilo. Amsterdam, 12/16/1999.
Boe, Vivia. Seattle, 5/25/2000.
Bohlen, Jim. Denman Island, British Columbia, 04/20/2000.
Coish, Calvin. Grand Falls-Windsor, Newfoundland, 06/24/2007.
Fineberg, Richard. Telephone interview and email correspondence, 09/18/2011.
Gannon, Bill. Vancouver, 5/2/2000.
Gannon, Linda. Vancouver 5/2/2000.
Hunter, Robert. Toronto, 7/24/2000.
Ingram, Nigel. Telephone interview, 5/14/2002.
Johnson. Al. Vancouver, 5/18/2000.
McDermott, Dan. Toronto, 7/25/2000.
McLeod, Dan. Vancouver, 4/10/2000.
McTaggart, David. Paciano, Italy, 10/15-16/1999.
Marining, Rod. North Vancouver, 4/16/2000.
Metcalfe, Ben. Shawnigan Lake, Vancouver Island, 5/9/2000.
Moore, Patrick. Vancouver, 4/14/2000.
Newborn, Susi. Interviewed by Rex Weyler. Used with permission.
Parmentier, Remi. Amsterdam, 12/14/1999.
Parmentier, Remi. Interviewed by Rex Weyler. Used with permission.
Sanger, Chesley, St. John's, Newfoundland, 06/23/2007.
Sawyer, Steve. Amsterdam, 10/15/1999.
Simmons, Terry. E-mail correspondence, August 2001.
Spong, Paul. Vancouver, 4/16/2000.
Stowe, Dorothy. Vancouver, 4/11/2000.
Taunt, Robert. Helena, Montana, 5/31/2000.
Treakle, Kay. Washington D.C., 2/20/2000.
Tussman, David. Berkeley, 6/6/2000.
Watson, Paul. Friday Harbor, Washington, 5/12/2000.
Weyler, Rex, Vancouver, 4/13/2000.

Sources

Allen, Jeremiah. "Anti Sealing as an Industry." *Journal of Political Economy* 87, no. 2 (1979): 424–28.
Altmann, Christian, and Marc Fritzler. *Greenpeace: Ist die Welt noch zu Retten?* Düsseldorf: ECON Taschenbuch Verlag, 1995.
American Friends Service Committee. *Speak Truth To Power: A Quaker Search for an Alternative to Violence*. N.p.: AFSC, 1955. Available at http://www.quaker.org/sttp.html.
Amory, Cleveland. *Man Kind?: Our Incredible War on Wildlife*. New York: Harper and Row, 1974.
Anderson, Terry H. *The Movement and the Sixties: Protest in America from Greensboro to Wounded Knee*. New York: Oxford University Press, 1995.

Armitage, Kevin C. *The Nature Study Movement: The Forgotten Popularizer of America's Conservation Ethic*. Lawrence: University Press of Kansas, 2009.

Au, Whitlow W. L. *The Sonar of Dolphins*. New York: Springer-Verlag, 1993.

Bacon, Margaret H. *The Quiet Rebels: The Story of the Quakers in America*. New York: Basic Books, 1969.

Barbour, Michael. "Ecological Fragmentation in the 1950s." In *Uncommon Ground: Toward Reinventing Nature*, ed. William Cronon. New York: W.W. Norton, 1995.

Barman, Jean. *The West Beyond the West: A History of British Columbia*. Toronto: University of Toronto Press, 1996.

Barry, Donald. *Icy Battleground: Canada, the International Fund for Animal Welfare, and the Seal Hunt*. St. John's, Newfoundland: Breakwater Books, 2005.

Bazell, Robert J. "Nuclear Tests: Big Amchitka Shot Target of Mounting Opposition." *Science* 172, no. 3989 (June 18, 1971): 1219–21.

Bennett, Scott H. *Radical Pacifism: The War Resisters League and Gandhian Nonviolence in America, 1915–1963*. Syracuse, NY: Syracuse University Press, 2003.

Bess, Michael. *The Light-Green Society: Ecology and Technological Modernity in France, 1960–2000*. Chicago: University of Chicago Press, 2003.

Bigelow, Albert. *The Voyage of the Golden Rule: An Experiment with Truth*. Garden City, NY: Doubleday, 1959.

Bocking, Stephen. *Ecologists and Environmental Politics: A History of Contemporary Ecology*. New Haven, CT: Yale University Press, 1997.

Bohlen, Jim. *Making Waves: The Origins and Future of Greenpeace*. Montreal: Black Rose Books, 2001.

Bölsche, Jochen. *Die Macht der Mütigen: Politik von Unten: Greenpeace, Amnesty und Co*. Hamburg: Spiegel Verlag, 1995.

Boyer, Paul. *By the Bomb's Early Light: American Thought and Culture at the Dawn of the Atomic Age*. New York: Pantheon, 1985.

Brand, Stewart. *Whole Earth Discipline: An Ecopragmatist Manifesto*. New York: Viking, 2009.

Braunstein, Peter, and Michael William Doyle. "Introduction: Historicizing the American Counterculture of the 1960s and '70s." In *Imagine Nation: The American Counterculture of the 1960s and '70s*, ed. Peter Braunstein and Michael William Doyle. New York: Routlege, 2002.

Brown, Cassie. *Death on the Ice: The Great Newfoundland Sealing Disaster of 1914*. Toronto: Doubleday Canada, 1972.

Brown, Michael, and John May. *The Greenpeace Story*. London: Dorling Kindersley, 1989.

Brown, Paul. *Greenpeace*. New York: New Discovery, 1994.

Bruemmer, Fred. *Life of the Harp Seal*. Montreal: Optimum, 1977.

Brulle, Robert. *Agency, Democracy, and Nature: The U.S. Environmental Movement from a Critical Theory Perspective*. Cambridge, MA: MIT Press, 2000.

Bryld, Mette, and Nina Lykke. *Cosmodolphins: Feminist Cultural Studies of Technology, Animals and the Sacred*. London: Zed Books, 1999.

Bryman, Alan. *The Disneyization of Society*. London: Sage, 2004.

Bunnell, Sterling. "The Evolution of Cetacean Intelligence." In *Mind in the Waters*, comp. Joan McIntyre.

Burnett, Alexander. *A Passion for Wildlife: The History of the Canadian Wildlife Service*. Vancouver: University of British Columbia Press, 2003.

Burnett, D. Graham. *The Sounding of the Whale: Science and Cetaceans in the Twentieth Century*. Chicago: University of Chicago Press, 2012.

Busch, Briton Cooper. *The War against the Seals: A History of the North American Seal Fishery*. Kingston and Montreal: McGill-Queen's University Press, 1985.

Carson, Rachel. *Silent Spring*. New York: Houghton-Mifflin, 1962.

Cassidy, Sean D. "Mind Bombs and Whale Songs: Greenpeace and the News." PhD diss., University of Oregon, 1992.

Chafer, Tony. "Politics and the Perception of Risk: A Study of the Anti-Nuclear Movements in Britain and France." *West European Politics* 8, no. 1 (1985): 10–27.

Chmielewski, Wendy. "Speak Truth to Power: Religion, Race, Sexuality, and Politics During the Cold War." Paper presented at Peace Movements in the Cold War and Beyond: An International Conference at the London School of Economics, UK, Feb. 1–2, 2008. Available at www.lse.ac.uk/Depts//global/PDFs/Peaceconference/Chmielewski.DOC.

Caron, Elsa, ed. *Fri Alert*. Dunedin, New Zealand: Caveman Press, 1974.

Clode, Danielle. *Killers in Eden: The Story of a Rare Partnership Between Men and Killer Whales*. Crows Nest, NSW, Australia: Allen and Unwin, 2002.

Cogan, Charles G. "From the Fall of France to the *Force de Frappe*: The Remaking of French Military Power, 1940–1962." In *The Fog of Peace and War Planning: Military Strategic Planning Under Uncertainty*, ed. Talbot C. Imlay and Monica Duffy Toft. New York: Routledge, 2006.

Cohen, Michael P. *The History of the Sierra Club, 1892–1970*. San Francisco: Sierra Club Books, 1988.

Coish, E. Calvin. *Season of the Seal: The International Storm over Canada's Seal Hunt*. St. John's, Newfoundland: Breakwater, 1979.

Commoner, Barry. *The Closing Circle: Nature, Man, and Technology*. New York: Knopf, 1971.

Cooney, Robert, and Helen Michalowski, eds. *The Power of the People: Active Nonviolence in the United States*. Philadelphia: New Society, 1987.

Coupland, Douglas. *Marshall McLuhan: You Know Nothing of My Work!* New York: Atlas, 2010.

Craige, Betty Jean. *Eugene Odum: Ecosystem Ecologist and Environmentalist*. Athens: University of Georgia Press, 2001.

———. *Laying Down the Ladder: The Emergence of Cultural Holism*. Amherst: University of Massachusetts Press, 1992.

Crane, Jeff, and Michael Egan, eds. *Natural Protest: Essays on the History of American Environmentalism*. New York: Routledge, 2009.

Dale, Stephen. *McLuhan's Children: The Greenpeace Message and the Media*. Toronto: Between the Lines, 1996.

Dallmayr, Fred. *Return to Nature?: An Ecological Counterhistory*. Lexington: University of Kentucky Press, 2011.

Dalton, Russell J. *The Green Rainbow: Environmental Groups in Western Europe*. New Haven, CT: Yale University Press, 1994.

Danielsson, Bengt, and Marie-Thérèse Danielsson. *Poisoned Reign: French Nuclear Colonialism in the Pacific*. Ringwood, Australia: Penguin Books, 1986.

Daston, Lorraine, and Gregg Mitman, eds. *Thinking with Animals: New Perspectives on Anthropomorphism*. New York: Columbia University Press, 2005.

Dauvergne, Paul, and Kate J. Neville. "Mindbombs of Right and Wrong: Cycles of Contention in the Activist Campaign to Stop Canada's Seal Hunt." *Environmental Politics* 20 (March 2011): 192–209.

Davies, Brian. *Savage Luxury: The Slaughter of Baby Seals*. New York: Taplinger, 1971.

Davis, L. E., R. E. Gallman, and K. Gleiter. *In Pursuit of Leviathan: Technology, Institutions, Productivity, and Profits in American Whaling, 1816–1906*. Chicago: University of Chicago Press, 1998.

Davis, Susan G. *Spectacular Nature: Corporate Culture and the Sea World Experience*. Berkeley: University of California Press, 1997.

Day, David. *The Whale War*. London and New York: Routledge and Kegan Paul, 1987.

DeBenedetti, Charles. *The Peace Reform in American History*. Bloomington and London: Indiana University Press, 1980.

Deese, R. S. "The New Ecology of Power: Julian and Aldous Huxley in the Cold War Era." In *Environmental Histories of the Cold War*, ed. J. R. McNeill and Corinna R. Unger. New York: Cambridge University Press, 2010.

Deloria, Philip J. "Counterculture Indians and the New Age." In *Imagine Nation: The American Counterculture of the 1960s and '70s*, ed. Peter Braunstein and Michael William Doyle. New York: Routlege, 2002.

———. *Playing Indian*. New Haven, CT: Yale University Press, 1998.

DeLuca, Kevin Michael. *Image Politics: The New Rhetoric of Environmental Activism*. New York and London: Guilford Press, 1999.

Devall, Bill. *Simple in Means, Rich in Ends: Practicing Deep Ecology*. Salt Lake City, UT: Peregrine Smith, 1988.

Devall, Bill and Sessions, George. *Deep Ecology: Living as if Nature Mattered*. Salt Lake City, UT: Peregrine Smith, 1985.

Dickerson, James. *North to Canada: Men and Women Against the Vietnam War*. Westport, CT: Praeger, 1999.

Disch, R. ed. *The Ecological Conscience*. Englewood Cliffs, NJ: Prentice-Hall, 1970.

Divine, Robert A. *Blowing on the Wind: The Nuclear Test Ban Debate, 1954–1960*. New York: Oxford University Press, 1978.

Donner, Frank J. *The Age of Surveillance: The Aims and Methods of America's Political Intelligence System*. New York: Knopf, 1980.

Dougan, J. L. "The Ice Lover: Biology of the Harp Seal." *Science* 215, no. 4535 (Feb. 19, 1982): 928–33.

Dowie, Mark. *Losing Ground: American Environmentalism at the Close of the Twentieth Century*. Cambridge, MA: MIT Press, 1995.

Doyle, Julie. "Picturing the Clima(c)tic: Greenpeace and the Representational Politics of Climate Change Communication." *Science as Culture* 16 (June 2007): 129–50.

Draper, Dianne Louise. "Eco-Activism: Issues and Strategies of Environmental Interest Groups in British Columbia." MA thesis, University of Victoria, 1972.

Dudzinski, Kathleen M., and Toni Frohoff. *Dolphin Mysteries: Unlocking the Secrets of Communication*. New Haven, CT: Yale University Press, 2008.

Dunlap, Thomas. *DDT: Scientists, Citizens, and Public Policy*. Princeton, NJ: Princeton University Press, 1981.

———. *Faith in Nature: Environmentalism as Religious Quest*. Seattle: University of Washington Press, 2004.

———. *Nature and the English Diaspora: Environment and History in the United States, Canada, Australia, and New Zealand*. Cambridge: Cambridge University Press, 1999.

Durland, Steven. "Witness: The Guerrilla Theater of Greenpeace." *High Performance* 40 (1987): 30–35.

Eayrs, James George. *Greenpeace and Her Enemies*. Toronto: Anansi Press, 1973.

Eder, Klaus. "The Rise of Counter-Culture Movements Against Modernity: Nature as a New Field of Class Struggle." *Theory, Culture and Society* 7 (1990): 13–35.

Editors of Ramparts, eds. *Eco-Catastrophe*. San Francisco: Canfield Press, 1970.

Edwards, Gordon. "Canada's Nuclear Industry and the Myth of the Peaceful Atom." In *Canada and the Nuclear Arms Race*, ed. Ernie Regehr and Simon Rosenblum. Toronto: James Lorimer, 1983.

Egan, Michael. *Barry Commoner and the Science of Survival: The Remaking of American Environmentalism*. Cambridge, MA: MIT Press, 2007.

Ellis, Richard. *The Empty Ocean: Plundering the World's Marine Life*. Washington, D.C.: Shearwater Books, 2003.

———. *Men and Whales*. New York: Alfred A. Knopf, 1991.

Ellul, Jacques. *The Technological Society*, trans. John Wilkinson. New York: Vintage, 1964.

Engels, Jens Ivo. "Modern Environmentalism." In *The Turning Points of Environmental History*, ed. Frank Uekoetter. Pittsburgh: University of Pittsburgh Press, 2010.

Environmental Action, eds. *Earth Day—The Beginning: A Guide for Survival*. New York: Bantam Books, 1970.

Epstein, Barbara. *Political Protest and Cultural Revolution: Nonviolent Direct Action in the 1970s and 1980s*. Berkeley: University of California Press, 1991.

Epstein, Charlotte. *The Power of Words in International Relations: Birth of an Anti-Whaling Discourse*. Cambridge, MA: MIT Press, 2008.

Farber, David. *Chicago '68*. Chicago: University of Chicago Press, 1988.

———. "The Intoxicated State/Illegal Nation: Drugs in the Sixties Counterculture." In *Imagine Nation: The American Counterculture of the 1960s and '70s*, ed. Peter Braunstein and Michael William Doyle. New York: Routlege, 2002.

Fichtelius, Karl-Erik, and Sverre Sjölander. *Smarter than Man? Intelligence in Whales, Dolphins, and Humans*. New York: Pantheon Books, 1972.

Fine, Gary Alan, and Randy Stoecker. "Can the Circle Be Unbroken? Small Groups and Social Movements." *Advances in Group Processes* 2 (1985): 1–28.

Firth, Stuart. *Nuclear Playground*. Honolulu: University of Hawaii Press, 1987.

Fitzpatrick, Elayne Wareing. *Doing It with the Cosmos: Henry Miller's Big Sur Struggle for Love Beyond Sex*. Bloomington, IN: Xlibris, 2001.

Fitzsimmons, Neal. "Brief History of American Civil Defense." In *Who Speaks of Civil Defense?* ed. Eugene P. Wigner. New York: Scribner, 1968.

Flader, Susan L. *Thinking Like a Mountain: Aldo Leopold and the Evolution of an Ecological Attitude toward Deer, Wolves, and Forests*. Madison: University of Wisconsin Press, 1974.

Flowers, Charles. "Between the Harpoon and the Whale." *New York Times Magazine*, Aug. 24, 1975.

Fortune Editors. *The Environment: A National Mission for the Seventies*. New York: Harper and Row, 1970.

Foster, Janet. *Working for Wildlife: The Beginning of Preservation in Canada*. Toronto: University of Toronto Press, 1978.

Fox, Stephen. *John Muir and His Legacy: The American Conservation Movement*. Boston: Little, Brown, 1981.

Fraser, Nicholas, Philip Jacobson, Mark Ottaway, and Lewis Chester. *Aristotle Onassis*. London: Weidenfeld and Nicholson, 1977.

Frost, Sidney. *The Whaling Question: The Inquiry by Sir Sidney Frost of Australia*. San Francisco: Friends of the Earth, 1979.

Fuller, R. Buckminster. *Operating Manual for Spaceship Earth*. New York: Dutton, 1968.

Gaskin, David. *The Ecology of Whales and Dolphins*. London and Exeter, NH: Heinemann, 1982.

Gerlach, Luther P. "The Structure of Social Movements: Environmental Activism and its Opponents." In *Waves of Protest: Social Movements Since the Sixties*, ed. Jo Freeman and Victoria Johnson. Lanham, MD: Rowman and Littlefield, 1999.

Gibson, James Williams. *A Reenchanted World: The Quest for a New Kinship with Nature*. New York: Holt Paperbacks, 2009.

Gidley, Isobelle, and Richard Shears. *The Rainbow Warrior Affair*. Sydney: Unwin Paperbacks, 1986.

Gillis, R. Peter, and Thomas R. Roach. *Lost Initiatives: Canada's Forest Industries, Forest Policy and Forest Conservation*. New York: Greenwood Press, 1986.

Gitlin, Todd. *The Sixties: Years of Hope, Days of Rage*. New York: Bantam, 1987.

Godlovitch, Stanley, Rosalind Godlovitch, and John Harris, eds. *Animals, Men and Morals: An Enquiry into the Maltreatment of Non-Humans*. New York: Taplinger, 1972.

Gofman, John, and Arthur Tamplin. *"Population Control" Through Nuclear Pollution*. Chicago: Nelson-Hall, 1970.

Golley, Frank. *A History of the Ecosystem Concept in Ecology: More than the Sum of the Parts*. New Haven, CT: Yale University Press, 1993.

Gottlieb, Robert. *Forcing the Spring: The Transformation of the American Environmental Movement*. Washington, D.C.: Island Press, 1993.

Granatstein, J. L. *Yankee Go Home: Canadians and Anti-Americanism*. Toronto: HarperCollins Canada, 1996.

Grant, George. *Lament for a Nation: The Defeat of Canadian Nationalism*. Toronto: McLelland and Stewart, 1965.

Gray, Earle. *The Great Uranium Cartel*. Toronto: McLelland and Stewart, 1982.

Gray, Timothy. *Gary Snyder and the Pacific Rim: Creating Counter-Cultural Community*. Iowa City: University of Iowa Press, 2006.

Greenpeace. *Greenpeace Worldwide*. http://www.greenpeace.org/international/en/about/worldwide/.

Greenpeace Deutschland. *Das Greenpeace Buch: Reflektionen und Aktionen*. Munich: Beck, 1997.

———. "30 Jahre, 1980–2010." Available at http://www.greenpeace.de/fileadmin/gpd/user_upload/wir_ueber_uns/erfolge_kampagnen/30JahreGP_Finale_inLitho.pdf.

———. *Volle Kraft voraus für die Umwelt: Greenpeace stellt sich vor—Aktionen, Themen, Lösungen*. Hamburg: Greenpeace e.V., 2001.

Gregg, Richard. *The Power of Nonviolence*. Philadelphia: J. B. Lippincott, 1934.

Griefahn, Monika. (Porträtiert von Jürgen Streich). *Monika Griefahn: Politik, Positionen, Perspektiven*. Cologne: Zebulon Verlag, 1997.

Grier, Katherine C. *Pets in America: A History*. Chapel Hill: University of North Carolina Press, 2006.

Guevara, Gloria Yolanda. "Assessing the Effectiveness of Transnational Activism: An Analysis of the Anti-Whaling and Anti-Sealing Campaigns." PhD diss., University of Southern California, 2008.

Guha, Ramachandra. *Environmentalism: A Global History*. New York: Longman, 2000.

Hagen, Joel. *An Entangled Bank: The Origins of Ecosystem Ecology*. New Brunswick, NJ: Rutgers University Press, 1992.

———. "Teaching Ecology During the Environmental Age, 1965–1980." *Environmental History* 13 (Oct. 2008): 704–23.

Hagen, John. *Northern Passage: American Vietnam War Resisters in Canada*. Cambridge, MA: Harvard University Press, 2001.

Haig-Brown, Allan. *Hell No We Won't Go: Vietnam Draft Resisters in Canada*. Vancouver: Raincoast Books, 1996.

Haig-Brown, Roderick. *Measure of the Year*. New York: Morrow, 1950.

Hansen, Anders. "Greenpeace and the Press Coverage of Environmental Issues." In *The Mass Media and Environmental Issues*, ed. Anders Hansen. Leicester: Leicester University Press, 1993.

Haq, Gary, and Alistair Paul. *Environmentalism Since 1945*. New York: Routlege, 2012.

Hardin, Garrett. "The Tragedy of the Commons." *Science* 162 (Dec. 13, 1968): 1244–49.

Harter, John-Henry. "Environmental Justice for Whom? Class, New Social Movements and the Environment: A Case Study of Greenpeace Canada, 1971–2000." *Labour/Le Travail* 54 (Fall 2004): 83–119.

Harvey, Mark W. T. *A Symbol of Wilderness: Echo Park and the American Conservation Movement*. Albuquerque: University of New Mexico Press, 1994.

Hayes, Peter. "France Bombs Mururoa, India Bombs India." *Not Man Apart*, June 1974.

Hays, Samuel P. *Beauty, Health, and Permanence: Environmental Politics in the United States, 1955–1985*. New York: Cambridge University Press, 1987.

———. *Conservation and the Gospel of Efficiency: The Progressive Conservation Movement, 1880–1920*. Cambridge, MA: Harvard University Press, 1959.

Hecht, Gabrielle. *The Radiance of France: Nuclear Power and National Identity after World War II*. Cambridge, MA: MIT Press, 1998.

Heimann, Rolf. *Knocking on Heaven's Door*. Melbourne: Friends of the Earth, 1979.

Heinz, Bettina, Hsin-I Cheng, and Ako Inuzuka, "Greenpeace Greenspeak: A Transcultural Discourse Analysis." *Language and Intercultural Communication* 7, no. 1 (2007): 16–36.

Helvarg, David. *The War Against the Greens: The "Wise Use" Movement, the New Right, and Anti-Environmental Violence*. San Francisco: Sierra Club Books, 1994.

Hemsing, Natalie Joan. "Production of Place: Community, Conflict and Belonging at Wreck Beach." MA thesis, University of British Columbia, 2005.

Henke, Janice Scott. "Canada and the WTO in the 21st Century." *Sustainable eNews*, Jan. 2008. Available at http://www.maninnature.com/MMammals/Seals/Seals1o.html.

———. *Seal Wars!: An American Viewpoint*. St. John's, Newfoundland: Breakwater Books, 1985.

Henningham, Stephen. *France and the South Pacific: A Contemporary History*. Sydney: Allen and Unwin, 1992.

Herron, Matt. "A Not-Altogether Quixotic Face-Off with Soviet Whale Killers in the Pacific." *Smithsonian* 7, no. 5 (Aug. 1976): 22-31.

Hodgins, Bruce W., John J. Eddy, Grant Shelagh, and James Atchison. "Dynamic Federalism: Continuity and Change since World War One." In *Federalism in Canada and Australia: Historical Perspectives, 1920–1988*, ed. Bruce W. Hodgins, John J. Eddy, Grant Shelagh, and James Struthers. Peterborough: Trent University, 1989.

Hoffman, Abbie. *Soon to Be a Major Motion Picture*. New York: G. P. Putnam's Sons, 1980.

Hook, Diana ffarington [*sic*]. *The I Ching and Mankind*. London: Routledge and Kegan Paul, 1975.

Horowitz, Gad. "Conservatism, Liberalism, Socialism: An Interpretation." *Canadian Journal of Economics and Political Science* 32, no. 2 (June 1966): 143–71.

———. "On the Fear of Nationalism." *Canadian Dimension* 5, no. 6 (1967): 8–9.

Horwood, Harold. "Tragedy on the Whelping Ice." *Canadian Audubon*, Mar.–Apr. 1960, 37–41.

Hunter, Robert. *The Enemies of Anarchy*. Toronto: McClelland and Stewart, 1970.

———. *Erebus*. Toronto: McClelland and Stewart, 1968.

———. "Fighting Back." *Vancouver Magazine*. June 1980, 57–66.

———. *Red Blood: One (Mostly) White Guy's Encounters with the Native World*. San Francisco: Sierra Club Books, 1999.

———. *The Storming of the Mind*. Toronto: McClelland and Stewart, 1971.

———. *Warriors of the Rainbow: A Chronicle of the Greenpeace Movement*. New York: Holt, Rinehart and Winston, 1979.

Hunter, Robert, with photographs by Robert Keziere. *The Greenpeace to Amchitka: An Environmental Odyssey*. Vancouver: Arsenal Pulp Press, 2004.

Hunter, Robert, and Rex Weyler. *To Save a Whale: The Voyages of Greenpeace*. San Francisco: Chronicle Books, 1978.

Inglehart, Ronald. *Culture Shift in Advanced Industrial Societies*. Princeton, NJ: Princeton University Press, 1990.

Isitt, Benjamin. *Militant Minorities: British Columbia Workers and the Rise of a New Left, 1948–1972*. Toronto: University of Toronto Press, 2011.

Ivakhiv, Adrian. *Claiming Sacred Ground: Pilgrims and Politics at Glastonbury and Sedona*. Bloomington: Indiana University Press, 2001.

Jacobs, Paul. "The Coming Atomic Blast in Alaska." *New York Review*, July 22, 1971, 34–35.

Jamieson, Dale, and Tom Regan. "Whales Are Not Cetacean Resources." In *Advances in Animal Welfare Science*, ed. M. W. Fox and L. D. Mickley. Boston: Martinus Nijhoff, 1984.

Janssen, Peter R. "The Age of Ecology." In *Ecotactics: The Sierra Club Handbook for Environmental Activists*, ed. John G. Mitchell. New York: Simon and Schuster, 1970.

Johnson, Christina S. "Jumbo Squid Following Low-Oxygen Zone." *Sea Grant California*, Jan. 28, 2008. Available at http://www-csgc.ucsd.edu/NEWSROOM/NEWSRELEASES/2008/jumbosquid.html.

Johnstone, Diana. "How the French Left Learned to Love the Bomb." *New Left Review* 146 (July/Aug., 1984): 5–36.

Jordan, Grant. *Shell, Greenpeace and the Brent Spar*. New York: Palgrave, 2001.

Jordan, Grant, and William A. Maloney. *The Protest Business?: Mobilizing Campaign Groups*. Manchester: Manchester University Press, 1997.

Joseph, Lawrence E. *Gaia: The Growth of an Idea*. New York: St. Martin's Press, 1990.

Kahn, Miriam. *Tahiti Beyond the Postcard: Power, Place, and Everyday Life*. Seattle: University of Washington Press, 2011.

Kaiser, Rudolf. "Chief Seattle's Speech(es): American Origins and European Reception." In *Recovering the Word: Essays on Native American Literature*, ed. Brian Swann and Arnold Krupat. Berkeley: University of California Press, 1987.

Kalland, Arne. *Unveiling the Whale: Discourses on Whales and Whaling*. New York: Berghahn, 2008.

Kalland, Arne, and Brian Moeran. *Japanese Whaling: End of an Era?* London: Curzon Press, 1992.

Keeling, Arn. "Ecological Ideas in the British Columbian Conservation Movement, 1945–1970." MA thesis, University of British Columbia, 1999.

———. "Sink or Swim: Water Pollution and Environmental Politics in Vancouver, 1889–1975." *BC Studies* 142–143 (Summer-Autumn 2004): 69–101.

Kellogg, Remington. "Whales, Giants of the Sea." *National Geographic* 77(Jan. 1940): 35–90.

Kelsey, Elin. *Watching Giants: The Secret Lives of Whales*. Berkeley: University of California Press, 2009.

Kesey, Ken. "The I Ching." *The Realist* 89 (Mar.–Apr., 1971). (Supplement in the *Whole Earth Catalog*). Available at: http://www.ep.tc/realist/89/11.html.

Keziere, Robert, and Robert Hunter. *Greenpeace*. Toronto: McClelland and Stewart, 1972.

Khatchadourian, Raffi. "Neptune's Navy: Paul Watson's Wild Crusade to Save the Oceans." *New Yorker*, Nov. 5, 2007.

King, Michael. *Death of the Rainbow Warrior*. Auckland: Penguin, 1986.

Kingsland, Sharon. *The Evolution of American Ecology, 1890–2000*. Baltimore, MD: Johns Hopkins University Press, 2005.

Kinkela, David. *DDT and the American Century: Global Health, Environmental Politics, and the Pesticide That Changed the World*. Chapel Hill: University of North Carolina Press, 2011.

Kirk, Andrew G. *Counterculture Green: The Whole Earth Catalog and American Environmentalism*. Lawrence: University Press of Kansas, 2007.

Kline, Benjamin. *First Along the River: A Brief History of the U.S. Environmental Movement*. 4th ed. Lanham, MD: Rowman and Littlefield, 2011.

Knappe, Bernhard. *Das Geheimnis von Greenpeace: Die Andere Seit Der Erfolgstory*. Vienna: Orac, 1993.

Kohlhoff, Dean. *Amchitka and the Bomb: Nuclear Testing in Alaska*. Seattle: University of Washington Press, 2002.

———. *When the Wind Was a River: Aleut Evacuation During World War II*. Seattle: University of Washington Press, 1995.

Kool, V. K. *The Psychology of Nonviolence and Aggression*. New York: Palgrave Macmillan, 2008.

Kosek, Joseph Kip. *Acts of Conscience: Christian Nonviolence and Modern American Democracy*. New York: Columbia University Press, 2009.

———. "Richard Gregg, Mohandas Gandhi, and the Strategy of Nonviolence." *Journal of American History* 91 (Mar. 2005): 1318–48.

Kostash, Myrna. *Long Way from Home: The Story of the Sixties Generation in Canada*. Toronto: James Lorimer, 1980.

Krech, Shepard III. *The Ecological Indian: Myth and History*. New York: Norton, 1999.

Kricher, John. *The Balance of Nature: Ecology's Enduring Myth*. Princeton, NJ: Princeton University Press, 2009.

Kriesi, Hanspeter, Ruud Koopmans, Jan Willem Dyvendak, and Marco G. Guigni. *New Social Movements in Western Europe: A Comparative Analysis*. Minneapolis: University of Minnesota Press, 1995.

Kripal, Jeffrey J. *Esalen: America and the Religion of No Religion*. Chicago: University of Chicago Press, 2007.

Krüger, Christian,and Mattias Müller-Henig, eds. *Greenpeace auf dem Wahrnehmungsmarkt: Studien zur Kommunikationspolitik und Medienresonanz*. Hamburg: Bureau für Publizistik, 2001.

Kurlansky, Mark. *Nonviolence: Twenty-Five Lessons from the History of a Dangerous Idea*. New York: Modern Library, 2006.

Lahusen, Christian. *The Rhetoric of Moral Protest: Public Campaigns, Celebrity Endorsement, and Political Mobilization*. Berlin and New York: Walter de Gruyter, 1996.

Lamb, David. "Animal Rights and Liberation Movements." *Environmental Ethics* 4 (Fall 1982): 215–33.

Lamb, Robert. *Promising the Earth*. London and New York: Routledge, 1996.

Laughlin, William S. *Aleuts: Survivors of the Bering Land Bridge*. New York: Rinehart and Winston, 1980.

Lavigne, David M. "Estimating Total Kill of Northwest Atlantic Harp Seals." *Marine Mammal Science* 15 (1999): 871–78.

———. "Harp Seal: *Pagophilus groenlandicus*." In *Encyclopedia of Marine Mammals*, eds. William F. Perrin, Bernd G. Würsig, and J. G. M. Thewissen. 2nd ed. New York: Academic Press, 2008.

———. "Life or Death for the Harp Seal." *National Geographic*, Jan. 1976, 128–42.

Lear, Linda J. *Rachel Carson: Witness for Nature*. New York: Henry Holt, 1997.

Lee, Martha F. *Earth First!: Environmental Apocalypse*. Syracuse, NY: Syracuse University Press, 1995.

Leopold, Aldo. *Sand County Almanac*. 1948; repr., New York: Ballantine, 1970.

———. "Some Fundamentals of Conservation in the Southwest" [1923] in *The River of the Mother of God and Other Essays by Aldo Leopold*, ed. Susan L. Flader and J. Baird Callicott. Madison: University of Wisconsin Press, 1991.

Leviatin, David. *Followers of the Trail: Jewish Working-Class Radicals in America*. New Haven, CT: Yale University Press, 1989.

Levitt, Kari. *Silent Surrender: The Multinational Corporation in Canada*. Toronto: Macmillan, 1970.

Lilly, John C. *Communication Between Man and Dolphin: The Possibilities of Talking with Other Species*. New York: Crown, 1978.

———. *The Deep Self: Profound Relaxation and the Tank Isolation Technique*. New York: Simon and Schuster, 1977.

———. "A Feeling of Weirdness." In *Mind in the Waters*, comp. Joan McIntyre.

———. *Man and Dolphin*. New York: Pyramid, 1961.

———. *The Mind of the Dolphin: A Non-Human Intelligence*. Garden City, NY: Doubleday , 1967.

Lilly, John C., and E. J. Gold. *Tanks for the Memories: Flotation Tank Talks*. Nevada City, CA: Gateways Books and Tapes, 1996.

Linton, R. M. *Terracide—America's Destruction of Her Living Environment*. Boston: Little, Brown, 1970.

Locke, Elsie. *Peace People: A History of Peace Activities in New Zealand*. Christchurch and Melbourne: Hazard Press, 1992.

Loo, Tina. *States of Nature: Conserving Canada's Wildlife in the Twentieth Century*. Vancouver: University of British Columbia Press, 2006.

Lovejoy, Arthur O. *The Revolt Against Dualism: An Inquiry Concerning the Existence of Ideas*. Chicago: Open Court, 1930.

Lowe, George D., and T.K. Pinhey. "Rural-Urban Differences in Support for Environmental Protection." *Rural Sociology* 47, no. 1 (1982): 114–28.

Luccioni, Xavier. *L'Affaire Greenpeace: Une Geurre des Médias*. Paris: Payot, 1986.

Lurz, Robert W. *Mindreading Animals: The Debate over What Animals Know about Other Minds*. Cambridge, MA: MIT Press, 2011.

Lust, Peter. *The Last Seal Pup: The Story of Canada's Seal Hunt*. Montreal: Harvest House, 1967.

Lutts, Ralph H. "Chemical Fallout: Rachel Carson's *Silent Spring*, Radioactive Fallout, and the Environmental Movement." *Environmental Review* 9 (Fall 1985): 210–25.

Lytle, Mark Hamilton. *The Gentle Subversive: Rachel Carson,* Silent Spring, *and the Rise of the Environmental Movement*. New York: Oxford University Press, 2007.

McBride, Arthur. "Evidence for Echolocation by Cetaceans." *Deep-Sea Research* 3 (1956): 153–54.

———. "Meet Mister Porpoise." *Natural History* 45(Jan. 1940): 16–29.

McCarthy, John D., and Mayer N. Zald. "Resource Mobilization and Social Movements: A Partial Theory." *American Journal of Sociology* 82 (May 1977): 1212–41.

McCloskey, J. Michael. *In the Thick of It: My Life in the Sierra Club*. Washington, D.C.: Island Press, 2005.

McCormick, John. *Reclaiming Paradise: The Global Environmental Movement*. Bloomington: Indiana University Press, 1989.

McEvilly, Wayne. "Synchronicity and the I Ching." *Philosophy East and West* 18 (July 1968): 137–49.

McGarity, Thomas O. "Our Science is Sound Science and Their Science is Junk Science: Science-Based Strategies for Avoiding Accountability and Responsibility for Risk-Producing Products and Activities." *Kansas Law Review* 52, no. 4 (2004): 897–937.

McIntyre, Joan. "An Invitation." In *Mind in the Waters.*

———. Comp. *Mind in the Waters: A Book to Celebrate the Consciousness of Whales and Dolphins.* New York and San Francisco: Charles Scribner's Sons and Sierra Club Books, 1974.

———. "Mind Play." In *Mind in the Waters.*

McLuhan, Marshall. *The Medium Is the Message.* New York: Bantam Books, 1967.

———. *Understanding Media.* New York: McGraw-Hill, 1964.

McTaggart, David. *Outrage!: The Ordeal of Greenpeace III.* Vancouver: J. J. Douglas, 1973.

McTaggart, David, with Robert Hunter. *Greenpeace III: Journey into the Bomb.* London: Collins, 1978.

McTaggart, David, with Helen Slinger. *Rainbow Warrior: Ein Leben gegen alle Regeln.* Munich: Riemann, 2001.

McTaggart, David, with Helen Slinger. *Shadow Warrior: The Autobiography of Greenpeace International Founder David McTaggart.* London: Orion, 2002.

McVay, Scott. "Can Leviathan Long Endure So Wide a Chase?" *Natural History* 80, no. 1 (1971): 36–40, 68–72.

———. "The Last of the Great Whales." *Scientific American* 215, no. 2 (Aug. 1966): 13–21.

———. "One Strand in the Rope of Concern." In *Mind in the Waters,* comp. Joan McIntyre.

———. "Reflections on the Management of Whaling." In *The Whale Problem,* ed. W. E. Schevill. Cambridge, MA: Harvard University Press, 1974.

———. "Stalking the Arctic Whale." *American Scientist* 61, no. 1 (1973): 24–37.

Macdonald, Dwight. *Politics Past: Essays in Political Criticism.* New York: Viking Press, 1970.

Macpherson, Kay, and Meg Sears. "The Voice of Women: A History." In *Women in the Canadian Mosaic,* ed. Gwen Matheson. Toronto: Peter Martin, 1976.

Macy, Michael. "Greenpeace: An Analysis." *Communication Arts,* Mar./Apr., 1980, 67–71.

Maddison, Sarah, and Sean Scalmer. *Activist Wisdom: Practical Knowledge and Creative Tension in Social Movements.* Sydney: University of New South Wales Press, 2006.

Manes, Christopher. *Green Rage: Radical Environmentalism and the Unmaking of Civilization.* Boston: Little, Brown, 1990.

Marchak, Patricia. "History of a Resource Industry." In *A History of British Columbia: Selected Readings,*ed. Patricia Roy. Toronto: Copp Clark Pitman, 1989.

Marchand, Philip. *Marshall McLuhan: The Medium and the Messenger.* New York: Ticknor and Fields, 1989.

Marcuse, Herbert. *One-Dimensional Man: Studies in the Ideology of Advanced Industrial Society.* Boston: Beacon Press, 1966.

Matusow, Allen J. *The Unraveling of America: A History of Liberalism in the 1960s.* New York: Harper and Row, 1984.

May, Elaine Tyler. *Homeward Bound: American Families in the Cold War Era.* New York: Basic Books, 1988.

Meine, Curt. *Aldo Leopold: His Life and Work.* Madison: University of Wisconsin Press, 2010.

Mendes-Flohr, Paul. *Divided Passions: Jewish Intellectuals and the Experience of Modernity.* Detroit, MI: Wayne State University Press, 1991.

Merle, Robert. *The Day of the Dolphin.* New York: Simon and Schuster, 1969.

Metcalfe, E. Bennett. *A Man of Some Importance: The Life of Roderick Langmere Haig-Brown.* Vancouver and Seattle: James W. Wood, 1985.

Meyer, David S., and Sidney Tarrow. "A Movement Society: Contentious Politics for a New Century." In *The Social Movement Society: Contentious Politics for a New Century,* ed. David S. Meyer and Sidney Tarrow. Lanham, MD: Rowman and Littlefield, 1998.

Miller, Char. *Gifford Pinchot and the Making of Modern Environmentalism.* Washington, D.C.: Island Press/Shearwater Books, 2001.

Miller, G. Tyler Jr. *Replenish the Earth: A Primer in Human Ecology.* Belmont, CA: Wadsworth, 1972.

Miller, Pam, and Norman Buske. *Nuclear Flashback: Report of a Greenpeace Scientific Expedition to Amchitka Island, Alaska—Site of the Largest Underground Nuclear Test in U.S. History.* Greenpeace USA: Oct. 30, 1996.

Mines, S. *The Last Days of Mankind*. New York: Simon and Schuster, 1971.

Mitcalfe, Barry. *Boy Roel: Voyage to Nowhere*. Martinborough, NZ: Alister Taylor, 1972.

———. "Why I Am Sailing into the Fallout Area." *New Zealand Monthly Review* 133 (Aug. 1972): 9–10.

Mitchell, Alanna. "The 'Mindbomb' That Was Greenpeace." *Literary Review of Canada* (Dec. 2004): 13–15.

Mitchell, David J. *W.A.C. Bennett and the Rise of British Columbia*. Vancouver: Douglas and MacIntyre, 1983.

Mitman, Gregg. *The State of Nature: Ecology, Community, and American Social Thought, 1900–1950*. Chicago: University of Chicago Press, 1992.

Mollin, Marian. *Radical Pacifism in Modern America: Egalitarianism and Protest*. Philadelphia: University of Pennsylvania Press, 2006.

Molotch, Harvey. "Santa Barbara: Oil in the Velvet Playground." In *Eco-Catastrophe*, ed. Editors of Ramparts.

Monbiot, George. *Heat: How to Stop the Planet from Burning*. New York: South End Press, 2007.

Moore, Patrick. "The Administration of Pollution Control in British Columbia: A Focus on the Mining Industry." PhD diss., University of British Columbia, 1973.

———. *Confessions of a Greenpeace Dropout: The Making of a Sensible Environmentalist*. Vancouver: Beatty Street, 2010.

Moore, Steve. "The I Ching in Time and Space." In *I Ching: An Annotated Bibliography*, ed. Edward A. Hacker, Steve Moore, and Lorraine Patsco. New York: Routledge, 2002.

Moreby, Christopher. "What Whaling Means to the Japanese." *New Scientist* 9 (Dec. 1982): 661–63.

Morris, Desmond. *The Human Zoo*. New York: McGraw-Hill, 1969.

Mowat, Farley. *A Whale for the Killing*. London: Heinemann, 1972.

Mowat, Farley, and David Blackwood. *Wake of the Great Sealers*. Toronto: McClelland and Stewart, 1973.

Naess, Arne. "The Shallow and the Deep, Long-Range Ecology Movements: A Summary." *Inquiry* 16 (Spring 1973): 95–100.

Nash, Roderick F. *The Rights of Nature: A History of Environmental Ethics*. Madison: University of Wisconsin Press, 1989.

———. *Wilderness and the American Mind*. 3rd. ed. New Haven, CT: Yale University Press, 1982.

Newborn, Susi. *A Bonfire in My Mouth: Life, Passion and the Rainbow Warrior*. Auckland: HarperCollins New Zealand, 2003.

Norton, Bryan. *Toward Unity Among Environmentalists*. New York: Oxford University Press, 1991.

Nutter, Kathleen Banks. "Jessie Lloyd O'Connor and Mary Metlay Kaufman: Professional Women Fighting for Social Justice." *Journal of Women's History* 14 (Summer 2002): 132–35.

O'Connor, Harvey. *Revolution in Seattle: A Memoir*. New York: Monthly Review Press: 1964.

Odum, Eugene. *The Fundamentals of Ecology*. Philadelphia, W. B. Saunders, 1953.

———. "The New Ecology." *BioScience* 14 (July 1964): 7–41.

Ohm, Viveca. "Greenpeace." *Northword*, July 1971.

Olafsson, Haraldur. "A True Environmental Parable: The Laxa-Myvatn Conflict in Iceland, 1965–1973." *Environmental Review* 5(Autumn 1981): 2–38.

Palmer, Bryan D. *Canada's 1960s: The Ironies of Identity in a Rebellious Era*. Toronto: University of Toronto Press, 2009.

Pash, Chris. *The Last Whale*. Fremantle, Australia: Fremantle Press, 2008.

Pauls, Naomi, and Charles Campbell, eds. *The Georgia Straight: What the Hell Happened?* Vancouver: Douglas and McIntyre, 1997.

Payne, Roger S., and Scott McVay. "Songs of the Humpback Whales." *Science* 173 (Aug. 13, 1971): 585–91.

Peace, Roger C. III. *A Just and Lasting Peace: The U.S. Peace Movement from the Cold War to Desert Storm*. Chicago: Noble Press, 1991.

Peck, Abe. *Uncovering the Sixties: The Life and Times of the Underground Press*. New York: Pantheon, 1985.

Pelligrini, Charles. *Greenpeace, la Manipulation*. Paris: Carriére, 1995.

Perls, Frederick S. *Gestalt Therapy Verbatim*. La Fayette, CA: Real People Press, 1969.

———. *In and Out of the Garbage Pail*. Lafayette, CA: Real People Press, 1969.

Perls, Frederick, Ralph Hefferline, and Paul Goodman. *Gestalt Therapy: Excitement and Growth in the Human Personality*. New York: Dell, 1951.

Peterson del Mar, David. *Environmentalism*. Harlow, UK: Pearson, 2006.

Phillips, Dana. *The Truth of Ecology: Nature, Culture, and Literature in America*. New York: Oxford University Press, 2003.

Phillips, Jim. *Raising Kane: The Fox Chronicles*. Rices Landing, PA: Kindred Spirits Press, 1999.

Phillipson, Anthony. "Dr. John Gofman, a Nuclear Researcher Who Refuses to Lie About Radiation Dangers." *Nuclear Guardianship Forum on the Responsible Care of Radioactive Materials*, 3 (Spring 1994). Available at http://www.ratical.org/radiation/NGP/DrJohnGofman.html.

Picaper, Jean-Paul, and Thibaud Dornier. *Greenpeace: l'écologie à l'an vert*. Paris: Editions Odilon Media, 1995.

Pierce, Christine. "Can Animals Be Liberated?" *Philosophical Studies* 36 (July 1979): 69–75.

Pinchot, Gifford. "Whale Culture—A Proposal." *Perspectives in Biology and Medicine* 10 (Autumn 1966): 33–43.

Poole, Robert. *Earthrise: How Man First Saw the Earth*. New Haven, CT: Yale University Press, 2010.

Porter, John. *The Vertical Mosaic: An Analysis of Social Class and Power in Canada*. Toronto: University of Toronto Press, 1965.

Prendiville, Brendan. *Environmental Politics in France*. Boulder, CO: Westview Press, 1994.

Princen, Thomas, and Matthias Finger. *Environmental NGOs and World Politics: Linking the Local and the Global*. New York: Routledge, 1994.

Qureshi, Yasmeen. "Environmental Issues in British Columbia: An Historical Geographical Perspective." MA thesis, University of British Columbia, 1991.

Rainbow, Stephen. "New Zealand's Values Party: The Rise and Fall of the First National Green Party." In *Environmental Politics in Australia and New Zealand*, ed. Peter Hay, Robyn Eckersley, and Geoff Holloway. Hobart: University of Tasmania Centre for Environmental Studies, 1989.

Read, Jennifer. "'Let Us Heed the Voice of Youth': Laundry Detergents, Phosphates and the Emergence of the Environmental Movement in Ontario." *Journal of the Canadian Historical Association* 9 (1996): 227–50.

Regan, Tomo. "Holism as Environmental Fascism." In *Contemporary Moral Problems*, ed. James E. White. St. Paul, MN: West, 1985.

Reich, Charles. *The Greening of America*. New York: Random House, 1970.

Reiss, Diana. *The Dolphin in the Mirror: Exploring Dolphin Minds and Saving Dolphin Lives*. New York: Houghton Mifflin Harcourt, 2011.

Reiss, Jochen. *Greenpeace: der Umweltmulti: Sein Apparat, Seine Aktionen*. Rheda-Wiedenbrück: Daedalus Verlag, 1988.

Resnick, Phillip. *The Land of Cain: Class and Nationalism in English Canada, 1945–1975*. Vancouver: New Star Books, 1977.

Reynolds, Earle. *The Forbidden Voyage*. New York: David McKay, 1961.

———. "Irradiation and Human Evolution." *Human Biology* 32 (Feb. 1960): 89–108.

Ritvo, Harriet. *The Animal Estate: The English and Other Creatures in the Victorian Age*. Cambridge, MA: Harvard University Press, 1987.

Robin, Martin. *The Company Province: The Rush for Spoils, 1871–1933*. Toronto: McClelland and Steward, 1972.

———. *The Company Province: Pillars of Profit, 1934–1972*. Toronto: McClelland and Stewart, 1973.

Rodman, John. "The Liberation of Nature?" *Inquiry* 20 (Spring 1977): 94–101.

Roman, Joe. *Whale*. London: Reaktion Books, 2006.

Rome, Adam. "The Genius of Earth Day." *Environmental History* 15 (Apr. 2010): 194–205.

———. "'Give Earth a Chance': The Environmental Movement and the Sixties." *Journal of American History* 90 (Sept. 2003): 525–54.

———. "Nature Wars, Culture Wars: Immigration and Environmental Reform in the Progressive Era." *Environmental History* 13 (July 2008): 432–53.

Ronald, K. and Dougan, J. L. "The Ice Lover: Biology of the Harp Seal *(Phoca groenlandica)*." *Science* 215 (Feb. 19, 1982): 928–33.

Roos, Leslie J. Jr., ed. *The Politics of Ecosuicide.* New York: Holt, Rinehart and Winston, 1971.

Ross, Nancy Wilson, ed. *The World of Zen: An East-West Anthology*. New York: Vintage, 1960.

Roszak, Theodore. *Where the Wasteland Ends: Politics and Transcendence in a Postindustrial Society*. Garden City, NY: Doubleday, 1972.

———. *The Making of a Counter Culture: Reflections on the Technocratic Society and Its Youthful Opposition.* Garden City, NY: Doubleday, 1969.

Rothenberg, David. *Thousand Mile Song: Whale Music in a Sea of Sound.* New York: Basic Books, 2008.

Rothman, Hal K. *The Greening of a Nation?: Environmentalism in the United States Since 1945.* Fort Worth, TX: Harcourt Brace, 1998.

———. *Saving the Planet: The American Response to the Environment in the Twentieth Century.* Chicago: Ivan R. Dee, 2000.

Rothman, Stanley, and S. Robert Lichter, eds. *Roots of Radicalism: Jews, Christians, and the Left.* New York: Oxford University Press, 1982.

Rubin, Jerry. *Do It!* New York: Ballantine Books, 1970.

Rucht, Dieter. "Ecological Protest as Calculated Law-Breaking: Greenpeace and Earth First! in Comparative Perspective." In *Green Politics Three*, ed. Wolfgang Rüdig. Edinburgh: Edinburgh University Press, 1995.

Ruddock, Joan. *CND Scrapbook.* London: Macdonald, 1987.

Ryan, Shannon, assisted by Martha Drake. *Seals and Sealers: A Pictorial History of the Newfoundland Seal Fishery*. St. John's, Newfoundland: Breakwater Books, 1987.

Sagoff, Mark. "Animal Liberation and Environmental Ethics: Bad Marriage, Quick Divorce." *Osgood Hall Law Journal* 22 (1984): 297–307.

———. "On Preserving the Natural Environment." *Yale Law Review* 84 (Dec. 1974): 220–69.

Sale, Kirkpatrick. *The Green Revolution: The American Environmental Movement, 1962–1992.* New York: Hill and Wang, 1993.

Saler, Michael. "Modernity and Enchantment: A Historiographic Review." *American Historical Review* 111 (June 2006): 692–716.

Sanders, Ed. *Tales of Beatnik Glory*. New York: Stonehill, 1975.

Scalmer, Sean. *Gandhi in the West: The Mahatma and the Rise of Radical Protest.* New York: Cambridge University Press, 2011.

Scarce, Rik. *Eco-Warriors: Understanding the Radical Environmental Movement.* Chicago: Noble Press, 1990.

Scheffer, Victor. "The Case for a World Moratorium on Whaling." In *Mind in the Waters,* comp. Joan McIntyre.

Schrader, Gene. "Atomic Doubletalk: A Review of Matters withheld from the Public by the Atomic Energy Commission." *The Center Magazine*, Jan./Feb. 1971, 29–52.

Schumacher, E. F. *Small Is Beautiful: Economics as if People Mattered.* New York: Harper, 1973.

Sears, Paul. "Ecology—A Subversive Subject." *BioScience* 14 (July 1964): 11–13.

Sergeant, D. E. "Harp Seals and the Sealing Industry." *Canadian Audubon* 25 (1963): 29–35.

Sessions, George. "Deep Ecology and the New Age." *Earth First!* Sept. 23, 1987, 27–30.

———. "The Deep Ecology Movement: A Review." *Environmental Review* 11 (Summer 1987): 105–25.

Shadbolt, Maurice. *Danger Zone*. London: Hodder and Stoughton, 1975.

Shaeffer, Francis. *Pollution and the Death of Man*. Wheaton, IL: Tyndale House, 1970.

Shaheen, Jack G., ed. *Nuclear War Films*. Carbondale: Southern Illinois University Press, 1978.

Shaiko, Ronald G. "Greenpeace U.S.A.: Something Old, New, Borrowed." *Annals of the American Academy of Political and Social Science* 528 (July 1993): 88–100.

Sharp, Gene. *The Politics of Nonviolent Action*. Boston: Extending Horizons Books, Porter Sargent, 1973.

Shepard, Paul. "Introduction: Ecology and Man—a Viewpoint." In *The Subversive Science: Essays Toward an Ecology of Man*, ed. Paul Shepard and Daniel McKinley. New York: Houghton-Mifflin, 1969.

Shoemaker, Nancy. "Whale Meat in American History." *Environmental History* 10 (April 2005): 269–94.

Sideris, Lisa H. "The Secular and Religious Sources of Rachel Carson's Sense of Wonder." In *Rachel Carson: Legacy and Challenge*, ed. Lisa H. Sideris and Kathleen Dean Moore. Albany: State University of New York Press, 2008.

Simmons, Terry. "Poverty of Plenty: Conservation in British Columbia." *Conifer* 16, no.1 (1970): 8–17.

Singer, Peter. "All Animals Are Equal." *Philosophical Exchange* 1 (Summer 1974): 103–16.

———. "Animal Liberation." *New York Review of Books*, April 5, 1973, 17–18.

———. *Animal Liberation: A New Ethics for Our Treatment of Animals*. New York: Random House, 1975.

———. *Pushing Time Away: My Grandfather and the Tragedy of Jewish Vienna*. New York: Harper-Collins, 2002.

Sluga, Glenda. "UNESCO and the (One) World of Julian Huxley." *Journal of World History* 21 (Sept. 2010): 251–74.

Smith, Bradley F. *The Shadow Warriors: OSS and the Origins of the CIA*. New York: Basic Books, 1983.

Snyder, Gary. *Earth House Hold: Technical Notes and Queries to Fellow Dharma Revolutionaries*. New York: New Directions, 1969.

Soderberg, Steffan, and Lennart Almkvist. *In Prospect of the Seal Hunt in Canada 1977*. Stockholm: Swedish Museum of Natural History, 1977.

SPEC, "SPEC's Beginnings." Available at http://www.vcn.bc.ca/spec/spec/Spectrum/spring1999/beginframeset.htm (accessed 11/23/11).

Spencer, Leslie, with Jan Bollwerk and Richard C. Morais. "The Not So Peaceful World of Greenpeace." *Forbes*, Nov. 11, 1991.

Spong, Paul. "Cortical Evoked Responses and Attention in Man." PhD diss., UCLA, 1966.

Spong, Paul. "The Whale Show." In *Mind in the Waters*, comp. Joan McIntyre.

Spong, Paul, Manfred Haider, and Donald B. Lindsley. "Selective Attentiveness and Cortical Evoked Responses to Visual and Auditory Stimuli." *Science* 48, no. 3668 (Apr. 16, 1965): 395–97.

Spong, Paul, and Don White. *Cetacean Research at the Vancouver Public Aquarium, 1967–1969*. Technical Report no. 2, Summer 1969. UBC Division of Neurological Sciences, Cetacean Research Laboratory.

Sprague, T. L. S. "Philosophers and Antivivisectionism." *Alternatives to Laboratory Animals* 13 (1985): 99–106.

Stephens, Julie. *Anti-Disciplinary Protest: Sixties Radicalism and Postmodernism*. Cambridge: Cambridge University Press, 1998.

Sternglass, Ernest J. "The Death of All Children." *Esquire* 72 (Sept. 1969):1a–1d.

———. "Has Nuclear Testing Caused Infant Death?" *New Scientist* 43 (July 24, 1969): 178–81.

———. "Infant Mortality and Nuclear Tests." *Bulletin of the Atomic Scientists*, Apr. 1969, 26–28.

———. *Low-Level Radiation*. New York: Ballantine Books, 1972.

Stevens, Jay. *Storming Heaven: LSD and the American Dream*. New York: Atlantic Monthly Press, 1987.

Stewart, Larry R. "Canada's Role in the International Uranium Cartel." *International Organization* 35, no. 4 (Autumn 1981): 657–89.

Stoll, Steven. *U.S. Environmentalism since 1945: A Brief History with Documents*. Boston: Bedford/St. Martin's, 2007.

Stone, Christopher. *Should Trees Have Standing? Toward Legal Rights for Natural Objects*. Los Altos, CA: W. Kaufmann, 1974.

Streich, Jürgen. *Betrifft: Greenpeace: Gewaltfrei gegen die Zerstörung*. Munich: Beck, 1986.

Strømsnes, Kristin, Per Selle, and Gunnar Grendstad. "Environmentalism Between State and Local Community: Why Greenpeace Has Failed in Norway." *Environmental Politics* 18 (May 2009): 391–407.

Sunday Times Insight Team. *Rainbow Warrior: The French Attempt to Sink Greenpeace*. London: Hutchinson, 1986.

Sutphen, John. "Body State Communication Among Cetaceans." In *Mind in the Waters*, comp. Joan McIntyre.

Suzuki, Daisetz Teitaro. *An Introduction to Zen Buddhism*. London: Rider, 1983.

Szabo, Michael. *Making Waves: The Greenpeace New Zealand Story*. Auckland: Reed Books, 1991.

Talbot, Lee. "The Great Whales and the International Whaling Commission." In *Mind in the Waters*, comp. Joan McIntyre.

Tarrow, Sidney. *Power in Movement: Social Movements, Collective Action, and Politics*. New York: Cambridge University Press, 1994.

Taylor, Bron. *Dark Green Religion: Nature Spirituality and the Planetary Future*. Berkeley: University of California Press, 2010.

Teilhard de Chardin, Pierre. *The Phenomenon of Man*, trans. Bernard Wall. New York: Harper, 1959.

Terpenning, J. G. "The B.C. Wildlife Federation and Government: A Comparative Study of Pressure Group and Government Interaction for Two Periods, 1947 to 1957 and 1958 to 1975." MA thesis, University of Victoria, 1982.

Thompson, John, and Stephen Randall. *Canada and the United States: Ambivalent Allies*. 2nd ed. Athens: University of Georgia Press, 1997.

Tønnessen, J. N., and A. O. Johnsen. *The History of Modern Whaling*. London and Canberra: C. Hurst, Australian National University Press, 1982.

Touraine, Alain. *Anti-nuclear Protest: The Opposition to Nuclear Energy in France*. New York: Cambridge University Press, 1983.

Tracey, David. *Guerrilla Gardening: A Manifesto*. Gabriola Island, BC: New Society, 2007.

Tremblay, K. R., and Riley E. Dunlap. "Rural-Urban Residence and a Concern with Environmental Quality: A Replication and Extension." *Rural Sociology* 43, no. 3 (1978): 474–91.

Tribe, Laurence. "Ways Not to Think about Plastic Trees: New Foundations for Environmental Law." *Yale Law Review* 83 (June 1974): 1340–46.

Turner, Fred. *From Counterculture to Cyberculture: Stewart Brand, the Whole Earth Network, and the Rise of Digital Utopianism*. Chicago: University of Chicago Press, 2006.

Turner, James Morton. "'The Specter of Environmentalism': Wilderness, Environmental Politics, and the Evolution of the New Right." *Journal of American History* 96 (June 2009): 123–48.

Turner, R. N., and K. S. Norris. "Discriminative Echolocation in a Porpoise." *Journal of the Experimental Analysis of Behavior* 9 (Sept. 1966): 535–44.

U.S. Congress. *Congressional Record*. "Planned Nuclear Bomb Tests in Alaska This Year." Vol. 117 (92-1). June 4, 1971.

U.S. Congress. House. Committee on Merchant Marine and Fisheries, Subcommittee on Fisheries and Wildlife Conservation Hearings. *Marine Mammals*. 91st Cong., 1st sess. Sept. 9, 13, 17, 23, 1971, 65–66.

Veysey, Lawrence. *The Communal Experience: Anarchist and Mystical Counter-Cultures in America*. New York: Harper and Row, 1973.

Wager, W. Warren. *Terminal Visions: The Literature of Last Things*. Bloomington: University of Indiana Press, 1982.

Walth, Brent. *Fire at Eden's Gate: Tom McCall and the Oregon Story*. Portland: Oregon Historical Society Press, 1994.

Wapner, Paul. *Environmentalism and World Civic Politics*. Albany: State University of New York Press, 1996.

Warford, Mark. Comp. *Greenpeace Witness: Twenty-five Years on the Environmental Front Line*. London: André Deutsch, 1996.

Watson, Paul. *Ocean Warrior: My Battle to End the Illegal Slaughter on the High Seas*. Toronto: Key Porter Books, 1994.

Watson, Paul. As told to Warren Rogers. *Sea Shepherd: My Fight for Whales and Seals*. New York: W.W. Norton, 1982.

Webb, Phyllis. "Protest in Paradise." *MaClean's*, June 1973.

Weissman, David. "Farewell to the Fox." *Chicago Wilderness Magazine*, Spring 2002.

Wenzel, George. *Animal Rights, Human Rights: Ecology, Economy and Ideology in the Canadian Arctic*. Toronto: University of Toronto Press, 1991.

Weyler, Rex. *Greenpeace: How a Group of Ecologists, Journalists, and Visionaries Changed the World*. Vancouver: Raincoast Books, 2004.

———. "Portfolio." *BC Photographer*, Fall 1975: 16–32.

———. *Song of the Whale*. Garden City, NY: Anchor Press/Doubleday, 1986.

White, Lynn Jr. "The Historical Roots of Our Ecological Crisis." *Science* 155 (Mar. 10, 1967): 1204–8.

White, Richard. "'Are You an Environmentalist or Do You Work for a Living?': Work and Nature." In *Uncommon Ground: Rethinking the Human Place in Nature*, ed. William Cronon. New York: W.W. Norton, 1996.

Widdowson, Frances Mary. "The Framing of Greenpeace in the Mass Media." MA thesis, University of Victoria, 1992.

Wilkinson, Pete, with Julia Schofield. *Warrior: One Man's Environmental Crusade*. Cambridge: Lutterworth Press, 1994.

Willoya, William, and Vinson Brown. *Warriors of the Rainbow: Strange and Prophetic Dreams of the Indians*. Healdsburg, CA: Naturegraph, 1962.

Winkler, Allen M. "A 40-Year History of Civil Defense." *Bulletin of Atomic Scientists*, June–July 1984, 16–22.

———. *Life Under a Cloud: American Anxiety About the Bomb*. New York: Oxford University Press, 1993.

Wittner, Lawrence S. *Rebels Against War: The American Peace Movement, 1933–1983*. Philadelphia: Temple University Press, 1984.

———. *Resisting the Bomb: A History of the World Nuclear Disarmament Movement, 1954–1970*. Stanford, CA: Stanford University Press, 1997.

———. *Toward Nuclear Abolition: A History of the World Nuclear Disarmament Movement, 1971–Present*. Stanford, CA: Stanford University Press, 2003.

Wöbse, Anna-Katherina. "Die Brent Spar-Kampagne: Plattform für diverse Wahrheiten." In *Wird Kassandra heiser? Die Geschichte falscher Ökoalarme*, ed. Frank Uekötter and Jens Hohensee. Stuttgart: Steiner, 2004.

Wood, Linda Sargent. *A More Perfect Union: Holistic Worldviews and the Transformation of American Culture after World War II*. New York: Oxford University Press, 2010.

Worster, Donald. *Nature's Economy: A History of Ecological Ideas*. 2nd ed. Cambridge: Cambridge University Press, 1994.

———. *A Passion for Nature: The Life of John Muir*. New York: Oxford University Press, 2008.

———. "Two Faces West: The Development Myth in Canada and the United States." In *Terra Pacifica: People and Place in the Northwest States and Western Canada*, ed. Paul Hirt. Pullman: Washington State University Press, 1998.

Wrightson, Kendall. "An Introduction to Acoustic Ecology." *Soundscape* 1 (Spring 2000): 10–13.

Wynne, Clive D. L. *Do Animals Think?* Princeton, NJ: Princeton University Press, 2004.

Yablokov, Alexey V. "On the Soviet Whaling Falsification, 1947–1972." *Whales Alive!* 6, no. 4 (1997). Available at http://csiwhalesalive.org/newsletters/csi97403.html (accessed Jan. 3, 2012).

Yavenditti, Michael J. "The American People and the Use of Atomic Bombs on Japan: The 1940s." *Historian* 36 (Feb. 1974): 224–25.

Zakin, Susan. *Coyotes and Town Dogs: Earth First! and the Environmental Movement*. Tucson: University of Arizona Press, 2002.

Zelko, Frank. "Challenging Modernity: The Origins of Postwar Environmental Protest in the United States." In *Shades of Green: Environmental Activism Around the Globe*, ed. Christof Mauch, Nathan Stoltzfus, and Douglas R. Weiner. Lanham, MD: Rowman and Littlefield, 2006.

Zunes, Stephen, Lester R. Kurtz, and Sarah Beth Asher, ed. *Nonviolent Social Movements: A Geographical Perspective*. Malden, MA: Blackwell, 1999.

Films and Broadcasts

All Things Bright and Beautiful. Toronto: Canadian Broadcasting Corporation, 1985.
Bitter Harvest. Vancouver: Northern Lights Films, 1981.
Desperate Measures. Greenpeace Communications, 1983.
Destination Iceland. Greenpeace Communications, 1980.
Ecology in Action. Northern Lights Films, 1985.
Silent Warrior. Soapbox Productions, Vancouver, 2000.
The Man in the Rainbow. Nordiskfilm, Denmark, 1992.
Whale Hunters. Produced by Jeremy Bristow. London: BBC, 2002.

INDEX